ASPECTS OF ARTIFICIAL I

STUDIES IN COGNITIVE SYSTEMS

James H. Fetzer
University of Minnesota, Duluth
Editor

ASPECTS OF ARTIFICIAL INTELLIGENCE

Edited by

JAMES H. FETZER

Dept. of Philosophy and Humanities, University of Minnesota, Duluth, U.S.A.

KLUWER ACADEMIC PUBLISHERS

DORDRECHT / BOSTON / LANCASTER / TOKYO

Library of Congress Cataloging-in-Publication Data

CIP-data will appear on a separate card.

ISBN 1-55608-037-9 (HB)
1-55608-038-7 (PB)

Published by Kluwer Academic Publishers,
P.O. Box 17, 3300 AA Dordrecht, Holland

Sold and distributed in the U.S.A. and Canada
by Kluwer Academic Publishers,
101 Philip Drive, Norwell, MA 02061, U.S.A.

In all other countries, sold and distributed
by Kluwer Academic Publishers,
P.O. Box 322, 3300 AH Dordrecht, Holland

2-1289-300-ts

printed on acid free paper

To
Chuck, Ken, Adam, and Joe

TABLE OF CONTENTS

SERIES PREFACE

This series will include monographs and collections of studies devoted to the investigation and exploration of knowledge, information and data-processing systems of all kinds, no matter whether human, (other) animal or machine. Its scope is intended to span the full range of interests from classical problems in the philosophy of mind and philosophical psychology through issues in cognitive psychology and sociobiology (concerning the mental capabilities of other species) to ideas related to artificial intelligence and to computer science. While primary emphasis will be placed upon theoretical, conceptual and epistemological aspects of these problems and domains, empirical, experimental and methodological studies will also appear from time to time.

The present volume illustrates the approach represented by this series. It addresses fundamental questions lying at the heart of artificial intelligence, including those of the relative virtues of computational and of non-computational conceptions of language and of mind, whether AI should be envisioned as a philosophical or as a scientific discipline, the theoretical character of patterns of inference and modes of argumentation (especially, defeasible and inductive reasoning), and the relations that may obtain between AI and epistemology. Alternative positions are developed in detail and subjected to vigorous debate in the justifiable expectation that — here as elsewhere — critical inquiry provides the most promising path to discovering the truth about ourselves and the world around us.

J. H. F.

FOREWORD

The contributions to this special symposium devoted to the theoretical foundations of artificial intelligence tend to fall into two broad categories: those dealing with AI in relation to issues in the philosophy of language and the philosophy of mind, which concern its ontology; and those dealing with AI in relation to issues in the theory of knowledge and the philosophy of science, which concern its methodology. Part I focuses upon these matters of ontology, anticipated by Smith's fascinating exploration of the semantics of clocks as a study of computational systems that participate in the world with us, where "participatory interpreted systems" are fundamentally different in kind from traditional Turing machines, with which we are already familiar. Moor takes up the challenge advanced by Searle in the form of his famous "Chinese room" argument, contending that Searle is both "right" and "wrong": right about a wrong conception of AI (since no program as "a mere formal structure" is sufficient to be a mind) but wrong about a right conception (since nothing has shown that "high-level semantics" cannot be generated by "low-level syntax"), whose truth is an open question. Maloney pursues a different issue, arguing in support of the role of representations — especially in the form of propositional attitudes — for understanding behavior, suggesting that the conception of "modularized information subsystems" can contribute toward explaining how certain cognitive systems have the capacity to adjust their cognitive states. Rapaport responds to Searle with a "Korean room" argument of his own, contending that the kind of semantics necessary to understand natural language is not the sort that Searle envisions but rather "mere symbol manipulation". He supports his position by elaborating a prototype "natural-language-understanding" system, which may or may not actually be one; but it is not essential to agree with Rapaport to recognize this study as a *tour de force*. If "mere symbol manipulation" were enough for mental activity, of course, contemporary AI might already qualify as a science of the mind. Fetzer, however, distinguishes between systems for which signs function as signs for those systems themselves and systems for which signs function as signs for the users

xi

James H. Fetzer (ed.), Aspects of Artificial Intelligence, xi—xiii.
© 1988 *by Kluwer Academic Publishers.*

of those systems, introducing a pragmatic theory of mind based upon Peirce's theory of signs. If his position is correct, then the symbol manipulation conception that lies at the core of computational interpretations may be adequate for machines but not for minds. Also inspired by Peirce, MacLennon contends that AI is moving into a new phase that accentuates "nonverbal reasoning". Since traditional AI technology rests upon idealized "verbal reasoning", the field confronts a dilemma: analysis can be shallow, yielding crude and rigid classifications, or deep, leading to a combinatorial "explosion of possibilities", because of which AI needs "a new logic".

Part II presents a collection of reflections upon the theoretical and experimental dimensions of artificial intelligence and the character of various types of reasoning that might be implemented therein. Glymour advances the thesis that artificial intelligence is perhaps best understood as (a branch of) philosophy, an intriguing contention that he supports by identifying various (implicit and explicit) philosophical commitments that can be discerned as underlying a broad range of AI programs. An alternative conception is offered by Buchanan in defense of the (somewhat more plausible) position that there are three broad stages in the development of an AI project, namely: (i) identifying the problem to be solved and designing (or otherwise finding) methods appropriate to its resolution; (ii) implementing that method in the form of a program and demonstrating the power of that program; and, (iii) analyzing the results of those demonstrations and generalizing their significance. From this perspective, AI tends to assume something of the classical hypothetico-deductive, conjectures-and-refutations, trial-and-error character of sciences that are both theoretical and experimental in general. Nute's impressive study not only seeks to clarify the nature of defeasible reasoning, in principle, but also attempts to illustrate the utilization of computer programs in the investigation of philosophical positions, a view that harmonizes well with the perspective presented by Glymour. Indeed, the methodological benefits that may be gleaned from this approach do not depend upon the more specific conception of defeasible reasoning that Nute endorses, where defeasible reasoning "proceeds from the assumption that we are dealing with the usual or normal case". Rankin's analysis of the features that tend to distinguish monotonic from non-monotonic reasoning, moreover, poses a subtle challenge to Nute's characterization, since Rankin's results suggest that monotonicity and non-monotonicity are necessary properties of (valid) deduction and of (proper) induction, respectively, where other modes of argumentative reasoning appear to be fallacious. Defeasible reasoning, in fact, appears to require satisfaction of the principle of total evidence, which Carnap advanced as a fundamental feature of inductive methodology. Kelly offers a study of the significance of discovery and of justification within the context of AI, suggesting that AI practice approximates "the study of epistemic norms for computational agents", which promotes "effective epistemology". Vaughan discusses some general aspects of providing expert systems with the capacity to reason inductively, illustrating the problems involved in implementing Bayesian conditionalization and likelihood principles. Scheines appropriately closes the symposium with a stimulating exploration of certain conditions under which "scientific creativity" could be realized by a machine. And the editor once again has the pleasure of thanking Jayne Moneysmith for her superb assistance in this effort.

J.H.F.

PROLOGUE

BRIAN CANTWELL SMITH

THE SEMANTICS OF CLOCKS

> *The inexorable ticking of the clock may have
> had more to do with the weakening of God's
> supremacy than all the treatises produced by
> the philosophers of the Enlightenment . . . .
> Perhaps Moses should have included another
> Commandment: Thou shalt not make mechan-
> ical representations of time.*
>
> — Neil Postman [1985, pp. 11—12]

1. INTRODUCTION

Clocks?

Yes, because they participate in their subject matter, and participation — at least so I will argue — is an important semantical phenomenon.

To start with, clocks are about time; they represent it.[1] Not only that, clocks themselves are temporal, as anyone knows who, wondering whether a watch is still working, has paused for a second or two, to see whether the second hand moves. In some sense everything is temporal, from the price of gold to the most passive rock, manifesting such properties as fluctuating wildly or being inert. But the temporal nature of clocks is essential to their semantic interpretation, more than for other time representations, such as calendars. The point is just the obvious one. As time goes by, we require a certain strict coördination. The time that a clock represents, at any given moment, is supposed to be the time that it is, at that moment. A clock should indicate 12 o'clock just in case it *is* 12 o'clock.

But that's not all. The time that a clock represents, at a given moment, is also a function of that moment, the very moment it is meant to represent. I.e., suppose that a clock does indicate 12 o'clock at noon. The time that it indicates a moment later will differ by an amount that is not only proportional to, but also dependent on, the intervening passage of time. It doesn't take God or angels to keep the clock coördinated; it does it on its own. This is where participation takes hold.

3

James H. Fetzer (ed.), Aspects of Artificial Intelligence, 3—31.
© 1988 *by Kluwer Academic Publishers.*

As well as representing the current time, clocks have to identify its "location" in the complex but familiar cycle of hours, minutes, etc. They have to measure it, that is, in terms of a predetermined set of temporal units, and they measure it by participating in it. And yet the connection between their participation and their content isn't absolute — clocks, after all, can be wrong. How it is that clocks can participate and still be wrong is something we will have to explain.

For clocks, participation involves being dynamic: constantly changing state, in virtue of internal temporal properties, in order to maintain the right semantic stance. This dynamic aspect is a substantial, additional, constraint. A passive disk inscribed with 'NOW' would have both temporal properties mentioned above (being about time, and having the time of interpretation relevant to content) and would even maintain perfect coördination. A rendering of this word in blinking lights, mounted on an chrome pedestal, might even deserve a place on California's Venice Boardwalk. But even though it would be the first time piece in history to be absolutely accurate, such a contraption wouldn't count as a genuine chronometer.

We humans participate in the subject matter of our thoughts, too, when we think about where to look for our glasses, notice that we're repeating ourselves, or pause to ask why a conversant is reacting strangely. Why? What is this participation? It's hard to say exactly, especially because we can't get outside it, but a sidelong glance suggests a thick and constant interaction between the contents of our thoughts, on the one hand, and both prior and subsequent non-representational activity, on the other, such as walking around, shutting up, or pouring a drink.

Take the glasses example. Suppose, after first noticing their absence, I get up and look on my dresser, asking myself "Are they here?" My asking the question will be a consequence of my wonder, but so will my (non-representational) standing in front of the dresser. Furthermore, the two are related; the word 'here' will depend for its interpretation on where I am standing. And who knows, to drive the example backwards in time, what caused the initial wonder — eye strain, perhaps, or maybe an explicit comment. The point is that the representational and non-representational states of participatory systems are inexorably inter-twined — they even rest on the same physical substrate. We can put it even more strongly: the physical states that realise our thoughts are caused by non-representational conditions, and engender non-representational consequences, in ways that must be coördinated with the contents of the very representational states they realise. Participation is something like that.

AI and general computational systems also participate — more and more, in fact, as they emerge from the laboratory and take up residence with us in life itself: landing airplanes, teaching children, launching nuclear weapons. Far from being abstract, computers are part of the world, use energy, affect the social fabric. This participation makes them quite a lot like us, quite unlike the abstract mathematical expression types on which familiar semantical techniques have been developed.

My real reason for studying clocks, therefore, can be spelled out as follows. First, issues of semantics, and of the relationship between semantics and mechanism, are crucial for AI and cognitive science (this much I take for granted). Second, it is terrifically important to recognise that computational systems participate in the world along with us. That's why they're useful. Third, as I hope this paper will show, participation has major consequences for semantical analyses: it forces us to develop new notions and new vocabulary in terms of which to understand interpretation and behaviour. Clocks are an extremely simple case, with very modest participation. Nonetheless, their simplicity makes them a good foil in terms of which to start the new development.

So they're really not such an unlikely subject matter, after all.

2. INFERENCE AND TIME-KEEPING

Let's start by reviewing the current state of the semantical art. Consider a familiar, paradigmatic case: a theorem-prover built according to the dictates of traditional mathematical logic. As suggested in Figure 1, two relatively independent aspects will be coördinated in such a system. First, there is activity or behaviour — what the system does — indicated as Ψ (for psychology). All systems, from car engines to biological mechanisms of photosynthesis, of course do something; what distinguishes theorem provers is the fact that their Ψ implements (some subset of) the proof-theoretic inference relation ($\vdash$). Second, there is the denotation or interpretation relation, indicated as Φ (for philosophy), which maps sentences or formulae onto model-theoretic structures of some sort, in terms of which the truth-values of the formulae are determined. In a computer system designed to prove theorems in abstract algebra, for example, the interpretation function would map states of the machine (or states of its language) onto groups, rings, or numbers — the subject matter of the algebraic axioms.

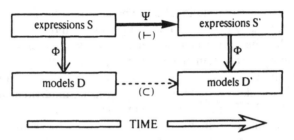

Fig. 1. Activity and semantics for a theorem prover

Four things about this situation are important. First, although proof theory's putative formality suggests that Ψ must be definable independent of Φ, you couldn't claim to have a proof-theoretic or inference relation except with reference to some underlying notion of semantic interpretation. Conceptually, at the very least, Ψ and Φ are inextricably linked (salesmen for inference systems without semantics should be reported to the Better Business Bureau). Furthermore, the two relations are coördinated in the well-known way, using notions of soundness and completeness: inferences (Ψ) should lead from one set of sentences to another only if the latter are true just in case the former are true ($\vdash$ should honour $\models$). And truth, as we've already said, is defined as in terms of Φ: the semantic relation to the world.

Second, even though the proof-theoretic derivability relation ($\vdash$) can be modelled as an abstract set-theoretic relation among sentences, I will view inference itself (Ψ) as fundamentally temporal — as an activity. 'Inference' is a derived noun; 'infer' first and foremost a verb, with an inherent asymmetry corresponding directly to the asymmetry of time itself. It might be possible to realise the provability relation non-temporally, for example by writing consequences of sentences down on a page, but you could hardly claim that the resulting piece of paper was doing inference.

Third, when its dynamic nature is recognised, inference is (quite properly) viewed as a temporal relation between sentences or states of the machine's memory, not as a function from times onto those corresponding sentences or states. Mathematically this may not seem like much of a difference, but conceptually it matters a lot. Thus, taking σ to range over interpretable states of the system, and t over times, Ψ is of type $\sigma \rightarrow \sigma$, not $t \rightarrow \sigma$. Of course it will be possible to define a temporal state function of the latter type, which I will call Σ; the point is that it is Ψ, not Σ, that we call inference. Details will come later, but the relation between the two is roughly as follows: if t' is one temporal unit past t, and $\Sigma(t) = \sigma$, then $\Sigma(t') = \Psi(\sigma)$. Inference, that is, has more to do with changes in state than with states themselves. To study inference is to study the dynamics of representational systems.

Fourth, of all the relations in Figure 1, only Ψ need be effective; *neither Φ nor Σ can be directly implemented or realised*, in the strong sense that there cannot be a procedure that uses these functions' inputs as a way of producing their outputs (the real reason to distinguish Ψ and Σ). This claim is obviously true for Φ. If I use the name 'Beantown' to refer to Boston, then the relation between my utterance and the town itself is established by all sorts of conventional and structural facts about me, about English, about the situation of my utterance, and so forth. The town itself, however, isn't the output of any mechanisable procedure, realised in me, in you, or in anyone else (fortunately — as it would be awfully heavy). It might require inference to understand my utterance, but that would only put you in some state σ with the same referent as my utterance, or state. You don't compute the referent of an utterance you hear, that is, in the sense of producing that referent as an output of a procedure. Nor is the reference relation directly mediated, at least in any immediate sense, by the physical substrate of the world. Not even the NSA could fabricate a sensor, to be deployed on route 128, that could detect Boston's participation as a referent in an inferential act.[2]

That Σ isn't computed is equally obvious, once you see what it means. The point is a strong metaphysical one: times themselves — metaphysical moments, slices through the *flux quo* — aren't causally efficacious constituents of activity; they don't have causal force. If they were, clocks wouldn't have been so hard to develop.[3] As it is, mechanisms, like all physical entities, manifest whatever temporal behaviour they do in virtue of momentum, forces acting on them, energy expended, etc., all of which operate in time, but don't convert time, compare it to anything else, or react with it. The only thing that's available, as a determiner of how a system is going to be, is how it was a moment before, plus any forces impinging on it. That, fundamentally, is why inference is of type $\sigma \rightarrow \sigma$, not $t \rightarrow \sigma$. *It could not be otherwise.* The inertness of gold, and the indifference of a neutrino, are nothing as compared with the imperturbability of a passing moment.

Given these properties of theorem provers, what can we say about clocks? Well, to start with, their situation certainly resembles that of Figure 1. As in the inference case, a clock's being in some state σ represents (Φ) it's being noon, or 7 : 15, or whatever; the interpretation function is what matters. Similarly, clocks, like theorem provers, change state (Ψ) in a simple but important way. Not only that; state change is what the clock designer has to work with; no mortal machinist, unfortunately, could build a device that would directly implement Σ. Furthermore, as in the case of the theorem prover, the change in state of the clock face is important *only because of its relation to its content.* Forget the Better Business Bureau; no one would buy a clock without a clue as to how its states represented time. Once again, systematic coördination between activity and interpretation is what matters.

But despite these similarities, there is a difference between clocks and theorem provers — suggested by the fact that many people (including me) would be reluctant to say that a clock was doing *inference.* To get at the difference, note that we haven't yet said what inference's coördinated pattern of events is for (on the face of it, going from truths to truths sounds a little boring). But the answer isn't hard to find: given a set of sentences or axioms that stand in (or enable you to stand in) a given semantical or informational relation to a subject matter, proofs or inference lead you to a new informational relation to the same, unchanged, subject matter. For example, the famous puzzle of Mr. S and Mr. P[4] focuses your attention on a pair of numbers under a peculiar description; a considerable amount of inference is required in order to give you access to those same numbers under a more tradi- tional description (or give you access to other more familiar properties of numbers — there are many ways to discharge the ontological facts). The numbers themselves, however, and their possession of all the relevant properties, are expected to stay put during the inferential process. None of this implies, of course, that the subject matter of inference cannot itself be temporal, as the situation calculus and temporal logics illustrate. The point is that the temporality of the inference process and the temporality of the subject domain aren't expected to interact.

The situation for clocks, on the other hand, is almost exactly the opposite. What changes, across the time slice mediated by Ψ, isn't the stance or attitude or property structure that clocks get at. What changes, rather, is the subject matter itself. Clocks never have a moment's rest; no sooner have they achieved the desired relationship to the current time than time slips out from under their fingers, as if God were constantly saying "It's later than you think!" Clocks should perhaps be viewed as the world's first truth maintenance systems: they do what they do merely in order to retain the validity of their single semantic claim. Like any other meter or measuring instrument, they must track the world.

We can summarise:

> *Inference, at least as traditionally construed, is a technique that enables you to change your relation to a fixed subject matter. Clocks, in contrast, maintain a fixed relationship to a changing subject matter.*

If reconstructing time-pieces were really my subject matter, rather than simply being a foil, I might stop here. But my real interest is in developing a single semantical framework so that we can not only handle both of these cases (mathematical inference and real-time clocks), but also locate everything in between. So let's spend a minute to see how clocks fit into the general case.

3. SEMANTICALLY COHERENT ACTIVITY

I will use the term 'representational system' to cover anything whose behaviour fits within the broad space of semantically constrained activity. To be a representational system, in other words, is to be an element of the natural order that acts in a semantically coherent way. Of all possible kinds of representational activity, inference will be analysed as a particular type. The representational space is large, of course, and certainly includes all of computation (more about that in a moment), but it's still a substantive notion: not everything is in it. Planets, for example, are excluded, because planets don't represent their orbits; they just have them. Clocks, on the other hand, do represent the time, just as I can represent to myself how the sunrise looked this morning, as I drove down from the mountains.

Clocks do however fall outside most traditional models of computation, including the "formal symbol manipulation" model so familiar in cognitive science.[5] First, clocks (their faces, and the clockworks that run them) are fully concrete, physical objects, part of the natural order; nothing abstract here. Furthermore, this concreteness is crucial to our understanding of them; for some purposes one might treat clocks at a level of description that abstracted away from their physical being, including their temporal being, but since our purpose is to show how participation in their subject matter influences their design, to do so would be to miss what matters most. Second, at least some clocks (especially electrical ones operating on alternating current) are analog, even though more and more recent ones are digital. Third, to the extent that clocks have representational ingredients, there is no obvious decoupling to be made between a set of structures that represent and an independent process that inspects and manipulates them according to the shapes it sees. In other words, whereas Fodor's characterisation of a computer's "standing in relation" to representational ingredients suggests a modular division between symbols and processor, no such division is to be found in the chronological case. Fourth, there is another separation that can't be maintained in the case of clocks: that between "internal" and "external" properties. Time (rather like neutrinos) permeates everything equally, being as much an influence on internal workings as on surrounding context. And of course it is one and the same time, inside and out — clock design depends on this. Fifth, clocks, especially analog clocks, aren't usually "programmed" in any sense; they are designed, but they aren't universal computers specialised by physical encodings of time-keeping instructions. Like so many other properties of clocks, this is important, and leads to the sixth salient difference. Even on the view that Turing machines are concrete, physical objects (of which abstract mathematical quadruples are merely set-theoretic models), there is still no guarantee, given a particular universal one, that *any set of instructions could make it be, or even simulate, an accurate time keeper* — because there need be no consistency or regularity as to how long its state changes take. Turing machines, *qua* Turing machines, don't really participate.

I have come to believe, however, that not one of these six properties — being abstract, being digital, exhibiting a process/structure dichotomy, having a clear boundary between inside and outside, being programmable, or being necessarily equivalent to any Turing machine — is essential to the notion of computation on which the economy of Silicon Valley is based, or to the notion that underlies AI's hunch that the mind is computational. Quite the contrary. In (Smith, forthcoming) I argue for a much stronger conclusion: that the only regularity essential to computation has to do with computation being a physically embodied representational process — an active system or process whose behaviour represents some part or aspect of the embedding world in which it participates. This has the consequence, needless to say, of defining computation squarely in terms of undischarged semantical predicates. My position on theoretical cartography is therefore the inverse of Newell's (1980): whereas he thinks that computer science has answered the question of what it is to be a symbol, I believe in contrast that the integrity of computation as a notion rests full-square on semantics. So we have lots of homework, but it's homework for another day.

In the meantime, clocks are a good test case for comprehensive semantical frameworks. They lack many important properties of more general computers: they don't act, for example, or have sensors. But since every semantical property they do exhibit is one that computers can exhibit too — including participation — they are a useful design study.

4. THREE POINTS ON TWO FACTORS

In the previous section I distinguished two aspects or factors of any representational system: its behaviour, activity, or causal connection with the world (which I'll call the *first factor*) and its interpretation, content, or relation to its subject matter (the *second factor*). I have previously used this two-factor framework to reconstruct the semantics of Lisp, the programming *lingua franca* of AI, and argued for its general utility in analysing knowledge representation systems (Smith, 1982, 1984, 1986). And I will use it here, to analyse clocks. But three points must be made clear.

First, the ordering of the two factors may seem odd. There is no doubt that having interpretation or content — standing in semantic relation to a subject matter — is what particularly distinguishes the systems we are interested in. Given this pride of place, it might seem that content should be called first. But this is a mistake. We theoreticians typically treat semantics as primary when we analyse both natural and artifactual languages (such as the predicate calculus). We typically define semantics over rather abstract entities — sentence types, for example — and then understandably define the other dimension (proof theory, inference) over the same domain. But this overall strategy, especially in conjunction with the formal-symbol manipulation view of computation, gives a very abstract feel to inference, leading such people as Searle to wonder how, or even whether, such a system could ever possess genuine semantical powers. In contrast, by calling activity the first factor I want to recognise that computational systems are, first and foremost, systems in the world. *Everything* has what I am calling a first factor; that's what gives a system the ability to participate. The second factor of representation or content, which enables a system (a thinker, a clock) to stand in relation to what isn't immediately accessible or discriminable, is a subsequent, more sophisticated capacity. It is the second factor, furthermore, that distinguishes the representational or interpretable systems from other natural systems, but it distinguishes them as a sub-type, not as a distinct class. First factor participation in the world ("being there", roughly) is always available — which is fortunate, since it is only with respect to the first factor that second factor content can ever be grounded. In sum, recognising the metaphysical primacy of the first factor is an important ingredient in the defense of naturalism.

Second, there is a natural (almost algebraic) tendency to think that, in accepting a two-factor stance, one is committed to thinking that the two factors, in any given system, will in some important sense be *independent*. This tendency is amplified by the fact that in standard first-order logic an almost total independence of factors is achieved — this is one of the many meanings of the ambiguous claim that first-order logic is *formal*. Truth, content, and interpretation in logic are thought to be relatively independent of proof-theoretic role, and provability or inferential manipulation analogously independent of content or interpretation. In fact it is only because of this conceptual independence that proofs of soundness and completeness, even the very notions of soundness and completeness, are conceptually coherent. In computer systems, however — and minds, and clocks — there is no reason to expect this total degree of disconnection or independence. We should expect something more like the relationship between the mass and velocity of a physical object, on the one hand, and the center of gravity or resonance of the system of which it is a part, on the other: a web of constraints and conditions tying the two factors together, piece-wise, incrementally, thereby giving rise to a comprehensive whole. The situation of a complete proof system defined on an abstract set of mathematical expression types is extreme: a global but locally un-mediated coherence, with no part of the proof or inferential system touching the semantic interpretation or content, except in the final analysis, when an outside theorist's proof grandly ties the whole thing together. For computers, and for us, it seems much more plausible to take a step or two apart from our subject matter, and then check in with it, to stay in "sync" — by taking a look, for example, or (following AT&T's recommendation) by reaching out and touching it. Participation is a resource, not a complication.

Third, as both the first two points make clear, it's a little hard to justify calling the two factors *semantical*, especially when the first is shared with every other participant in the natural order. It's not just that the first should be viewed as syntax, the second as semantics (as application of this more general framework to the predicate calculus would suggest). Rather, it's not clear what, if anything, the terms "syntax" and "semantics" should mean in a context where the coupling between factors is so much richer and more complex than in the traditional idealised case — if indeed they mean anything at all. Clockworks are mechanisms that enable first-factor behaviour — that much seems innocuous enough; calling the momentum of a clock's pendulum *semantic* is more difficult. First and second factors aren't distinct objects that somehow coöperate in engendering semantical activity; rather, one and the same causal constituents of a semantic system play both first and second factor roles.

This whole question is complicated by the use of the word 'semantics' (especially in AI) to describe inferential and structural relations among ingredients within a computational system. In (Smith, 1986) I attempt to resolve some of these issues, but instead of reconstructing that argument here I'll simply use the two-factor terminology without prejudice as to what does and doesn't have legitimate claim to the overloaded term.

5. THEORETIC MACHINERY AND ASSUMPTIONS

Let's look, then, at how clocks represent time, starting with some basic assumptions. As suggested in Figure 2, *qua* theorists we need accounts of four things:

1.   States of the clock itself, including the face (σ);

2.   The time or passage of time that the clock represents (τ);
3.   The first factor movement or state change between clock states (Ψ);   and
4.   The second factor representation relation (Φ) between clock states and times.

All four of these are shared with standard semantical analysis: the first two would be the syntactic and semantic domains; the third, inference or proof theory; the fourth, semantics or interpretation.

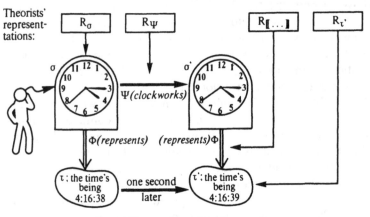

Fig. 2. The typology of clock semantics

I will adopt what I will call a *direct* rather than model-theoretic approach to these analytic tasks. Typically, when doing semantics, instead of talking directly about clock faces, orientations of hands, etc., you model them. For example, the state of a three-hand analog clock might be modelled as a triple, consisting of the orientations of the hour-hand, minute-hand, and second-hand, respectively, measured clock-wise from the vertical, in degrees. Thus the clock face shown in Figure 2 would be modelled as follows:

(S1) M_σ: $\langle 128.31666\ldots, 99.8, 228 \rangle$

The problem with this technique, however, as suggested in Figure 3, is that a model M of a situation S is itself a representation of S, since modelling is a particular species of representation (M_σ, for example, represents the clock face; it isn't the clock face, since for example it has

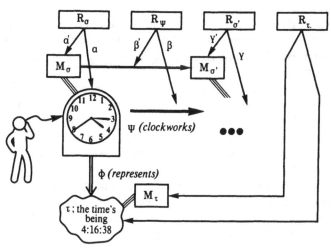

Fig. 3. The model-theoretic approach

a length of 3). The general character and complexity of the model—clock relation M_σ—σ, therefore, is the same as that between the clock and the time it represents (σ—τ). It is therefore very hard to know whether what is crucial about σ—τ will be revealed or hidden in its M_σ—M_τ form. For example, using simple numbers to represent the orientations of hands presumes an absolute accuracy on the clock face, counter to fact. When studying something like natural language, which makes use of a much more complex representation relation than a model, the problems of indiscriminate theoretic modelling may be minor, or (more likely) go unnoticed. In our case, however, the representation relation we are studying, between clock faces and periodic times, is essentially an isomorphism. In this situation indiscriminate modelling would be much more theoretically distracting.

This direct semantical stance will have consequences, of two main sorts. First, we will need some machinery for talking precisely about the world without modelling it; for this I will use an informal "pocket situation theory", based unapologetically on Barwise and Perry (Barwise and Perry, 1983; Barwise, 1986a). Second, in the analog case it will be tempting to use some elementary calculus, which is problematic because a situation-theoretic reconstruction of continuity hasn't been developed yet. On the other hand, since the continuities underlying the integrity of the calculus presumably derive, ultimately, from the fundamental continuity of the physical phenomena that the mathematics was developed to describe, and since exactly those continuous phenomena will be our subject matter here, I'll take the liberty of applying its insights anyway. We're not really going to do any mathematics, so we won't get into trouble.

The direct semantical stance also highlights a question: how as theorists are we going to describe or *register* the phenomena we are going to study — i.e., in terms of what concepts, categories, and constraints are we going to explicate its regularity? When giving semantical analyses of linguistic or syntactic objects (sentences, expression types, etc.), tradition provides standard registrations in terms of constituent terms, predicate letters, etc. Similarly, purely abstract objects are typically categorised in advance in terms of a defining set of properties or relations. Clocks, on the other hand, are neither traditional nor abstract, so the question remains.

My metaphysical bias is to treat the world as infinitely rich, not only in the sense that there is more to everything than anything we can say, but also in that there is both more uniformity and structure, and more heterogeneity and individual difference, than theory or language can ever encompass. So I will say that clock faces, being actual, have enough structure so that one can be wrong about them, but still don't come labelled in advance by God, like plant slips at a nursery identified with a white plastic tag. Since every clock face, furthermore, exemplifies an infinite number of properties and relations (such as the property of being the subject matter of this paragraph), even after a basic registration scheme has been settled on, we have considerable latitude in making our choice.

None of this is intended to be problematic, or new; it's worth mentioning only because we need to make room for there being a difference between how we theorists do it, and how clocks do it, for themselves or (more likely, in the case of clocks) for their users. The problem is particularly acute for time itself, especially the periodic cycle of hours, minutes, and seconds that I keep referring to without explanation. If this were a paper on the semantics of time, not just of clocks, that explanation would have to be given, which would raise the incestuous fact that clocks themselves are probably largely responsible for the temporal registration (hours, mintues, seconds, etc.) of the times they represent, as argued for example by Lewis Mumford (1934). In this paper, however, I will merely adopt the periodic cycle without analysis, taking its explanation as a debt that should ultimately be paid.

Given these preliminaries, we should set out the ontological type structure, as summarised in Figure 4. Variables ranging over objects will be spelled with lower-case italic letters; over properties and relations, in lower-case Greek; over functions, in upper-case Greek. Thus c and c' will range over clocks; t, t', etc., over full-blooded times, which are taken to be instantaneous slices through the metaphysical flux. Times are meant to include the time Kennedy was shot, the referent of "now", the point when the ship passed out of sight behind the island — that sort of thing. Intervals — intuitively, temporal durations between times — will be indicated by Δt, $\Delta t'$, etc. I will extend the use of '+' to allow adding intervals to times (overloading '+', as computer scientists say); thus $t + \Delta t$ will be of type t.

Objects and Properties

$c, c', \ldots$	— clocks
$t, t', \ldots$	— times (instantaneous moments: slices through the flux quo)
$\Delta t, \Delta t', \ldots$	— temporal intervals
$\tau, \tau', \ldots$	— o'clock properties: being midnight, being $4:01:23, \ldots$
	τ_t — the o'clock property that holds of time t
$\sigma, \sigma', \ldots$	— states of clock faces (both hands pointing upwards, . . .)
	$\sigma_{c,t}$ — the state of clock c at time t

Primary Theoretic Functions

$\Psi : \sigma, \Delta t \to \sigma$	— clockworks (from clock states and intervals onto clock states)
$\Sigma : c, t \to \sigma$	— state function (from clocks and times onto clock states)
$[\ldots] : \sigma \to \tau$	— content function (from clock states onto o'clock properties)

Overloaded Addition

$t + \Delta t : t$	— times plus intervals are times
$\tau + \Delta t : \tau$	— o'clock properties plus intervals are o'clock properties

---

Fig. 4. Theoretic type structure

As opposed to times themselves, I will assume that times are located on the periodic cycle by what I will call the *o'clock properties*, such as that of "being $4:01:23$", "being midnight", etc. The idea is not so much to license a continuum of distinct properties, but rather to assume that they arise out of a continuous relation between times and the abstract locations on the periodic time cycle to which they correspond

("4 : 00", etc.). Various possible explanations of this relation are possible, but since the intent of this paper is not to present an independently justified metaphysical account of time, but only to relate clocks to such a thing, I will employ a notation that simply picks up o'clock properties, whatever they are, from times that have them. Thus I will use τ_t to refer to the particular o'clock property that actually holds of time t. Also, I'll take differences between o'clock properties to be intervals (e.g., the difference between 5 : 00 and 3 : 00 will be two hours). Thus the sentence $\tau_t(t')$ says of time t' that it has o'clock property τ_t — i.e., that it has whatever o'clock property t has. The term $\tau_t - \tau_{t'}$ denotes an interval, of type Δt.[6]

In an analogous way, σ, σ', etc. will range over a continuous (in the analog case) set of states of clock faces. For traditional circular analog clocks, a σ representing 4 : 30 might be "having the hour hand at 135°, the minute hand at 180°, and the second hand at 0°, all measured clockwise from the 'XII'."

Given this framework we can type the various semantical functions already encountered. As suggested in the previous section, Σ will be a (non-computed!) function of type $t \rightarrow \sigma$, from times onto clock states; Ψ, a function of type σ, $\Delta t \rightarrow \sigma$, from clock states and temporal intervals onto clock states; and Φ, a function of type $\sigma \rightarrow \tau$, from clock states onto o'clock properties. The important typological point for general semantic analysis is that both factors (Ψ and Φ) are defined as functions between states objects can be in, not between objects that are in them. This is as you would expect for scientific laws.

Two more theoretical points, before we take up the analysis itself. First, as just mentioned, I claimed in Section 2 that times t weren't causal agents — that they couldn't be in the domain of a strongly effective realisable function. It is probably more important to the life of clock designers that the o'clock properties (τ) are equally impotent. Even if it's 4 : 00 all around you, there's nothing that it's being 4 : 00 can cause to happen — like serving tea and crumpets. With respect to engendering behaviour, a moment's being midnight is more like Boston's being a referent than it is like ice-cream's being sticky: it just isn't the sort of thing that a sensor could detect. So functions of the form $\tau \rightarrow x$ are as unrealisable (in the strong sense discussed earlier) as those of type $t \rightarrow x$, for arbitrary x. Such is life.

Second, I mentioned earlier that using numbers to represent the orientations of the hands of clocks presumes an accuracy that outstrips physical plausibility. Even if quantum physics would theoretically support there being a fact of the matter as to where a hand points within $\pm 10^{-50}$ degrees, say (which it won't), there are also pragmatic realities of producing a macroscopically observable clock subject to the forces of gravity, anomalies of manufacture, etc. Furthermore, if the hour-hand were anything like this accurate, then at least for theoretical purposes the minute and second hands would be redundant: a perfect observer could gaze at a clock and read off a time of, say, $4:15:38:17.$[7] One might object, of course, that human users wouldn't be able to register the hour-hand more accurately than, say, $\pm$ 1° or 2°, and therefore, even with internal calculation, wouldn't be able to determine the time on a single-handed clock more accurately than to within about 5 minutes, no matter how much more accurately than that the time was actually signified. In fact casual observation suggests that hour hands on modern analog clocks are much more accurately positioned than necessary merely to determine which hour the minute hand signifies time with respect to.

These issues again raise the question of the relation between how we as theorists register clock faces and the times they represent, and how clock faces themselves register those represented times.[8] But I won't answer this question here, since we will primarily be dealing with semantic constraints on clock and time registrations, rather than with individual registrations themselves.

6. TEMPORAL REPRESENTATION: THE SECOND FACTOR

Given these premises and caveats, let's look at how times are represented. Intuitively, we are aiming for something like the following:

(S2)   [ ] = the property of being 4 : 16

To do this, we start with Φ, of type $\sigma \rightarrow \tau$ from (representing) states of clock faces onto (represented) states of times — i.e., onto o'clock properties. Instead of the name 'Φ', however, I will use so-called semantic brackets ('[...]'), in the following way: $[\ldots \sigma_{c,t} \ldots]$ will be the o'clock property signified by the state '$\ldots \sigma_{c,t} \ldots$', assuming that $\sigma_{c,t}$ is the state σ of clock c at time t. For example, the sentence $[\sigma_{c,t'}](t)$

claims of time t that it has the o'clock property that clock c indicates at time t'; $[\Psi(\sigma_{c,t'}, \Delta t)](t)$ claims of time t that is has the o'clock property that clock c would indicate Δt later than time t', since $\Psi(\sigma_{c,t'}, \Delta t)$ is the state it would then be in.

Using this terminology, we can say that clock c is chronologically *correct* at time t just in case t is of the type that the clock then indicates:

(S3) $\text{Correct}(c, t) \equiv_{df} [\sigma_{c,t}](t)$

So far, of course, this is a constraint on possible interpretation functions [. . .], since we haven't defined any specific instances. Longer-term notions of correctness (over extended intervals, for example) could be defined by quantifying over times; similarly, approximate degrees of correctness could be characterised in terms of the difference between what time it actually was and what time was indicated.

7. CLOCKWORKS: THE FIRST FACTOR

With respect to operation, the basic point is this: if at time t a clock is so-and-so (σ), then at some point Δt later it will be such-and-such (σ'), where σ' is $\Psi(\sigma, \Delta t)$. The function Ψ, which takes a clock into the future in this way, must be realised by the underlying physical machine — must be implemented, that is, by the clockworks. The important constraint on this relation, which I will call the *realisability* constraint, is that $\Psi(\sigma, \Delta t)$ can depend on σ and on Δt, but not on the time t that is "happening" when the clock is in state σ.

In symbol manipulation or semantical contexts, where time and symbols are both digital, we often view Ψ as a state-transition function (such as for a Turing machine controller). In such cases Δt drops out, being assumed to be a single time "click". For example, suppose $\mathbb{S}$ is a (discrete) function from states to states ($\sigma \rightarrow \sigma$). The equation for a single state change, of the sort one would expect in a digital world, would be something like $\sigma' = \mathbb{S}(\sigma)$, or generalised to Δt's of n tick's duration, $\sigma' = \mathbb{S}^n(\sigma)$. In the continuous world of physical mechanics, on the other hand, Ψ is merely "what the world does", explained in terms of velocities, accelerations, etc. From this perspective, the calculus can be viewed as a theoretical vehicle with which to explain first factor futures for continuous systems, where the state σ of some system in an amount of time Δt after it is in a starting state σ_0, assumed to depend on the continuity of the underlying phenomena, can be expressed in the familiar equation:

$$\text{(S4)} \qquad \sigma = \sigma_0 + \frac{d\sigma}{dt}\,\Delta t + \frac{1}{2}\,\frac{d^2\sigma}{dt_2}\,\Delta t^2 + \ldots$$

My aim isn't to contrast the discrete and continuous cases (I want to develop results applicable to both analog and digital clocks), but rather to highlight the common focus on state change, represented computationally by state transition functions, and physically by temporal derivatives. There is, however, this apparent difference: the theoretic notions employed in physics (force, acceleration, etc.) are essentially relative; they describe how the new state will differ from the old one. The real identity of the new state — what state the system will actually arrive in — is obtained, as if it were conceptually subsidiary, by altering the previous state in the prescribed manner. State transition tables, in contrast, are typically absolute. They still describe state *change*, of course — they aren't temporal state functions like Σ. The point is that the new state is specific *de novo*, so to speak, not as a modification of the old one, though of course the extent to which the new state differs from the old can be calculated as a difference between the two.

This difference in theoretic stance, however, is superficial, since in actual use (in describing programs, operations on memory, etc.) state transition functions are defined with explicit reference to how the new state differs from the old. In giving environment transition functions, for example, showing the consequence of binding a variable, the requisite function from total environments onto total environments is defined as modifying the value of the given variable in question, and *otherwise being just like the prior one*. Practice suggests, in other words, that in the computational case, as in the physical case, state change is conceptually prior, new total state conceptually dependent. Thus there is general support for our specific focus on Ψ.

Intuitively, a proper Ψ for a clock will specify that it runs at the right speed. It is easy enough to calculate, in the case of circular analog clocks, that this amounts to having the hour hand, minute hand, and second hand rotate at $0.008333°/\text{sec}$, $0.1°/\text{sec}$, and $6°/\text{sec}$, respectively. But to *characterise* correctness this way is exactly like characterising the correctness of a proof procedure by pointing to the syntactic inference rules. It may indeed be true that, if this condition is met, the clock will be running at the correct speed, but that doesn't mean that this condition expresses what it is to be running correctly. Rather, we want to say that if at time t (say, $12:00$) a clock designates o'clock property $\tau_{t'}$ (say, $3:11$), then at time $t + \Delta t$ ($12:01$, for a one minute Δt) it should indicate the property of being Δt later, i.e., $\tau_{t'+\Delta t}$ ($3:12$). We can do this as follows:

(S5)   Right-speed$(c, t, \Delta t) \equiv_{df} [\sigma_{c, t+\Delta t}] = [\sigma_{c, t}] + \Delta t$

which has the consequence, given the definition of Ψ, that

(S6) $[\Psi(\sigma_{c, t}, \Delta t)] = [\sigma_{c, t}] + \Delta t$

Properly, we should state something stronger: that a clock runs correctly throughout the interval from t to $t + \Delta t$ if and only if it advances at the right speed for the whole time (note that the following is neutral as to whether this is a continuous or discrete interval — i.e., as to whether $\forall$ is a discrete or continuous quantifier):

(S7)   Right-speed$(c, t, \Delta t) \equiv_{df} \forall \Delta t' \mid 0 \leqslant \Delta t' \leqslant \Delta t$
$[\sigma_{c, t+\Delta t'}] = [\sigma_{c, t}] + \Delta t'$

again directly yielding

(S8) $\forall \Delta t' \mid 0 \leqslant \Delta t' \leqslant \Delta t \ [\Psi(\sigma_{c, t}, \Delta t')] = [\sigma_{c, t}] + \Delta t'$

These equations involve a property identity, but I defer any questions on that issue to situation theory. Note also that in each version the two instances of '+' are of different types: the first takes a time and an interval onto a time, the second an o'clock property and an interval onto an o'clock property. No problem.

Given (S3) and (S7), the temporal analogues of soundness and completeness can be proved: if a clock is correct at time t, and runs at the right speed during the interval from t to t', then it will be correct during that interval, and conversely if it is correct throughout the interval it must be running at the right speed. But it is more fun to do this in the continuous case, so let's turn to that.

Very simply, we want to talk of an analog clock's running at the right speed *instantaneously*, which means, intuitively, that we should differentiate the temporal state function Σ — or, what is equivalent, take the limit of Σ as Δt approaches 0, in the standard way:

(S9) $\lim_{\Delta t \to 0} \dfrac{(([\sigma_{c, t}] + \Delta t) - [\sigma_{c, t}])}{\Delta t} = \lim_{\Delta t \to 0} \dfrac{([\sigma_{c, t+\Delta t}] - [\sigma_{c, t}])}{\Delta t}$

Since, as we've already said, differences between o'clock properties are intervals, the left side of this reduces to $\lim_{\Delta t \to 0} (\Delta t / \Delta t)$, which is identically 1, yielding:

(S10) $1 = \lim\limits_{\Delta t \to 0} \dfrac{([\sigma_{c,\,t+\Delta t}] - [\sigma_{c,\,t}])}{\Delta t}$

The right hand side, however, is merely the derivative, with respect to time, of the interpretation of the state. We can't differentiate σ directly, its not being a function of time (in fact it's not a function at all), but we can rewrite (S10) in terms of Σ:

(S11) $1 = \lim\limits_{\Delta t \to 0} \dfrac{([\Sigma(c, t + \Delta t)] - [\Sigma(c, t)])}{\Delta t}$

This enables us to take the limit (Σ is continuous by assumption), since the right hand side is the derivative of a function that is essentially the composition of the second and first factors ($[\ldots] \circ \Sigma$).[9] I will abbreviate this $[\Sigma]$, giving us:

(S12) $\text{Right-speed}_{\text{analog}}(c, t) \equiv_{df} \dfrac{d}{dt} [\Sigma] = 1$

If the derivative (with respect to time) of a function is unity, of course, it follows that the function is of the form $\lambda t \cdot t + k$ — or rather, in our case, $\lambda t \cdot \tau_t + k$, as dictated by our type constraints — where k is a constant of type Δt. This is exactly what we would expect: the constant represents the error in the clock's setting — the difference between the actual and indicated times. The equation, predictably, says that if a clock is running at the right speed the error will (instantaneously) remain constant. Furthermore, since (S3) implies that

(S13) $\text{Correct}(c, t) \Leftrightarrow [\Sigma(c, t)](t)$

it follows that the constant would be 0 for a correctly set clock, as expected.

We can summarise these results as follows:

(S14) $\text{Correct}(c, t) \qquad \equiv_{df} [\Sigma(c, t)](t)$

(S15) $\text{Right-speed}(c, t, t') \equiv_{df} \forall \Delta t \,|\, 0 \leqslant \Delta t \leqslant (t' - t)$
$\qquad\qquad\qquad\qquad\qquad [\sigma_{c,\,t+\Delta t}] = [\sigma_{c,\,t}] + \Delta t$
$\qquad\quad$*implying that*$\qquad \forall \Delta t \,|\, 0 \leqslant \Delta t \leqslant (t' - t)$
$\qquad\qquad\qquad\qquad\qquad [\Psi(\sigma_{c,\,t}, \Delta t)] = [\sigma_{c,\,t}] + \Delta t$
$\qquad\quad$*implying that*$\qquad \forall \Delta t \,|\, 0 \leqslant \Delta t \leqslant (t' - t)$
$\qquad\qquad\qquad\qquad\qquad [\Sigma(c, t + \Delta t)] = [\Sigma(c, t)] + \Delta t$

(S16) Right-speed$_{\text{analog}}(c, t) \equiv_{\text{df}} \dfrac{d}{dt} [\Sigma] = 1$

and in their terms define what it is for a clock to be "working" properly from time t to $t + \Delta t$:

(S17) Working$(c, t, \Delta t) \equiv_{\text{df}}$ Correct$(c, t) \wedge$ Right-speed$(c, t, \Delta t)$

(S18) Working$_{\text{analog}}(c, t) \equiv_{\text{df}}$ Correct$(c, t) \wedge$ Right-speed$_{\text{analog}}(c, t)$

For either version, the constraint can be shown to be satisfied (over the interval, or instantaneously, depending) in exactly the following condition:

(S19) $[\Sigma(c, t)] = \lambda t \cdot \tau_t$

Given the abbreviation adopted above, we can state this even more simply:

(S20) $[\Sigma] = \lambda t \cdot \tau_t$

I would be the first to admit that (S20) is obvious — at least retroactively, in the sense that, once stated, it is hard to imagine thinking anything else. In English, it says that the state function and the interpretation function should be proportional inverses; given a clock that (so to speak) maps time onto some sort of complex motion, the appropriate interpretation function is merely that function that maps that motion back onto the o'clock properties of the linear progression of time that was started with. So the putative clock of Figure 5, for example, with a million-mile pendulum and a 24 hour period, would have a pointer position (σ) proportional to $\sin(t)$, and an interpretation function analogously proportional to $\sin^{-1}(\sigma)$.[10]

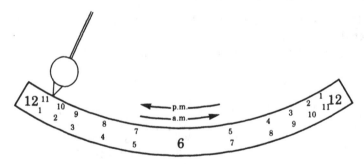

Fig. 5. The million mile clock

Still, (S20) isn't trivial, for a reason that shows exactly why clocks were hard to build. It says that working clocks map all times onto their o'clock properties. The problem for clockmakers is that Σ isn't directly computable, since, to repeat, neither times nor o'clock properties enter into causally efficacious behaviour. What can be implemented is Ψ, not Σ, and Ψ is essentially the temporal derivative of Σ.

In sum, we have determined the function of clockworks: to integrate the derivative of time. When you set the hands on the clock's face, you are supplying the integration constant.

8. MORALS AND CONCLUSIONS

What have we learned? Four things, other than some fun facts to tell our friends.

The first has to do with the interaction among notions of participation, realisation, and formality. Clocks' participation in their subject matter (being temporal, as a way of measuring time), which depends on their physical realisation, might seem to violate the formality constraint that is claimed to hold of computational systems more generally. In fact, however, clocks' temporality doesn't relieve them of much of the structure that characterises more traditional systems: separable Ψ and Φ, the possibility of being wrong, etc. This similarity of clocks to symbol manipulation systems arises from the fact that the particular aspect of times that clocks represent — the o'clock properties — aren't within immediate causal reach of a clockwork mechanism (or of much else, for that matter). In (Smith, forthcoming) I argue that this is a manifestation of a deep truth: the limitations of causal reach are the real constraints on representational systems. Formality, as a notion, is merely a cloudy and approximate projection of these limitations into a particular construal of the symbolic realm.

The second moral has to do with the impact, for theoretical analysis, of the relation between Ψ and Φ. The function Ψ, realised in clockwork, is what the engineers must implement; without an analysis of it, effective clocks couldn't be designed. But theories of clocks must go much further. Our characterisation of what it was for a clock to work properly, for example, had to reach beyond the immediate or causally accessible aspects of the underlying clockwork mechanism. Whatever one might think about more complex cases, methodological solipsism doesn't work in this particular instance.

Third, the similarity between the state transition functions of computer science and the temporal derivatives of mechanics, both of which focus not on time itself but on temporal change, suggest the possibility of a more unified treatment of representational dynamics in general. So far most of what we have to say deals with specific cases. So, for example, in Section 2 we characterised inference as a particular species of representational activity, having to do with changing content relations to a fixed subject matter. Inference was contrasted with clock's maintenance of a fixed content relation to a changing subject matter. Remembering what is perceived, to take quite a third sort of representational behaviour, is a form of retaining a fixed relation to a fixed subject matter in ways that make it immune to changes in the agent's circumstances. It doesn't seem impossible that a common framework could be uncovered.

Fourth, and finally, by occupying a place very different from that of either Turing machines or traditional theorem provers, clocks help illuminate the fundamental constraints governing computers and representational systems in general. As Figure 6 suggests, there are two basic kinds of constraint — causal relations and content relations — that a representational system must coördinate as it moves through the world.

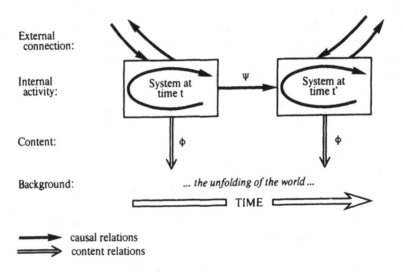

Fig. 6. C5: Coordinated constraints on content and causal connection.

Both kinds, in general, will be complex — much more so than we have seen in the case of clocks. Two aspects of content we haven't encountered, for example, are its "situational" dependence on surrounding circumstances, as discussed for example in (Barwise, 1986b; Perry, 1986), and the three-way semantic interactions among language, mind, and world that arise in cases of communication. Causal connections are similarly complex, and can be broken down into three main groups:

1. *Internal activity or behaviour*: the relation between a system at some time and the same system shortly thereafter. This is what we've called Ψ.
2. *External connection*: actions the system takes that affect the world, and effects on the system of the world around it — the results, that is, of sensors and effectors. (Clocks have none of this, but other systems are clearly not so limited.)
3. *Background dynamics*: the progress or flow of the surrounding situation. The passage of time would be counted as one instance, as would one's conversant's behaviour, or the passing visual scene.

In the traditional case of pure mathematical inference, there is no connection (action or sensation), and the background situation, as we saw, is presumed to stay fixed. Barwise's construal of "formal inference" (the "non-situated" reading), (Barwise, 1986b, p. 331) strengthens this constraint by assuming that the content relation is also independent of surrounding situation. The clock example gives us a different point in the space: again no connection, an essentially unchanging (and relatively situation-independent) content relation, but an evolving background situation, mirrored in the internal activity or behaviour. Finally, semantic theories of action, involving everything from intentionally eating supper to making a promise, must deal with cases where the connection aspect makes a contribution. They must therefore deal with cases where the surrounding situation is affected not only by its own background dynamics, but as a result of internal activity on the part of the representational agent. But simpler systems will require an analysis of external connection, as well: computerised (ABS) brakes on late model cars, for example, are directly connected (even vulnerable) to the content of their representations, in a way that seems to free them from the need to have their representational states externally interpreted.

In the end, however, the similarity among these systems is far more important than the variance. We can put it this way. Causal participation in the world is ultimately a two-edged sword. On the one hand, it is absolutely enabling. Not only could a system not exist without it, but in a certain sense it's total: everything the system is and does arises out of its causally supported existence. There are no angels. On the other hand, causal connection on its own — unless further structured — limits a system's total participation in the world to those things within immediate causal reach.

Representation, on this view, is a mechanism that honours the limits of causal participation, but at the same time stands a system in a content relation to aspects of the world beyond its causal reach. The trick that the system must solve is to live within the limits (and exploit the freedoms!) of the causal laws in just such a way as to preserve its representational stance to what is distal. This much is in common between an inference system and a clock.

ACKNOWLEDGEMENTS

This paper grew out of a bet made with Richard Weyhrauch during a discussion late one night in a bar in Alghero, Sardinia about what was involved in reading one's watch. Specifically, I promised to develop a semantical analysis of the familiar behaviour cited in the third paragraph of the paper: waiting a second to see whether a watch moves before reading the time. This paper is part one of the answer; interpreting a clock will come later. My thanks to him and other members of the Cost 13 Workshop on Reflection and Meta-Level Architectures, especially including Jim des Rivières and John Batali. Thanks also to Jon Barwise, John Etchemendy, David Israel, John Perry, Susan Stucky, and the other members of the situation theory and situation semantics (STASS) group at CSLI, to Pat Hayes for discussions of measurement, and to John Lamping for his help on celestial mechanics. The research was supported by Xerox Corporation and the System Development Foundation, through their mutual support of the Center for the Study of Language and Information.

NOTES

[1] Clocks represent time for us, as it happens, not for themselves, but that will count, at least here. I'm sympathetic to the distinction between original and derivative semantics (in fact I'm interested in participation for just such reasons), but I am very much against relativising representation to an observer at the outset, especially to a human observer

(Winograd and Flores, 1986). To do that would be to abandon any hope of explaining how the human mind might itself be representational, my ultimate goal. See (Smith, forthcoming).

[2] In computer science the claim that reference isn't computed is viewed suspiciously, for a very interesting reason. To see it, consider why the claim is true. Suppose in a room of 100 people some person A is the average height. Then suppose a new person enters the room. Suddenly, *and without any computation*, a different person B will be the average height. No work needs to be done to lift the property from A and settle it on B; no energy expended, no symbols massaged. The new state just *comes to be*, automatically, in virtue of the maze of conditions and constraints that hold. Reference, I take it, is something like that; conditions and constraints hold so that, when a word is uttered or a thought entertained, some object becomes the referent. (Nor is it possible to reply "Well, the room computed it"; on that recourse everything that happens would be computed, which would make the word 'compute' vacuous.)

How could computer scientists object to this? For the following reason. Note that the way that B becomes the person of average height is by participating in the situation at hand: he enters the room. Participation, in other words, is what enables relationship to exist. Computers, on the other hand, are traditionally viewed in purely abstract terms, and abstractions, whatever they are, presumably don't participate. The closest an abstraction comes to the property of average height — or indeed to anything at all — is by designating it. And so, because of this abstract conception of computers, one gets lulled into thinking that *everything* has to come into being in this disconnected, putatively "computational" way.

Needless to say, I don't believe the abstract conception of computers is right. More strongly, I am arguing that participation — the opposite of abstraction — is exactly what allows you to connect to the world in other ways than through explicit symbol manipulation. See Section 8, and (Smith, forthcoming).

[3] For accurately measuring distances on roads, one attaches a "fifth wheel" to a car and reads off the passing miles. Maybe, if time had been causally efficacious, we could have built clocks the same way, running a wheel against time and reading off the passing seconds.

[4] There are two numbers between 1 and 100. Mr. P knows their product, and Mr. S their sum. They have the following conversation:

> Mr. P.   *I don't know the numbers.*
> Mr. S.   *I knew you didn't. Neither do I.*
> Mr. P.   *Now I do.*
> Mr. S.   *Now I do too.*

What are the numbers?

The earliest publication of this problem I am aware of is by H. Freudenthal in the Dutch periodical *Nieuw Archief Voor Wiskunde*, series 3, **17**, 1969, p. 152 (a solution by J. Boersma appears in the same series, **18**, 1970, pp. 102–106). It was subsequently submitted by David J. Sprows to *Mathematics Magazine* **49**(2), March 1976, p. 96 (solution in **50**(5), Nov. 1977, p. 268). Perhaps the most widely read version appears in Martin Gardner's 'Mathematical Games' column in *Scientific American* **241**(6), Dec. 1979, pp. 22–30, with subsequent discussions and slight variations in 1980: **242**(3), March, p. 38; **242**(5), May, pp. 24–28; and **242**(6), June, p. 32.

[5] The two other primary models, conceptually distinct from the formal symbol manipulation idea, are the automata-theoretic notion of a digital or discrete system and the related idea of a machine whose behaviour is equivalent to that of some Turing machine. Although the formal symbol manipulation view seems to go virtually unchallenged in cognitive science, the other two have much more currency in modern computer science. See (Smith, forthcoming).

[6] A more detached theoretic viewpoint should point out that o'clock properties τ_t are in fact two-place relations between times and places (a time that is midnight in London will be 7 : 00 p.m. in New York). More generally, whereas I assume throughout that activity (Ψ) and interpretation (Φ) are functions, they should properly be viewed as more complex relations between agents and their embedding circumstances.

[7] "**third**, n. . . . 5. The sixtieth part of a second of time or arc." — *Webster's New International Dictionary*, Second Edition. New York: G. & C. Merriam, Co. 1934.

[8] Clock faces, and representations in general, don't need to register themselves, in order to represent.

[9] Strictly speaking this isn't quite accurate, since both $[\ldots]$ and Σ should depend on c and t: the function we are differentiating should really be $\lambda c, t. [\Sigma(c, t)]$. But being strict would add only complexity, not insight.

[10] This clock would be even harder to build than you might suppose. At first blush, it might seem as if the equation of motion for a pendulum would imply that a very large bob, swinging in an arc at the surface of the earth (an arc, say, 100 feet in length), whose mass completely dominated the mass of a long string by which it was suspended from a geosynchronous point 1 150 000 miles above the surface of the earth, would have a period of 24 hours. Unfortunately, however, such a device would have a period of slightly less than an hour and a half. Why this is so, and how to modify the design appropriately are left as an exercise for the reader (hint: the result would be difficult to read).

REFERENCES

Barwise, Jon, and Perry, John: 1983, *Situations and Attitudes*, Bradford Books, Cambridge, MA.

Barwise, Jon: 1986a, 'The Situation in Logic — III: Situations, Sets and the Axiom of Foundation', in Alex Wilkie (ed.) *Logic Colloquium 84*, North Holland, Amsterdam. Also available as CSLI Technical Report CSLI-85-26 from the Center for the Study of Language and Information, Stanford University, 1985.

Barwise, Jon: 1986b, 'Information and Circumstance', *Notre Dame Journal of Formal Logic,* **27**(3) 324—338.

Brachman, Ronald J., and Levesque, Hector J. (eds.): 1985, *Readings in Knowledge Representation*, Morgan Kaufmann, Los Altos, CA.

Fodor, Jerry: 1975, *The Language of Thought*, Thomas Y. Crowell Co., New York. Paperback version, Harvard University Press Cambridge, MA, 1979.

Fodor, Jerry: 1980, 'Methodological Solipsism Considered as a Research Strategy in Cognitive Psychology', *The Behavioral and Brain Sciences*, **3**(1), pp. 63—73. Reprinted in Fodor, Jerry, *Representations*, Bradford, Cambridge, MA, 1981.

Mumford, Lewis: 1934, *Technics and Civilization*, Harcourt, Brace & Co., New York. Reprinted 1943.

Newell, Allen: 1980, 'Physical Symbol Systems', *Cognitive Science* **4**, 135—183.

Perry, John: 1986, 'Circumstantial Attitudes and Benevolent Cognition', in J. Butterfield (ed.), *Language, Mind and Logic*, pp. 123—134, Cambridge University Press Cambridge.

Postman, Neil: 1985, *Amusing Ourselves to Death: Public Discourse in the Age of Show Business*, Penguin Books, New York.

Smith, Brian C.: 1982, *Reflection and Semantics in a Procedural Language*, Technical Report MIT/LCS/TR-272, M.I.T., Cambridge, MA, 495 pp. Prologue reprinted in (Brachman and Levesque, 1985), pp. 31—39.

Smith, Brian C.: 1984, 'Reflection and Semantics in Lisp', Conference Record of 11th Principles of Programming Languages Conference, pp. 23—35, Salt Lake City, Utah. Also available as Xerox PARC Intelligent Systems Laboratory Technical Report ISL-5, Palo Alto, California, 1984.

Smith, Brian C.: 1986, 'The Correspondence Continuum', appeared with the Proceedings of the Sixth Canadian AI Conference, Montreal, Canada, May 21—23. Available as CSLI Technical Report CSLI-87-71 from the Center for the Study of Language and Information, Stanford University, 1987. To appear in *Artificial Intelligence* (forthcoming).

Smith, Brian C.: *Is Computation Formal?* MIT Press/A Bradford Book Cambridge, MA. (Forthcoming.)

Winograd, Terry, and Flores, Fernando: 1986, *Understanding Computers and Cognition: A New Foundation for Design*, Ablex, Norwood, New Jersey.

Intelligent Systems Laboratory, and
Center for the Study of Language and Information
Xerox Palo Alto Research Center
Palo Alto, CA 94304, U.S.A.

PART I

ONTOLOGICAL FOUNDATIONS

JAMES H. MOOR

THE PSEUDOREALIZATION FALLACY AND THE CHINESE ROOM ARGUMENT

SEARLE'S CHALLENGE TO AI

John Searle with his now-famous Chinese room argument (1980, 1982, 1984) challenges the basis for a strong version of Artificial Intelligence (AI). Searle's argument has generated diverse and often strong reactions. Roland Puccetti says, "On the grounds he has staked out, which are considerable, Searle seems to me completely victorious," (1980, p. 441). Douglas Hofstadter remarks, "This religious diatribe against AI, masquerading as a serious scientific argument, is one of the wrongest, most infuriating articles I have ever read in my life" (1980, p. 433). My reaction to Searle's argument is that it is dead right and dead wrong. That is, his argument is right about a wrong conception of AI and wrong about a right conception of AI. But regardless of one's position on the Chinese room argument, there is no doubt that the argument raises critical issues about the nature and foundation of AI.

Searle offers us the following thought experiment:

> Suppose that I'm locked in a room and given a large batch of Chinese writing. Suppose furthermore (as is indeed the case) that I know no Chinese, either written or spoken, and that I'm not even confident that I could recognize Chinese writing as Chinese writing distinct from, say, Japanese writing or meaningless squiggles. ... Now suppose further that after this first batch of Chinese writing I am given a second batch of Chinese script together with a set of rules for correlating the second batch with the first batch. The rules are in English, and I understand these rules as well as any other native speaker of English. They enable me to correlate one set of formal symbols with another set of formal symbols, and all that "formal" means here is that I can identify the symbols entirely by their shapes. Now suppose also that I am given a third batch of Chinese symbols together with some instructions, again in English, that enable me to correlate elements of this third batch with the first two batches, and these rules instruct me to give back certain Chinese symbols with certain sorts of shapes in response to certain sorts of shapes given me in the third batch. (1980, pp. 417—418)

Searle's setup in the Chinese room is meant to imitate, if not parody, the approach of Roger Schank (Schank and Abelson, 1977). Schank's view is that computers can understand stories if they are given scripts (Searle's first batch of symbols). Scripts provide linked lists of informa-

James H. Fetzer (ed.), Aspects of Artificial Intelligence, 35—53.
© 1988 *by Kluwer Academic Publishers.*

tion about what usually happens in the world in particular situations. For example, in a restaurant one normally enters, orders, eats and leaves. Of course, there are many subactivities within each of these activities which are treated by further refining the script. Scripts give a computer a conceptual framework for organizing information which it receives in a story. So when the computer is given a story (Searle's second batch of symbols) the computer can put the details into a general framework provided by the appropriate script. Finally, when the computer is asked questions about the story (Searle's third batch of symbols), the computer can answer questions about the story even if it has to infer some of the information by using the script. The computer program (Searle's rules of manipulating symbols) guides the computer in a formal way from the questions about the story to the answers.

Searle proceeds with his argument as follows: He imagines that after a while he becomes so good at following the instructions for manipulating the Chinese symbols and the programmers get so good at writing the programs that, from the point of view of somebody outside the room in which he is locked, his answers to the questions are absolutely indistinguishable from those of native Chinese speakers. His answers in Chinese to Chinese questions are as good as his answers in English to English questions. The big difference for him is that he really understands English but he doesn't understand any Chinese, because he can produce answers in Chinese only by manipulating uninterpreted formal symbols. For Chinese he is the instantiation of a computer program (1980, p. 418).

Searle concludes that for the same reason he doesn't understand Chinese in the Chinese room Schank's computer doesn't understand English or any other language. Moreover, the computer and its program don't provide sufficient conditions for understanding because the computer and program can function without any understanding. What is sufficient for understanding, according to Searle, is the human brain. Searle believes that mental states are both caused by the operations of the brain and realized in the structure of the brain (1983, pp. 262–272). Searle is less clear about whether neurology is necessary for mentality. For instance, Searle says, "Whatever else intentionality is, it is a biological phenomenon, and it is as likely to be as causally dependent on the specific biochemistry of its origins as lactation, photosynthesis, or any other biological phenomena" (1980, p. 424). Some commentators take Searle to be advocating neural chauvinism (Cuda, 1985) or at least a claim that biology is essential for mental states (Bynum, 1985). But I think that, in the light of other passages, Searle holds a somewhat weaker claim, *viz.*, only brains or machines having the same causal powers as brains can have mental states. For instance, he says, "I offer no *a priori* proof that a system of integrated circuit chips couldn't have intentionality. That is, as I say repeatedly, an empirical question" (1980, p. 453). Thus, I take it that Searle's main objection to AI is not that computers are made of the wrong stuff, but that computers operate in the wrong way, i.e., mental states, like understanding, do not arise by simply instantiating a formal program.

THE MODEL AND METAPHOR DEFENSE

Searle's real target is strong AI and not weak AI. Searle tells us that in weak AI the computer is viewed as a powerful tool. In weak AI the principal value of the computer in the study of the mind is that it enables the formulation and testing of hypotheses in a more rigorous and precise fashion. "But according to strong AI," Searle says, "the computer is not merely a tool in the study of the mind; rather, the appropriately programmed computer really *is* a mind, in the sense that computers given the right programs can be literally said to *understand* and have other cognitive states" (1980, p. 417). Searle flatly rejects the claim that appropriately programmed computers can have cognitive states and that such programs can explain human cognition.

Searle's distinction between weak and strong AI is a little misleading in that it invites an image of two conflicting camps doing research within the AI community — one doing weak AI and the other doing strong AI. In fact, the strong AI camp is inhabited by philosophers and researchers in their philosophical moments, whereas actual scientific research in AI is done in the weak AI camp. Thus, one effective way of defending AI against Searle is to agree with him that the computer is only a tool. Because scientific AI really is weak AI, an attack on strong AI is irrelevant to the discipline at least as it currently exists and will exist for the foreseeable future.

Of course, the rhetorical force of the phrase "strong AI" invites a stronger defense. Who wants to work or get a Ph.D. in a weak field? Indeed, some do insist that strong AI really is being done. For example, Zenon Pylyshyn maintains in an argument against weak AI, ". . . computation is a literal model of mental activity, not a *simulation* of behavior, as was sometimes claimed in the early years of cognitive modeling" (1984, p. 43). But such zealous claims, though perhaps philosophically interesting, are scientifically dubious. If rhetoric is ignored and actual AI projects are examined objectively, then AI *as practiced* is weak in Searle's sense. Although AI researchers frequently use cognitive vocabulary in describing their projects, no AI project in the short history of the field would have to be classified or justified in terms of strong AI.

AI is such a diverse enterprise that it is difficult to characterize, but two working conceptions of AI are popular. Neither requires adherence to strong AI. One working conception is the operational account that AI is the study of making computers perform tasks which would be considered intelligent if done by humans, and the other working conception is the model building account that AI is the attempt to build computational models of cognitive processes (Bundy 1980, p. ix). These two conceptions reflect different emphases in the field. Some researchers are interested in the field of *Artificial* Intelligence. These researchers want machines to do intelligent tasks. If a robot welds well, who cares whether it simulates what a human welder thinks while welding? Other researchers are interested in Artificial *Intelligence*. These are the theoreticians who are interested in intelligence in general and who use a computer to model and investigate cognitive processes. The difference between *Artificial* Intelligence and Artificial *Intelligence* is not one of kind but one of degree. Many projects may fall between the two extremes. Consider PROSPECTOR, an expert system for making judgments about mineral deposits, as an example. This project is Artificial *Intelligence* in that it models inference patterns of various human experts, and it is *Artificial* Intelligence in that the goal of the project is to produce an artifact which really makes expert decisions about mineral deposits (Campbell, Hollister, Duda, and Hart, 1982).

What this brief survey of AI suggests is that nothing occurring in scientific AI really requires a defense of strong AI. Both *Artificial* Intelligence and Artificial *Intelligence* can occur within weak AI. Admittedly, researchers employ cognitive language to describe their projects and write books with tantalizing titles like *The Thinking Computer — Mind Inside Matter* (Raphael, 1976). But such talk is metaphor or hyperbole; any alleged cognitive process on a computer can be safely regarded as a model of the real thing. As Roger Schank and Robert Abelson put it, "[W]e need computers as the metaphor in terms of which we create our theories and as the arbiter of the plausibility of our theories" (Schank and Abelson, 1977, Preface). AI researchers may see strong AI as the ultimate goal, but nothing about their ongoing research requires that they defend the claims of strong AI. Again, as Schank describes his own position, "According to the distinction between weak and strong AI, I would have to place myself in the weak AI camp with a will to move to the strong side" (Schank 1980, p. 446).

The upshot of this defense against Searle is that research in AI is relatively immune to the Chinese room argument. Scientific AI past, present and foreseeable future does not depend on a strong conception of the enterprise. If the Chinese room argument were successful, it would certainly alter a popular philosophical conception of the field; but it would not close any projects or force AI researchers into unemployment lines.

Though strong AI is not essential for current scientific AI, it is, nevertheless, a possible conceptual foundation for scientific AI of the distant future. I believe strong AI — roughly the view that cognition is computation — is plausible and worthy of defense. The position isn't new. Thomas Hobbes suggested three centuries ago that reasoning is computation (Hungerland and Vick, 1981, p. 177). More recent versions of strong AI emerge from discussions of functionalism and cognitive science (Fodor, 1968, 1975, 1981; Pylyshyn, 1984; Haugeland 1985). Before I offer my defense of strong AI, some clarification of the details of the Chinese room argument is necessary.

THE STRANGE INHABITANT OF THE CHINESE ROOM

The thought experiment, as Searle describes it, is not coherent. We are asked to imagine that Searle-in-the-room gets good enough at identifying symbols that he executes the program so that his external behavior of speaking Chinese is indistinguishable from a native speaker of Chinese. What would this require? The first batch of symbols, the script, would be enormous. The script would be far more than a script for restaurants. The script would be the script for knowledge of the world possessed by a typical native Chinese speaker. The rules for manipulating the symbols would constitute a massive and extremely complex set of rules for adequately manipulating this body of world knowledge. Because no program like this really exists, it is difficult to judge its size, but it would not be surprising if millions of operations per second were required in order to execute such a program to conduct a normal conversation in real time. But for the sake of argument, suppose that only one hundred operations per second were required. This is still far beyond what any human being could do given human reaction times.

Because Searle is giving us only a thought experiment, we need not be limited by human nature. We might imagine that Searle-in-the-room can visually discriminate millions of forms per second, can look up the appropriate rules in English and execute them in a fraction of a second so that the linguistic behavior on the outside of the room is indistinguishable from that of a native speaker. With his enhanced visual system Searle-in-the-room will now be able to watch his own rapid hand movements which are performing millions of operations per second. Of course, his hand movements are not voluntary. Because Searle-in-the-room is an instantiation of a running computer program, there are causal as well as logical connections among the operations. Searle-in-the-room cannot freely choose, for example, not to follow a rule which applies. The millions of operations required to conduct a normal conversation on the outside are performed involuntarily. Searle-in-the-room does not have desires which can override his commitment to the task before him. For Searle-in-the-room there are no coffee breaks.

If we begin with Searle's assumption that we know no Chinese and believe we could be the person in the room, then his thought experiment may seem persuasive, for we still know no Chinese after receiving the batches of symbols. But, once the thought experiment is made coherent we realize that we could not be the person in the room even in principle. Searle-in-the-room couldn't be Searle or any other human being. Searle-in-the-room is an alien creature with an imagined mix of superhuman capacities and subhuman abilities, needs and desires. Searle-in-the-room is a kind of organic central processing unit. Therefore, since neither we nor Searle could be Searle-in-the-room, even in principle, the fact that we or Searle know no Chinese is irrelevant. We may believe that the alien organic central processing unit lacks cognitive states as well, but that requires further argument.

Unfortunately, making the Chinese room argument more coherent does not get to the heart of the argument. Searle's central thesis that supports his argument is somewhat obscured by the rhetoric of the Chinese room. Searle's central thesis is that whatever purely formal principles are put into a computer they will not be sufficient, and may not even be necessary, for understanding. Thus, Searle would argue, I am sure, that the alien organic central processing unit, which I contend is the real inhabitant of the Chinese room, does not understand Chinese because it operates in a syntactic manner on purely formal elements.

This is the general pattern of argument between Searle and his critics. Searle gives the Chinese room argument, a critic challenges the argument offering some kind of modification and Searle responds by emphasizing his central thesis. For example, Searle believes the Systems Reply, that is, the claim the whole system understands is incorrect because even if the inhabitant of the room were to incorporate the whole system, e.g., memorizes the rules and the lists of Chinese symbols, the whole system would still be operating in a purely formal manner and therefore would not understand. Searle believes the Robot Reply is inadequate for the same reason. Even if the inhabitant of the room were part of a robot with perceptual and motor systems, no understanding would occur if the information to and from these added systems were handled in a purely formal way (1980, pp. 419–420). Therefore, the philosophical discussion cannot be advanced by simply enhancing replies like the Systems Reply or the Robot Reply, because Searle will respond with his central thesis that formal programs are not sufficient for minds. Progress in the discussion requires a close look at Searle's reasons for his central thesis.

WEAK SEARLE VS. STRONG SEARLE

Why does Searle believe that his central thesis is true? Why is it that instantiating a computer program cannot be a sufficient condition for cognitive states like understanding? Searle says, "Because the formal symbol manipulations by themselves don't have any intentionality; they are quite meaningless; they aren't even *symbol* manipulations, since the symbols don't symbolize anything. In linguistic jargon, they have only a syntax but no semantics." (1980, p. 422) In his recent Reith Lectures Searle organizes these reasons into an argument for his central thesis. (1984, p. 39) I will call this "the foundational argument" because it presents a foundation for the Chinese room and Searle's various conclusions about the nature of minds.

1.   Brains cause minds.

2.   Syntax is not sufficient for semantics.
3.   Computer programs are entirely defined by their formal, or syntactical, structure.
4.   Minds have mental contents; specifically, they have semantic contents.

No computer program by itself is sufficient to give a system a mind. Programs, in short, are not minds, and they are not by themselves sufficient for having minds.

Searle draws other conclusions from the premises, but this conclusion is his central thesis and his primary challenge to strong AI. The advantage of looking at the foundational argument is that Searle's position can be examined without the rhetoric of the Chinese room. When the argument is examined carefully, however, it is difficult to determine what its premises and conclusion really claim. For example, how is the word "program" used? A program can be many things: (i) a program can be an abstract mathematical object such as a program for a Turing machine; (ii) a program can be a description of an algorithm as found in a flow chart; (iii) a program can be a set of instructions for a computer such as a program in Pascal; (iv) a program can be a physical object which encodes computer instructions such as that found on a disk or magnetic tape; (v) a program can be a routine running on a computer. I think Searle's foundational argument plays on such ambiguities. Consider two possible interpretations of the conclusion of his argument.

Weak Thesis: No program as a mere formal structure is a mind or sufficient for having a mind.

Strong Thesis: No computational operations of an electronic computer which can be described properly and adequately on a low level in terms of syntax or a formal program can produce a system which can be described properly and adequately on a high level in terms of semantics or mental processes.

I think the weak thesis is true. Formal structures as such don't cause anything, let alone minds. Even if the formal structure is instantiated, e.g. as a stack of punch cards on a filing cabinet, it is not a mind or by itself sufficient for having a mind. But, clearly the weak thesis doesn't challenge strong AI. I assume Searle believes he has proved something stronger, something like the strong thesis, for he says about his conclusion, "Now, that is a very powerful conclusion, because it means that the project of trying to create minds solely by designing programs is doomed from the start" (1984, p. 39). The problem for Searle is that the strong thesis doesn't follow from the premises of his argument. This conclusion is drawn from the second, third and fourth premises. Searle takes the second premise, syntax is not sufficient for semantics, as "a conceptual truth", the third premise, computer programs are entirely defined by their formal or syntactical structure, as "true by definition", and the fourth premise, minds have mental contents, as "an obvious fact about how our minds work" (1984, p. 39). Though the weak thesis may follow from these premises, thus construed, the strong thesis does not. There is no way that the strong thesis, an open and quite interesting empirical conjecture, follows validly from a conceptual truth, a definition, and an obvious fact. The dilemma for Searle is this: either his foundational argument is valid but has a weak conclusion which doesn't challenge strong AI or the foundational argument has a strong conclusion which challenges strong AI but is invalid.

THE EMPIRICIST REPLY

Searle's Chinese room argument and his replies to critics rest on one central claim: computational processes can't be mental processes. But this claim, the strong thesis, is not proven by Searle's foundational argument. On the contrary, whether computation is sufficient for cognition is largely an open empirical question. It may turn out empirically that the internal causal powers of brains responsible for mentality are just computational powers. Searle's Chinese room argument and his replies to critics depend on Searle's insistence that a particular answer be given to this open empirical issue. Therefore, the way to challenge Searle is to show that the issue is open and might be decided in the other direction.

Strong AI need not deny that there is a logical distinction between syntax and semantics. The thesis of strong AI is that it may be possible to construct *high*-level semantics from *low*-level syntax. In other words, although at a low level of analysis a digital computer's operation is syntactical, at higher levels of organization semantic structures may emerge. These semantic structures are composed of nothing but syntactic units but are semantic in that they are causally connected to the world in the right way. It is not a conceptual truth that this view is false. Rather it is an empirical matter whether AI can succeed in the endeavor of creating high-level semantics from low-level syntax. My strategy is to offer a series of examples which indicate the possibility of getting (high-level) semantics from (low-level) syntax.

Consider first the example of card flashers at a football game. This is a kind of real-life Chinese room. The members of the card section operate according to formal rules. On a given cue each puts up a card of a specific color. Although this activity involves parallel processing with many people flipping cards at once, it has a limited syntax of a digital nature. Each flipper either does or does not flip a card of a particular color. From a local point of view, e.g., sitting in the card section, one cannot tell what the cards say. But from a global point of view, e.g., across the stadium, the different-colored cards, the individual syntactic components, may spell out a word like "Touchdown", revealing semantic content beyond the individual cards. Now, of course, I am cheating in this example since arguably the system of cards takes on meaning because we ascribe meaning to the system. We, not the system of cards flashing, know what the word "Touchdown" means. Searle, I think quite correctly, would classify this as a case of derived not intrinsic intentionality. (Cf. Searle, 1983, pp. 26—29.) Nevertheless, the example is useful because it shows that a system may have high-level properties as well as low-level properties, and the scale of observation may determine which properties we notice. The best place to observe the card section is not located in the card section, and the best place to observe the functionality of a computer program is not located in the Chinese room.

Now consider another case in which the semantic component is inside a real computer. We need not imagine fancy AI applications but only everyday garden-variety computing. Consider a bar code reader in a supermarket. A machine optically scans merchandise as a customer checks out and records this information in a computer. Let's suppose it recognizes the bar code for chicken noodle soup. The price of this item will be added to the total bill for the customer and will be noted in a computer for inventory and marketing considerations later. Although this information processing on one level is completely syntactic, it is also the case that internal states in the computer, complexes of syntax, are semantic, for they contain information about the sales of this kind of soup. I grant that the semantics in this example is limited. First, the bar code reading system is extremely stupid., for it reads only bar codes. If a potato is wrapped with a bar code label from a can of chicken noodle soup, then the machine will treat the potato like a can of soup. The machine can discriminate bar codes but can't discriminate potatoes from soup cans, so one could argue that its semantics is at best about bar codes, not soup cans. Second, the computer has been directed by a programmer. Some human decided to let a variable in a program represent the sales of chicken noodle soup. Thus, Searle would be quick to point out that this is a case of derived and not intrinsic intentionality. While all of this is true, an important feature of this example remains. For a state of a computer to be considered semantic, even in a derived sense, it should play the right causal role both internally and externally. In this example the internal state of the computer which represents soup cans (or maybe strictly speaking bar codes for soup cans) has the right connections to other parts of the program and to the outside. Even if we are not the programmer, we can learn that a given variable in the program represents soup cans because that variable increases as soup cans are scanned and the program uses the variable later to report them as soup cans.

Finally, consider an example from AI of the distant future. Imagine a computer which has a learning program so that it generates its own scripts. These scripts are much like the scripts Schank gives his computer except that the computer produces its own scripts as it learns about its environment. We can further imagine that this computer is a robot with sensory and motor devices. It learns about the world by interacting with the world in much the way a human child does (cf. Turing, 1950). Gradually, the computer establishes, refines, tests, and modifies semantic networks as it interacts with the world. Such networks would allow the computer to make reasonable inferences about situations it has experienced even if some information is only implicit. I see no nonarbitrary way of denying intrinsic intentionality to such a computer without also denying it to other humans. The computer generates its own semantic networks as we do. The semantics are not derived from the programmer; the programmer has little idea what the semantics will be. The particular semantic categories are determined by the environment in which the computer learns about the world. Still, at the low level, of course, the computer is operating in a completely syntactic fashion.

This train of examples is meant to show how high-level semantics might emerge from low-level syntax. Semantic indicators would be built out of syntactic components and causally connected with each other and the world to produce what we normally consider to be intentional behavior. This is only a philosophical idea; no computer like the one in the last example exists. No current AI project is building one. But, the idea is coherent and an open empirical possibility. Searle has not given us any reason to think it is impossible. The real issue is not the conceptual distinction between semantics and syntax but the empirical question whether high-level semantics can be generated from low-level syntax when that low-level syntax constitutes the operation of a sophisticated computer program.

The empiricist's reply to Searle can be put as follows: Not only is it an empirical question whether a system of integrated circuit chips can have intentionality, as Searle says (1980, p. 453), but also it is an empirical question whether such intentionality has a purely syntactic basis on lower levels of analysis. That is, a machine with the causal powers equivalent to the human brain may turn out empirically to be nothing more than a machine which executes a certain kind of sophisticated formal program. We can imagine that this sophisticated program for learning and generating scripts can be run on various computers which configure the program somewhat differently. So it is the program, not a particular piece of hardward, that is crucial. We can even go so far as to imagine we discover that similar programs are carried out in human brains. Given all of this potential empirical evidence, it would be reasonable to conclude that semantics and cognition have a computational basis and that Searle's Chinese room argument and his replies to critics are based upon a mistaken thesis.

How can Searle respond to the empiricist reply? Searle can't appeal to the problem of other minds, which he claims isn't the issue (1980, pp. 421–422). Searle can't retreat to dualism for he is a materialist — brains cause minds. Searle can't insist on a biological basis for minds, because he thinks a nonbiological machine with the right causal powers would have mental states. Searle's rejoinder, I assume, would be that we have managed to fool ourselves in this empirical investigation. We might assign mental states to an advanced machine, but we would withdraw our attribution of intentionality once we learned it had a formal program. (Searle, 1980, p. 421; *cf*. Stalker, 1978). Searle's view is that once the formal program is discovered the machine would be unmasked for what it is. But that's the issue: what is it? Would we withdraw our attribution of intentionality to other humans or to ourselves if human cognition turns out to be essentially a formal program? There may be many levels of explanation in an intentional system (Dennett, 1971). Lower-level syntactical explanations might be impractical to use but not incompatible with higher-level cognitive accounts.

CAUSAL COMPUTATIONALISM

In what sense might minds be programs? To answer this I need to say more about the nature of a computational theory of the mind. A computational theory of the mind gives laws and principles, stated as algorithms, which account for mental processes. Visual processes, for example, would be described in terms of a complex computational network. A theory of mind is either true or false, depending on whether the computational processes described by the theory actually occur in the brain or computer and are properly connected to other computational systems and the appropriate input and output. Programs in a computational theory of mind are general terms which designate classes of computational systems. These computational systems must carry out the appropriate computation and function in the causally appropriate way within the overall system. For example, an actual computational system for vision must be sensitive to light radiation, analyze the object in the field of vision, and pass the information along so that other computational systems, e.g., the speech system, can generate the appropriate output.

I will call this view "causal computationalism". Mental processes are computational processes. But minds are not programs abstractly conceived, for only those computational structures which are adequate to their causal role in the overall system will be picked out by the computational theory. In other words, if a computational theory of mind is an actual scientific theory, not just a philosophical position, then it must be able to explain real cognitive behavior. To do this the scientific theory must have some empirical content. The computational theory must describe real structures which carry out computations in the right way to causally produce the right output for a given input at the right time and place. The empirical content of the computational theory will be determined in large measure by the requirements for overall behavior of the system. A real cognitive system will have both an internal and an external functionality. The external functionality described by the theory specifies the performance characteristics and thereby puts empirical constraints on potential realizations of the internal computational processes.

For example, human beings exhibit cognitive behavior presumably mediated by causal processes in the nervous system. If a scientific computational theory of mind is correct, then these mediating causal processes must be accounted for in computational terms. If the theory is correct, then it is the computational aspects of the processes that are crucial features so that any other system that carried out the appropriate computations and met the causal constraints of the theory would count as a functional equivalent. For instance, if the theory is correct, then if an area of a human brain is injured, another region of the brain may assume the computational duties. If the theory is correct, then in the future neurosurgeons might replace regions of the brain with electronic devices which carry out the appropriate computations and causally give the right output in the right place and time for any given input. Both the computational and causal features described by the theory are crucial. The electronic replacement must not only make the right computations but accept the appropriate inputs and give the right outputs at the right places at the right time.

On this view minds are computational structures which could occur in biological or nonbiological entities. However, there is no basis for dualism as Searle suggests.

Unless you believe that the mind is separable from the brain both conceptually and empirically — dualism in a strong form — you cannot hope to reproduce the mental by writing and running programs since programs must be independent of brains or any other particular forms of instantiation (1980, p. 424)

The charge of dualism is without foundation. Strong AI, properly understood, is not committed to the view that programs, as algorithms independent of any realization, are minds. The only dualism in strong AI is the harmless distinction between a theory and its subject matter. Computational theories are abstract enough that they may describe different kinds of entities including brains and electronic computers, just as theories in other sciences are abstract enough that they may describe different kinds of entities (Rapaport, 1986).

THE PSEUDOREALIZATION FALLACY

The empirical content of computational theories of the mind should be stressed to block a fallacy which occurs repeatedly in Searle's criticisms of strong AI. The fallacy takes the following form: Suppose minds are programs. Now imagine some fantastic realization of such a program. This fantastic realization can't possibly have or be a mind. Therefore, minds can't be programs. Searle says, "For example, there is a level of description at which my stomach does information processing, and it instantiates any number of computer programs, but I take it we do not want to say that it has any understanding" (1980, p. 420). Of course, strong AI does not contend that just any instantiation of any program is a mind. But, Searle's concern here is that there is no principled way to distinguish cognitive from noncognitive processes if the standard is only a formal program. Searle believes that the mental-nonmental distinction must be intrinsic to the system and not merely in the eye of the beholder. Otherwise, one can treat people as nonmental and hurricanes as mental (1980, p. 420). Nevertheless, he commits the fallacy of appealing to a pseudo-realization.

Picking appropriate realizations is not nearly as arbitrary as Searle suggests. The reason the pseudorealization fallacy is a fallacy is that the imagined fantastic realization is not a realization of the appropriate theory at all. The empirical content of the theory, especially its external functionality, will limit the kind of realizations possible. Let's suppose that processes of the stomach can be thought of as information processing. First, it must be shown that they really are processes which are specified by the internal functionality of the theory. The interpretation cannot be *ad hoc* but must survive testing. It must be shown that given the initial interpretations from the primitives of the computational theory, the stomach really does carry out the appropriate computations. Obviously, constantly reinterpreting the data to fit the theory is not a legitimate testing procedure. One cannot pick out a new section of the stomach in an *ad hoc* manner each time a new computational state is required. I predict that empirically the stomach system will fail to capture the internal functionality given in the theory. Second, the stomach interpretation will surely fail to capture the external functionality of the system. Stomachs may growl a little, but they don't talk or exhibit any real cognitive behavior. Searle's fear about making the cognitive/noncognitive distinction is unfounded, for the theory will give performance standards which make the distinction more than something in the eye of the beholder. It's a brute fact that humans exhibit intelligent behavior and hurricanes don't.

I think it is a mistake to take the external functionality of a cognitive theory for granted. It may be tempting to do this, for cognitive behavior seems so obvious. But then cognitive behavior is easily regarded as a set of abstract inputs and outputs in some logical space and time. From here it is a short step to viewing the computational processes in the theory as nothing but formal algorithms, and the door is left wide open to pseudorealizations. The way to close the door is to insist that the proposed realizations be realizations of the full empirical theory.

What I have been suggesting as requirements for a good computational theory of mind is no different than requirements for any theory of a functional system. A good theory about the heart system should describe how it works internally and how it performs externally. Something will count as an artificial heart only if its meets the minimum performance conditions of the heart system. A device which worked something like a heart, but which pumped only gases or which was not able to receive blood from a pulmonary vein or which pumped only one cubic centimeter of blood per day, would not be a realization of a heart.

Consider another fantastic realization suggested by Searle:

On this view, any physical system whatever that had the right program with the right inputs and outputs would have a mind in exactly the sense that you and I have minds. So, for example, if you made a computer out of old beer cans powered by windmills; if it had the right program, it would have to have a mind. (1984, pp. 28—29).

Elsewhere Searle imagines that millions or even billions of beer cans are rigged up to levers and powered by windmills to simulate neuron firings which occur when we are thirsty. The beer cans bang together as neurons would fire, so that at the end of the sequence of banging beer cans a can pops up on which it is written "I am thirsty" (1982, p. 4). Such fantastic realizations are fun to describe but hard to take seriously. Such fantastic realizations are pseudorealizations; they are not realizations at all. Even if the internal functionality of a computational theory were captured by such a realization, which I think is extremely unlikely, the external functionality which is required by the theory would not be. The beer can/windmill contraption would not have the performance characteristics of a cognitive system. Real cognitive systems accept real inputs and give real outputs in real time and situations. The appearance of a beer can, which has "I am thirsty" written on it, doesn't count as adequate behavior. Empirically, the pseudorealization would be no counterexample to an actual computational theory. Of course, one can throw empiricism to the winds and tell a story about a beer can/windmill system that actually drinks beer, gets drunk, complains about hangovers, etc. But such a story is only a fairy tale. Such a story threatens strong AI no more than a story about spinning straw into gold threatens chemistry.

The clearest case of the pseudorealization fallacy is, of course, the Chinese room example itself. If a human is in the room and realizes the internal functionality of a computational theory of strong AI, then the external functionality will not be properly realized. The system will not speak Chinese in a manner indistinguishable from a native speaker. If the external performance is realized, then the internal functionality will not be realized, for humans cannot function within such constraints. The Chinese room example is truly fantastic, for it cannot be a realization of an adequate computational theory of the mind at all.

I believe that Searle is right about a wrong conception of AI. If the notion of a program is taken in an abstract enough way, e.g. as a machine table of a Turing machine (cf. Putnam 1964), then Searle is right to deny that programs in this sense are sufficient for minds, for all sorts of fantastic realizations will be instantiations of such programs, but won't be minds. But Searle is wrong about a right conception of AI. Scientific theories which fit the requirements of causal computationalism will have empirical conditions that will rule out the fantastic realizations for, on examination, they will be seen to be pseudorealizations. Causal computationalism permits a reasonable course between neural chauvinism and panpsychism. All nonbiological entities we know about lack minds. But one day, far in the future, the appropriately programmed computer may have one.

REFERENCES

Bynum, T. W.: 1985, 'Artificial Intelligence, Biology, and Intentional States', *Metaphilosophy* **16**, 355—377.
Bundy, A. (ed.): 1980, *Artificial Intelligence: An Introductory Course*, Edinburgh University Press, Edinburgh.

Campbell, A. N., Hollister, V. F., Duda, R. O. and Hart, P. E.: 1982, 'Recognition of a Hidden Mineral Deposit by an Artificial Intelligence Program', *Science* **217**, 927—929.

Cuda, T.: 1985, 'Against Neural Chauvinism', *Philosophical Studies* **48**, 111—127.

Dennett, D. C.: 1971, 'Intentional Systems', *The Journal of Philosophy* **68**, 87—102.

Dennett, D. C.: 1980, 'The Milk of Human Intentionality', *The Behavioral and Brain Sciences* **3**, 428—430.

Fodor, J. A.: 1968, *Psychological Explanation: An Introduction to the Philosophy of Psychology*, Random House, New York.

Fodor, J. A.: 1975, *The Language of Thought*, Thomas Y. Crowell Co. New York.

Fodor, J. A.: 1981, *Representations*, MIT Press, Cambridge.

Haugeland, J.: 1985, *Artificial Intelligence*, MIT Press, Cambridge.

Hofstadter, D. R.: 1980, 'Reductionism and Religion', *The Behavioral and Brain Sciences* **3**, 433—434.

Hofstadter, D. R. and Dennett, D. C.: 1981, 'Reflections', in *The Mind's I*, Basic Books, New York, pp. 373—382.

Hungerland, I. C. and Vick, G. R.: 1981, *Thomas Hobbes: Part I of De Corpore*, Abaris Books, Inc., New York.

Moor, J. H.: 1976, 'An Analysis of the Turing Test', *Philosophical Studies* **30**, 249—257.

Moor, J. H.: 1978, 'Three Myths of Computer Science', *British Journal of the Philosophy of Science* **29**, 213—222.

Moor, J. H.: 1979, 'Are There Decisions Computers Should Never Make?' *Nature and System* **1**, 217—229.

Moor, J. H.: 1981, 'AI and Cargo Cult Science', *The Behavioral and Brain Sciences* **4**, 544—545.

Pylyshyn, Z. W.: 1984, *Computation and Cognition*, MIT Press, Cambridge.

Puccetti, R.: 1980, 'The Chess Room: Further Demythologizing of Strong AI', *The Behavioral and Brain Sciences* **3**, 441—442.

Putnam, H.: 1964, 'Minds and Machines', *Minds and Machines*, in A. R. Anderson (ed.), Prentice-Hall, Englewood Cliffs, N.J. pp. 72—97.

Rapaport, W. J.: 1986, 'Philosophy, Artificial Intelligence, and the Chinese-Room Argument', *Abacus* **3**, 7—17.

Raphael, B.: 1976, *The Thinking Computer — Mind Inside Matter*, W. H. Freeman and Company, San Francisco.

Schank, R. C.: 1980, 'Understanding Searle', *The Behavioral and Brain Sciences* **3**, 446—447.

Schank, R. C. and Abelson, R. P.: 1977, *Scripts Plans Goals and Understanding*, Lawrence Erlbaum Associates, Hillsdale, N.J.

Searle, J. R.: 1980, 'Minds, Brains, and Programs', *The Behavioral and Brain Sciences* **3**, 417—424, 450—457.

Searle, J. R. (1981), "Analytic Philosophy and Mental Phenomena", *Midwest Studies in Philosophy*, **6**, 405—423.

Searle, J. R.: (1982), 'The Myth of the Computer', *New York Review of Books* April 29, 3—6.

Searle, J. R.: 1983, *Intentionality*, Cambridge University Press, Cambridge.

Searle, J. R.: 1984, *Minds, Brains and Science*, Harvard University Press, Cambridge.
Stalker, D. F.: 1978, 'Why Machines Can't Think: A Reply to James Moor', *Philosophical Studies* **34**, 317—320.
Turing, A. M.: 1950, 'Computing Machinery and Intelligence', *Mind* **59**, 433—460.

Department of Philosophy
Dartmouth College
Hanover, N.H. 03755, U.S.A.

J. CHRISTOPHER MALONEY

IN PRAISE OF NARROW MINDS:
THE FRAME PROBLEM*

I

If you have a taste for realist doctrines, suppose that the mind is a store of real, efficacious beliefs, desires and propositional attitudes generally. Why should anyone agree that propositional attitudes exist? For much the same reasons that lead us to endorse other scientifically reputable entities. Our behavior is largely explicable by reference to the propositional attitudes we have, variation in behavior devolving from variation in propositional attitudes. This leads to two questions. First, how is it that if behavior is driven by propositional attitudes, it is typically appropriate to the circumstances of its production? And second, if variation in behavior falls to variation in propositional attitudes, what accounts for variation among propositional attitudes?

Both questions are answered by the computational theory of the mind. According to this proposal, propositional attitudes, beliefs in particular, are *representations*. An agent's (true) beliefs symbolize or represent her situation, including her external environment and internal states. As representations, beliefs can refer to, and predicate properties of, objects. When the object to which a belief refers itself exists and has the property ascribed to it by the belief, the belief is true. Should an agent's behavior result from true beliefs, the behavior has a fair chance of being appropriate to the agent's circumstances. This since behavior is caused by belief in a surprising way. Beliefs so cause behavior that the cluster of beliefs implicated in causing a bit of behavior is, as a sequence of representations, a *derivation* of the behavior. That is, the bit of behavior caused by a sequence of beliefs is, in the standard case, the rational consequence of what the beliefs in question collectively represent. Put differently, behavior is typically appropriate to an agent's situation because it is normally both the causal result, and inferential consequence, of true beliefs. Behavior, then, is a matter of inference,

* This essay was written with the support of a National Endowment for the Humanities Fellowship for College Teachers, for which I am grateful.

James H. Fetzer (ed.), Aspects of Artificial Intelligence, 55—80.
© 1988 *by Kluwer Academic Publishers.*

and what better than inference could secure the appropriateness of an agent's behavior to her situation?

So much for the first question. The second question is the one that will occupy us. But it takes its cue from the answer to the first. If behavior is inferentially powered, then variation among propositional attitudes is nothing more than the sort of propagation displayed within inferences. Beliefs arise in the manner of theorems; they are the inferential consequences of prior beliefs. A sequence of beliefs may, as a physical configuration, cause a new belief to occur. And what insures that the internal effect of a sequence of beliefs is itself a belief is that the elements in its collective cause constitute a proof. The effect of such a sequence, therefore, represents whatever the proof suffices to establish. Of course, proofs must have first lines. And here computationalism may adopt variants of either nativism or empiricism. On the one alternative, agents simply come originally equipped with beliefs sufficient to generate others. On the other alternative, agents' sensory systems somehow inscribe primordial beliefs from which all others are derived.

Let's review what we have here. Computationalism considers the mind to be an inferential system. That clearly presupposes the existence of structures over which the inferences are defined. Certainly, the mind is thoroughly physical, and, hence, the structures that carry the mind's computational load must themselves be physical. Now the only things that seem physically fit to function as elements in material inferences are sentences (Fodor, 1975.)[1] Thus, we had better take it to heart that computationalism is wed to the idea that propositional attitudes are attitudes towards sentences, sentences actually encoded in — well — a cryptic mental language within the brain.

It is easy to see why computationalism goes hand in glove with Artificial Intelligence. For according to computationalism the mind is exactly like a programmable computer. Both are systems whose internal activities are traced in terms of computationally characterized transformations. Both, then, essentially rely upon alterations of their theoretically relevant states occurring in accordance with certain principles insuring that their outputs can be viewed as rational relative to assignments of content to their internal states. Indeed, it is primarily because computationalism is plausible as a theory of mental processes that Artificial Intelligence holds the interest it does as branch, if not the trunk, of cognitive science. Since Artificial Intelligence is an empirical science with the task of producing programs that, minimally, simulate genuinely cognitive processes, it is only to be expected that its practice will unearth difficulties in implementation. Importantly, these problems, if especially recalcitrant, may ring of theoretically fundamental failings within the computational doctrine itself.

This is precisely the nature of the frame problem (McCarthy and Hayes, 1969; Raphael, 1971; Boden, 1977; Dennett, 1978c; Fodor, 1983, 1985; Pylyshyn, 1984; Haugeland, 1985). Any program attempting to simulate, if not emulate, genuine cognition will necessarily face the task of accounting for belief propagation, including belief revision as a special case. An agent is, among other things, a store of beliefs. And if these beliefs are appropriately to direct her behavior, they normally must, as we have seen, accurately reflect or represent the situations in which she finds herself. Naturally, as her situation changes, so too must at least some of her beliefs. Otherwise, the agent, besides being unable to buy life insurance, would not manifest the plasticity characteristic of intelligent action. So, both artificially and naturally intelligent agents must constitute solutions to the frame problem. Importantly, if successive attempts to solve the frame problem within Artificial Intelligence go awry, then computationalism itself, as a theory of naturally intelligent minds must certainly become suspect. For if we cannot see our way through the frame problem in the case of artificially intelligent devices, we must speculate that nature has anticipated the problem and not selected the computational model in realizing natural intelligence.

We can quickly appreciate the difficulty of the frame problem within the confines of computationalism by looking at an apparently simple case of belief revision. Eloise glances out her window and sees Abelard standing in the courtyard, wearing a hat, leaning against the chestnut tree and speaking with her father. She registers this information and returns her attention to the poem she is composing. Soon one of her brothers enters and tells her that Abelard has departed. Now Eloise has to reassess her beliefs. Minimally, she must delete her belief that Abelard is in the courtyard. But what of his hat? She likely will want to erase the belief she evidently has to the effect that Abelard's hat is in the courtyard. But not everything has changed. The chestnut tree remains rooted in place, and Eloise had better continue believing it so.

And now that Eloise's father is no longer speaking with Abelard in the courtyard, he too might have left the scene. So, Eloise will reduce her estimate of her father's presence in the courtyard.

All this is natural enough. The problem is artificially to simulate it. For what we want is a program that will judiciously segregate those of Eloise's beliefs that require revision from those that do not. This, given computationalism, requires that those sentences that encode Eloise's beliefs be erased or retained depending upon whether the beliefs they encode require revision. What computational principles dictate the solution to Eloise's epistemic problem? We might suppose that she deletes her belief that Abelard is in the courtyard because that contradicts the information supplied by her brother. But from this we cannot conclude that Eloise's cognitive processes are governed by a rule to delete any mental sentence that contradicts 'Abelard has departed'. The reason for this is simple. Eloise has lots and lots of beliefs besides those we have attributed to her in describing her situation. Surely, she believes that Paris is in France, that London is in England, that Rome is in Italy and — well, you get the idea. Thus, if she were to alter her beliefs in accordance with the principle of noncontradiction, she would be at the task a very long time. For, with respect to each of her beliefs, she would need consider the mental sentence encoding it in order to assess its consistency with 'Abelard has departed'. Indeed, unless she luckily happens immediately to hit upon the inconsistency of 'Abelard is in the courtyard' and 'Abelard has departed', it could take her a very long time to drop her belief that Abelard is in the courtyard. But, of course, the fact of the matter is that Eloise is no dolt and instantly realizes that Abelard is now not in the courtyard. Thus, the problem becomes more difficult. What we see right away is that the frame problem calls for a solution that explains how an agent manages efficiently to isolate for consideration just those information bearing states that bear on the subject under cognitive consideration.

Indeed, the spectre of combinatorial explosion threatens. For notice that 'Abelard is in the courtyard' is not a straightforward syntactic contradiction of 'Abelard has departed'. The inconsistency falls to the fact that 'Abelard has departed' entails 'Abelard is not in the courtyard', which in turn directly contradicts 'Abelard is the courtyard'. But if belief revision requires that all, or even very many, of Eloise's beliefs be revised by way of gauging the consistency of their consequences with the consequences of her new information, the demands of the search for inconsistency will quickly outstrip whatever cognitive resources Eloise may have.

The complications continue. Remember the hat. Being intelligent, Eloise will realize that since Abelard is no longer in the courtyard, (probably) neither is his hat. But this is not a matter of "pure logic." Evidently, what Eloise antecedently knows about the practice of wearing hats contributes to her coming to believe that Abelard's hat is no longer in the courtyard. Thus, it appears as if any adequate computational simulation of natural belief change will necessarily include an itemization of what the simulated agent can be expected to know or otherwise believe. When the variety of topics about which a natural agent exhibits understanding is appreciated, the task of simulating the knowledge necessary to move belief revision along seems stunningly complicated. For it looks as if the process of belief revision must advert to an internally encoded encyclopedia, an encyclopedia with exhaustive cross references.

Of course, even this is not enough. An inert encyclopedia will not be of much use unless it comes with instructions for its use. An artificially intelligent clone of Eloise will need to rely on a reliable procedure that will enable her to refer to only the relevant entries in her internal encyclopedia. Otherwise, for each occasion of belief change she would stupidly need to read her entire encyclopedia, looking for information determining which beliefs she should add, revise and delete.

One last hurdle. Naturally intelligent Eloise can add to her encyclopedia; she can make discoveries and formulate and test hypotheses. She can abductively transcend her evidence, settling upon what she takes to be the best hypothesis to account for what awaits explanation. Some people call this common sense. Sherlock Holmes called it the science of deduction. Philosophers of science get promoted by showing that, regardless of what we call it, we do not know nearly enough about theory formation and confirmation to expect it soon to be mimicked by a formal program (Fodor 1983, 1985).

All this is sobering. For it indicates that belief revision, if it is to be compatible with computationalism, depends on some very clever heuristics. Happily, there are some very clever people engaged in simulating, if not creating, cognitive systems of belief. Perhaps one of the more promising lines of research starts with Winograd (1971), leads through Minsky (1975) and continues on to Schank and Abelson

(1977) and Schank (1982).[2] Their idea is, in essence, to compartmentalize the mind according to topics or subjects of belief so as to limit the beliefs up for reassessment when revision is in order. The idea, if not its implementation, *seems* simple and plausible enough, yet it is not without its detractors (Fodor, 1983; 1985; Haugeland, 1985). In what follows, I will discuss some, certainly not all, of the difficulties in the approach and argue that these objections are not as compelling as they may seem.

II

The picture of the mind that I want to paint here depicts the mind as a series of integrated modules (Chomsky, 1980; Fodor, 1983; Anderson, 1983). Modules are, somehow, encapsulated. That is simply to say that the data structures available to a module are limited in size and are relatively immune to the influence of information available to the system containing the module. And modules may not tell their containing system all that they themselves know. Suppose that a module contains 'Abelard is in the courtyard'; additionally assume that the cognitive system housing that module, but not the module itself, has access to 'If Abelard is in the courtyard, then Abelard is near the chestnut tree'. Given just this, 'Abelard is near the chestnut tree' need not be generated and encoded within the module, even if the module abides by *modus ponens*. Moreover, the system itself might fail to encode 'Abelard is near the chestnut tree' if 'Abelard is in the garden' is not part of the output of its contained module.

There is some striking, though not novel, evidence that we ourselves, as cognitive systems, are modular to some degree. Fodor (1983) argues quite persuasively that the familiar facts of perceptual illusion confirm the hypothesis that our perceptual processors are modular. We all know that what we see on the road ahead is not a puddle, but it persists in so appearing. In order to explain this and similar phenomena it is natural to conjecture that, under certain conditions, our visual system is *bound* to supply to cognitive central the representation to the effect that there is a puddle ahead. Cognitive central, with its impressive store of information, evaluates the information from the visual system and concludes that there is no puddle in the distance. If the etiology of a familiar illusion is something like this, then the visual system, in promoting its hypothesis regarding the presence of a puddle, could not be privy to what cognitive central knows. And that amounts to saying that the visual system is an encapsulated module. Much the same might be said, and Fodor says it, on behalf of the encapsulation of our speech decoders.

Yes, encapsulated systems are subject to characteristic errors, though that is definitely not to say that they characteristically err. But while they may persist in mistakes of certain sorts, they certainly are fast, and that can be a great advantage when the price of error is not abnormally high and quick decisions are at a premium. Presumably, the speed of a module is at least partially explained by the module's encapsulation. The less information a module must address in processing, the quicker the module can complete its computing. Established, then, is this: we do instantiate some encapsulated cognitive modules, the output of which is available for processing elsewhere within our cognitive systems. Very probably evolution has something to do with the modularity of perceptual and linguistic processors. But that is not to say that these modules are thoroughly independent of contributions from cognitive central or, for that matter, other modules. Presumably, we *learn* what puddles are. I do not know, and neither does anyone else, how we manage this, but this is significant for the encapsulation of perception. For it indicates that at least some of the concepts deployed in perceptual categorization are imported into the perceptual module. In other words, if a perceptual processor is, as computationalism supposes, an inference device, some of the concepts featured in the premises it applies, and perhaps even some of the premises themselves, are contributed by either some other module or cognitive central. I do not want to insist that this *must* be so for perceptual processors. It will do that some of the data utilized by an encapsulated subsystem can, on rare occasions, be contributed by another system. Of course, the encapsulation of the subsystem will not tolerate frequent or wholesale importation of externally provided information. But the encapsulation of a system is consistent with at least some of the system's aboriginal information being provided by the grace of another system. Principle: It's okay to give an encapsulated system a few axioms with which to start, but after that, don't interfere much, or you'll ruin its encapsulation and eventually make it slow and circumspect.

Our visual system is topic sensitive. It can detect and classify red balls and shimmering puddles, but it cannot discriminate the cube root of 793^{23965}, Toad's not having been considerate of Ratty, or what it's like to be a bat (Nagel, 1974).[3] The point is that, because of the poverty of its own informational store, the visual system is not prepared to react to and classify everything that might be present. This means that the visual system may be pretty stupid when it comes, say, to mathematics. But that does not compromise its ability intelligently to do what it is meant, by nature, to do. What wants emphasis is that encapsulated modules are topical and that, by being so, they are blind to certain facts that might be obvious to other cognitive systems. Of course, one might wonder whether for every topic that cognitive central might concoct there is an encapsulated module to attend to it. But, and this will be of some use later, there is certainly no reason to suppose that cognitive central cannot be served by lots of different topical modules.

Modules are, to trade on understatement, a bit easier to simulate than cognitive central. While very little is known about the actual cognitive processes involved in belief fixation,[4] some progress has been made on the computational nature of the visual system (Marr, 1982). And we do now know something about constructing well behaved parsers (Marcus, 1977). We have, then, a beginning of an explanation as to how, for example, the visual system manages to classify distal stimuli and how it updates its analysis of the changing visual scene. So, we have a hint as to how the visual system formulates and tests hypotheses about the ambient array. That, written very, very small, is a start on an account of the frame problem, relative to the visual module. If Caesar was so successful with his method, perhaps we might emulate him and try to solve the frame problem by dividing in order to conquer. While the frame problem may resist solution at the level of cognitive central, it may more readily yield in the case of encapsulated cognitive modules. This is because the inherent limitations and rigidity of modules insures that revision is computationally less expensive. Modules are satisfied with epistemically imperfect principles of belief modification, and will tolerate certain local epistemic failures so long as the function of the module itself is not compromised. Maybe, then, Eloise is as gifted as she is in revising her beliefs about Abelard because she possesses some interacting modules, modules perhaps respectively dedicated to Abelard, hats, courtyards and, well, yes, the list does get embarrassingly long. But that, as Fodor (1985) has remarked in a different context, may well be "the way that Nature likes to operate: 'I'll have some of each' — one damned thing piled on top of another, and nothing in moderation, ever."

The proposal, then, is that an agent is as good as she is at belief fixation and revision because she relies on highly topical conceptual modules. A conceptual module is pretty much what Schank and Abelson call a script. It marshalls a fair amount of information about some topic, perhaps about Abelard, in order to function as a question answering device. Just as the visual module has available to it encapsulated information that it relies upon to process information, so too for the conjectured Abelardean module. It contains a stock of information about Abelard, apparently including, for example, that he usually keeps his hat on his head and always keeps his head on his shoulders. Supposing that the Abelardean module is sufficiently information poor and inferentially hobbled so as to be able quickly to answer the questions it can, it will, subsequent to Eloise's being told by her brother that Abelard has departed the courtyard, speedily inform Eloise that Abelard's hat has gone with him.

Nature has done some hardwiring to insure that Eloise has a visual module, but nature could hardly anticipate that Eloise will need an Abelardean module. So, how is it that Eloise comes to have an Abelardean module, whereas Margaret Thatcher does not? The answer is both easy and hard. The answer is easy if we notice that Eloise's Abelardean module is pretty much what we used to call her *concept* of Abelard, concepts being organized bits of information on a subject. Thus, to ask why Eloise, but not Margaret Thatcher, has an Abelardean module is much like asking how Eloise learned the concept of Abelard though Margaret Thatcher did not. And that makes the question hard. Hard, that is, because its answer trades on a satisfactory theory of learning. And this, as earlier remarked, is a theory on which no one has a lock. But what we can venture here is that the explanation of learning is not in any evident way predestined to be incompatible with computationalism. And, so, we can provisionally suppose that the contingencies of her situation serve to modify Eloise's cognitive superstructure so as to carve out a module for Abelard. We do not know much about how this happens, but we do know that it does not happen to Margaret Thatcher because the contingencies under which she operates differ from those that shape Eloise.

Now if Eloise's Abelardean module is incorporated in her cognitive system as a result of learning, nature not having done any special wiring to insure its realization, we can appreciate two features of this module. First, the data store that it encapsulates is largely a matter of the selection that learning subserves. So, if, as seems natural, learning somehow selects for importance, Eloise's Abelardean module will encapsulate information central to Eloise's notion of Abelard. Indeed, it approaches an obvious, perhaps trivial, truth that concepts, modules, will generally encapsulate what is cognitively important if they result from the needs that fuel learning. Second, since Eloise's Abelardean module arises from learning, it will be relatively stable, though certainly not set in cement. Over time, learning can modify the store of information encapsulated within the module.[5] The stability of the module enables it to function throughout long periods in shaping and revising beliefs about Abelard. And the malleability of an acquired module enables it to be a bit smarter than we expect a hardwired module to be. For variations in an encapsulated module's information store may result in a reduction in the number or kinds of errors to which it is subject. Additionally, that learned modules are subject to the contingencies peculiar to the agents that house them should lead us to anticipate that, across agents, modules developed to handle information on the same topic will differ among themselves more radically than do hardwired, species specific modules such as visual processors.

This is not without consequence. For if, across naturally intelligent agents, we allow for wide variation in the reach of the same modules, we should expect that simulations of these modules are not as smart as those they emulate. This for the mundane reason that artificial cognitive modules do not enjoy the continuous and inexorable refining to which learning subjects natural cognitive processors. To see the point here, compare Eloise's Abelardean model with that edition of the same module as it occurs within Abelard's teacher, William of Champeaux. Eloise and William will have developed their Abelardean modules in reaction to different confrontations with Abelard. So, while Eloise and William may agree in many ways about Abelard, one of the axioms featured in Eloise's Abelardean module might be that Abelard is lovable, loyal and lucid, whereas William's module may replace it with one characterizing Abelard as vexing, vain and vulgar. Given that their respective Abelardean modules differ in detail, Eloise and William might well differently respond to the same novel information about Abelard. Told that Abelard is on his way to the University, Eloise might come to believe that he is pursuing his desire for knowledge. Informed of the same, William might suspect that Abelard is bent on another vicious argument on behalf of nominalism. And this, of course, is exactly what we expect in the world of cognitive agents, though, of course, we also expect some agreement. But once we focus on the variation that natural cognitive modules of the same topic tolerate, we cannot fail to realize that agents in whom a module is young and evolving will often be puzzled when directly questioned about the module's topic. Eloise's little sister, asked whether Abelard is trustworthy as a tutor, may plead ignorance, whereas Eloise and William would answer definitely, even if differently. Thus, if it turns out, as it does, that artificially composed scripts are frequently unable to answer questions posed about topics they are presumed knowledgeable, questions that any naturally intelligent, informed adult could answer, that alone does not demonstrate that representational modules could not possibly stand as answers to the frame problem. For failures of an artificial module, like that of Eloise's sister, might just as well be ascribed to ignorance, as opposed to any absolute lack of intelligence.

Modules are topical, but that does not mean that they must be woefully narrow-minded. Eloise's Abelardean module encapsulates information useful for recognizing and characterizing Abelard. Therefore, this module, in order to be about Abelard, will necessarily acknowledge some part of the network of relations in which Abelard operates. We can expect it to indicate that Abelard is a philosopher and a nominalist to boot. Naturally, since the module refers to philosophers and nominalists, it will very probably allow for consequences to follow from Abelard's being a philosophical nominalist. This it might accomplish either by simply containing some principles about philosophy and nominalism or giving directions to other modules, presumably the modules for philosophy and nominalism, along with instructions as to where to channel the output of those modules. The visual module is known to function in just this way. It encapsulates information about a variety of subjects — colors, lines, angles, shapes and the like — and signals other cognitive centers when additional information is required. That Eloise's Abelardean module should exploit similar interconnections is, then, no stain on its modularity.

But while Eloise's Abelardean module might refer to philosophy and nominalism, it need not mention everything Eloise knows or even everything Eloise knows about Abelard. Encapsulation insures that every module is, to an important degree, narrow-minded. And although topicality provides for modules being about certain subjects, it does not mandate that every bit of information about a subject that an agent may be able to exploit is encapsulated within the module, if any, for that subject. For example, Eloise may know that the wart on the end of her father's nose is not Abelard's maternal grandmother. This knowledge is, in some sense, *about* Abelard. But that it is does not necessitate that it be encapsulated in Eloise's Abelardean module. Of course, discerning the laws that dictate what information is encapsulated in a module is crucial to any cognitive theory of modules. But if modules are shaped by the pressures of learning in a relatively hostile environment, there is every reason to suspect that modules, though topical, will feature useful, even if not all available, information on topics of consequence.

The sketch of the mind we have been tracing might be viewed as presenting an agent's cognitive system as a disjointed conglomeration. But that would be most mistaken. Modules, like good old-fashioned concepts, are coordinated into schemes in naturally intelligent agents. The outputs of some qualify as the inputs to others, and nothing in principle precludes symmetrical arrangements. Once we recognize that modules are interacting elements within conceptual schemes, we must be wary of resurrecting the frame problem. For certainly, under pain of combinatorial explosion, we cannot allow for the many modules that are evidently required within the schemes sufficient for the demands of normally intelligent people while at the same time blithely supposing that they interact without constraint. Otherwise, the computational demands on the system as a whole will be impossible to satisfy. So, we must take a cue from the activity of the modules about which we have some understanding. The visual module, in identifying something as, say, a red ball, might pass this information on to selected modules, perhaps the red and ball modules (if such there be). But this output of the visual module cannot *in general* be required to visit the module for differential equations. True, this may occasionally cost the agent useful information, but that is a simple artifact of the kind of quick cognitive processing that wants explanation. And yes, we sorely need to know exactly what principles dictate on which modules the output of a selected module calls. But these principles, like the true principles of molecular biology, are susceptible to discovery only by arduous empirical investigation. In all likelihood, selected modules are sensitive to the same properties of bits of ambient information. Eloise's modules for Abelard and philosophers may both be activated by an utterance of 'Abelard is a philosopher', and that they are so activated on an occasion may determine that they interact on that, though not necessarily all, occasions. But, again, whether this is so is a thoroughly empirical matter. What demands emphasis here is that while modules must be designed so as to interact, they cannot interact with all other modules all the time if they are ever to get anything done.

We ourselves rely upon perceptual processing for no small portion of the information we exploit. This is, most likely, fundamentally important to the manner in which we solve the frame problem. Think, for the moment of the visual system acting as an oracle. It surveys a scene and proceeds to inscribe on its mental slate a series of sentences descriptive of the scene. This list is then sent to memory. Later it revisits the same scene, again compiles a list of descriptive sentences and sends this list to memory. Suppose that the first and second lists are subsequently compared, with the effect that any sentence appearing on the first but not the second list is deleted from memory. We would thereby have effected a crude solution to the frame problem as it applies to modifying those visual beliefs immediately tied to the assayed scene. This, of course, is an overly simplistic account of the functioning of the visual system. Still, there are two ideas here meriting attention. First, the perceptual systems may be able to contribute to the revision of perceptually encoded beliefs by way of a process as relatively simple as list comparison. And second, if this is how at least some of our beliefs undergo revision, then belief revision may, in part, be independent of knowledge of general laws governing the interactions of objects and events. These two points, together with the supposition that perception is the conduit of a great deal of the information that is subject to reassessment, suggest that perceptually active creatures may have, consistent with computationalism, a relatively simple way of revising a great deal of the information they possess. Now, this is certainly not to say that the frame problem is thereby solved in the general case. For much of what we believe is not a direct function of what perceptual oracles might inscribe. Nevertheless, if the perceptual systems do participate in belief modulation in something like the way here indicated, then we can be excused for anticipating that the frame problem poses less of a threat to computationally powered and perceptually equipped agents then might at first appear.

It may also pay some dividends to note that, apart from Oxford dons, cognitive agents are under no necessity to revise their beliefs in a manner that respects thoroughgoing consistency. Eloise originally believes Abelard to be in the courtyard and wearing a hat. Later she is

told that he has departed. Presumably, she manages to revise her belief
that Abelard is in the courtyard and, in actual fact, comes to believe that
he is not now in the courtyard. But what of his hat? Must she, as we
have previously assumed, also reassess that belief? She need not if, as it
happens, during the interval determined by the occasion of its encoding
and the receipt of the information about Abelard's departure that belief
has been lost to memory. That is, if belief revision is a matter of
reevaluating stored beliefs against novel information, the (putative) fact
that information literally decays insures that belief revision may not be
as computationally expensive as it might otherwise be. And while this
plainly does not trivialize the frame problem, it should make us
appreciate that the problem's solution in no case requires anything like
the total reevaluation of any and all the information an agent may have
ever acquired. In other words, forgetfulness mitigates the difficulty of
the frame problem, and if we only had a decent explanation of why we
just plain forget some of what we learn, we might more quickly see our
way through the frame problem.

This leads to another observation that may soften the severity of the
frame problem. Some beliefs are more central to an agent's behavior
than are others. Eloise knows that her brother is not the wife of
Aristotle's mother, but the belief, if any, encoding in her this information is, in some sense, inferentially inert. As a matter of presumed fact,
it does not enter into the mental calculations that determine any of her
behavior. An agent's inert beliefs normally need not, then, be updated
in order to enable the agent to behave intelligently. Certainly, failing to
update an inert belief *might* have debilitating consequences for an
agent. But the low probability of penalty for such a failure and the
rewards in the corresponding savings of computational time and energy,
may encourage the policy of ignoring inert beliefs when tidying the
belief store.

While this may be true, it is of little help in solving the frame
problem unless inert beliefs are somehow indicated as such. The sheer
content of a belief is surely no indication of its importance in shaping
an agent's behavior. For how, simply by attending to the content of a
belief, are an agent's computational processes to appreciate that the
belief is behaviorally inert? Nevertheless, this is not to say that a
computational system could not stumble upon a relatively reliable way
of discriminating inert from active beliefs. For suppose that an agent's
computational system were, like a gunslinger, to notch a belief each

time that belief occurred as part of the cause of a bit of behavior. If beliefs were also dated so as to recond the time of their fixation, the number of notches a belief exhibited would, when compared to the date of its fixation, serve as a simple, even if simplistic, measure of its utility to an agent. Aging beliefs with no notches would be *prima facie* inert, and an agent's computational system might well be excused from considering them when occupied with belief revision. The converse may hold as well. Beliefs loaded with notches are frequently called upon to direct behavior. So, they probably represent relatively constant features of the environment and, thus, do not need regularly to be updated. They too, then, might be ignored when belief modification is underway. Of course, ignoring inert and highly active beliefs when surveying candidates for revision can occasionally induce costly errors. But such may be a fixed cost of the cognitive enterprise.

That cognitive systems engaged in doxastic modulation actually do ignore inert and highly active beliefs is suggested by at least anecdotal evidence. You normally wear your waterproof, shock resistant watch, even when you wear nothing else. But, as the present unusual occasion happens to dictate, you remove you watch, perhaps to have it repaired. Later, you find yourself staring at your bare wrist, trying to learn the time. Evidently, you have failed to modify your belief to the effect that your watch is on your wrist even though you quite consciously and intentionally took it off only a short while ago. Here we have a classic failure of the processes involved in belief modification, and one hypothesis for explaining it is that the implicated processes rely on ignoring those beliefs that are frequently deployed in launching behavior, beliefs with many notches. But be all this as it may, if an agent's computational system is equipped with a way of registering inert and constantly called beliefs, it may be positioned to ignore them, for better or worse, when revising beliefs. If so, this may greatly reduce the difficulty a natural computational system has in solving the frame problem.

Let's pause to survey what we have so far. We are hypothesizing that cognitive agents are flush with encapsulated modules. Some, probably few, are hardwired. Most result from that miasma called learning. We recognize that possibly — but certainly not necessarily — a yet to be articulated and confirmed theory of learning may falsify the many module hypothesis. Different modules are, for the most part, differentially activated by different bits of incoming information, though, depending on the nature of the received information, the same bit of information may trigger several selected modules for analysis and subsequent interaction. Although modules are topical, they need not feature every bit of information on a topic available to an agent. The encapsulation of modules insures that, once they begin processing some information, they will not be annoyed by other modules or cognitive central. And the same allows modules to function without fear of the frame problem. For encapsulated modules need address only limited banks of information while processing any bit of received information. Thus they avoid bursting their encompassing system's combinatorial seams. Still, modules, at least those entrenched through learning, are not totally impenetrable (Pylyshyn, 1984, pp. 220—221). Learned, they can be shaped by further learning. Hence, across different agents at any given time and through time within any given agent, a module dedicated to specific topic can wax and wane, thereby exhibiting variation in its contribution to belief fixation and revision. Modules are willing to sacrifice some intelligence for speed, but those modules established through learning can, within limits, return some speed for a little intelligence. In an important sense, a system of interacting modules is more intelligent than its modules. For the output of one module may induce others to act, the net result being information that no single module could have generated on its own. There is no reason to suppose that all modules are equally gifted. Some may contain more information than others, whereas some may be faster or more frequently activated than their associates. In the teeth of the frame problem, we take some comfort in the idea that perceptual processors may make belief modulation somewhat easier than might at first appear. And that some beliefs naturally atrophy while others are either behaviorally inert or extraordinarily active may lighten the computational burden of any natural cognitive agent bent on belief revision.

III

Now that modularity has been unbridled, we had better turn to some of the more pressing philosophical objections, mindful that undiscussed empirically based difficulties might well undermine the modular enterprise.

Haugeland (1985, pp. 199 f.) lays down three poignant complaints against the idea that modules might play a fundamental role in extricating computationalism from the frame problem.[6] Modules, in essence, function as topically organized files of information. If they are to be of any use whatsoever, they must be cued on just the proper occasions. When Eloise needs to modulate her beliefs as they reflect on Abelard, she will require a way of determining that, of the many modules on which she can draw, it is her Abelardean module which awaits selection. It is Haugeland's first worry that there may be no generally reliable way for addressing modules in any way that simulates or emulates genuine intelligence.

How, then, might Eloise manage to call upon her Abelardean module? No doubt, her module should be brought on line when she receives information pertaining to Abelard. So, how might incoming information incite a module? Considering the visual module is instructive. Information about the colors of ambient objects presumably activates this module. Of course, not all color information will activate the visual module; only such information delivered up from certain sensory receptors will suffice. That is to say, the information must be so encoded as to *cause* the visual module to come into play. An utterance of 'red' will not activate the visual module, though a presentation of a red ball might. Both the utterance and the ball convey information about color, but only the ball impinges upon the ocular system so as to effect the activation of the visual module.

It is not difficult to see how this type of connection could stimulate Eloise's Abelardean module. It *might* be so designed as to be aroused by any utterance or inscription of 'Abelard'. Certainly, Eloise's Abelardean module will need be a bit more selectively sensitive than this if it is intelligently to contribute to her cognitive economy, but the idea is clear enough. Modules must be processors guaranteed to react to information encoded in certain ways. This hypothesis fits well with at least some familiar patterns of error exhibited by cognitive agents. Suppose that, besides the man Eloise knows and loves, there is another logician named 'Abelard', whom we will call, for the sake of clarity, The Pretender. The Pretender lives in Rome and has achieved some fame for his understanding of logic and his staunch nominalism. Assume, communications being what they were in the twelfth century, that Eloise, like everyone else in Paris, has no knowledge whatsoever of The Pretender. One day, however, word arrives from Rome about this logician. While on a walk that day, Eloise, still ignorant of the Pretender, happens to overhear a messenger from Rome say in conversation, "Abelard is in Rome." Of course, Eloise will mistakenly suppose that this information pertains to her lover and will quickly modify some of her beliefs about him. All this is to be expected if her Abelardean module is cued by an utterance of 'Abelard'.[7] Just as the visual system can be activated by illusory color stimuli, so too can Eloise's Abelardean module be roused by misleading information. This, in both cases, if the modules are sensitive to certain properties of stimuli that normally, though not invariably, coincide with the delivery of information pertaining to designated topics.

Conversely, Eloise occasionally will fail properly to react to information that actually is about Abelard if that information is encoded in a way to which her Abelardean module is blind. Suppose that Abelard happens to be the man praying in the chapel, that Eloise does not know this and that her father tells her, "The man praying in the chapel is unworthy of you." Eloise might believe her father, even though this flies in the face of everything she believes about Abelard. This too is explicable if her Abelardean module is sensitive to only certain coding properties of information, that is, if it is activated by 'Abelard' but not by 'the man praying in the chapel'.

The reply, then, to Haugeland's first objection is that modules are selected according to specified causal properties of information-bearing states. This entails that modules will sometimes be notified when they should not be and will sometimes fail to be notified when they should be. But that is quite consistent with the ways of natural, fallible intelligence, and we should expect the same in artificial simulations of natural cognitive modules.

Haugeland's (1985, p. 202) second reservation regarding the utility of modules is that there are not enough naturally available topics awaiting modularization to suffice to explain natural intelligence. The example Haugeland provides is neat. I tell you, "I left my raincoat in the bathtub, because it was still wet." You automatically realize that what was still wet, what 'it' represents, is my raincoat and not the bathtub. But in order to explain this, a computationalist committed to modularity would need to rely upon a module, or modules, that would encapsulate information about raincoats and bathtubs to the effect that wet raincoats might reasonably be left in bathtubs, whereas dry raincoats are not normally placed in wet bathtubs. Haugeland's claim is that there just is not any natural, efficient way of encapsulating this information. Put differently, it is easy to understand the quoted sentence about raincoats and bathtubs. But how could the mind ever have been so foresighted as to realize how handy it would be to encapsulate information about raincoats and bathtubs?

Of course, Haugeland is right. It is unreasonable to suppose that for every problem we solve there is some antecedently prepared module that does the job for us. But no one seriously supposes that this is so. What modularists do suspect is that the mind is so compartmentalized that its various encapsulated departments *cooperate* in solving problems. We do not need a module for *raincoats-in-bathtubs* if we have one each for raincoats and bathtubs that contribute to answering the question as to whether the raincoat or the bathtub was wet. And actual modules for raincoats and bathtubs are unnecessary if their functions are subserved by more general modules, so long as generality is not purchased at the price of encapsulation.[8]

But besides this, it is not at all evident that we do not have exactly the modules that Haugeland fears we lack. The example in the objection involves the problem of pronomial reference. There is fair reason to suppose that our linguistic system is an encapsulated module adept at determining the possible range of antecedents for pronouns, depending upon such factors as the nesting of clauses. In other words, we know in a flash that the only possible antecedents of 'it' in 'I left my raincoat in the bathtub because it was wet' are 'raincoat' and 'bathtub', this information apparently being the output of a linguistic analyzer. But — and this is what is important — choosing between 'raincoat' and 'bathtub' evidently presupposes knowing quite a bit about both raincoats and bathtubs. As a matter of plain fact, most of us do know all about these things. Aside from those, like very young children, who have not had the requisite learning opportunities, almost all of us have concepts of raincoats and bathtubs. And concepts, recall, are prime candidates for encapsulated modules. Anyone familiar with raincoats knows that they get wet and can be hung to dry, and survival in Seattle requires a near innate supposition that the bathtub is *the* proper place to hang wet things to dry. Why cannot this information, if encapsulated in cooperating modules, resolve the posed problem of pronomial reference with all the dispatch we expect of naturally fluent speakers of English?

The third objection Haugeland levies against cognitive modules is that their encapsulation effectively prevents them from displaying the intelligence minimally necessary for the kind of belief modulation they are to accommodate. Again, Haugeland's example isolates the issue.

Present an intelligent agent with these two "stories" in terms of which questions are to be answered:

(1)   When Daddy drove up, the boys quit playing cops and robbers. They put away their guns and ran out back to the car.

(2)   When the cops drove up, the boys quit their robbery attempt. They put away their guns and ran out back to the car.

Understanding these stories and administering one's beliefs accordingly is not, Haugeland would have it, a process that could depend on encapsulated modules. For such, he fears, could never account for the fact that anyone who understands (1) will thereby realize that it tells a charming tale of happy children glad to see their father, whereas (2) reports on an ominous scene threatening the happiness of all involved. Haugeland's complaint is that the sort of local constraints imposed on these stories by the modules activated to understand them could not possibly explain the appreciation we have for the tone of these tales.

I, for one, confess not to see why modularity could not take center stage in the intelligent comprehension of these little stories. Presumably, somewhere in the right account of the understanding of (1) will be reference to what is known about children playing. And appreciating the thrust of (2) will certainly revolve around knowing something about robbery and cops. A module devoted to child-play could easily specify that such play is a happy pastime, and modules respectively dedicated to robbery and cops had better say something about the dangers of each. So, it remains quite unclear as to why, finally, Haugeland thinks that modules suffer the sort of tunnel vision that would render them blind to the obvious implications of the stories.

It is worth pointing out that the tone a story is found to have is apparently a partial function of what the interpreting agent happens to believe. If you are convinced that playing cops and robbers with toy guns is bound to corrupt children, ruin their lives and ultimately threaten your own, most likely you will not attribute to (1) the tone that others not blessed with your background of beliefs discern. Yet, surely, you understand the story as well as anyone else, even if, in some extended sense tutored by your peculiar beliefs, you understand it differently. The point is that it is not evident that there is a uniquely correct way to interpret and react to an arbitrarily selected story. If this is so, we might surmise that the intelligent comprehension of stories can exhibit various forms. And given this, it is not so plain that the type of comprehension supplied by an artificially constructed script or module could not qualify as intelligent. For if comprehension of stories is somehow tied to the set of beliefs an agent brings to the interpretative task, it remains to be argued that the beliefs brought to bear by a cluster of cooperating modules must fall short of the goal of plausible, even if attenuated, interpretation.

IV

Fodor (1983, 1985), though a champion of modularity, assumes that modularity is too isolated a cognitive phenomenon to contribute to solving the frame problem. While he is happy to recognize modules for vision and speech, he is loath to suppose that there could be a module given over to worry about Abelard. According to Fodor, an agent's private, internal system of belief, like the public, external corpus of science, is, as he says, isotropic and Quinean. A system of belief is isotropic if any belief within the system can affect the confirmation of any other belief. And such a system is Quinean if its manner of belief confirmation is sensitive to the structure of the whole system by way of favoring beliefs that, if added to the system, preserve such systematic properties as simplicity and conservatism. In contrast to isotropic and Quinean information systems are encapsulated, modular systems in which only differentially selected bits of information are deployed in confirming any new bit of information. Now Fodor takes the natural process of belief fixation, and hence modulation, to be nondemonstrative inference that results in an isotropic and Quinean system of belief. And this, if Fodor is right, is what makes the frame problem so very intractable. For any module that marks off as relevant for processing a subset of beliefs is bound to be implausible as a solution to the frame problem because, by the nature of modules, it must ignore an arbitrarily large number of beliefs within the system, beliefs that partially determine how modification of any belief within the system is to occur. The idea, then, is that the frame problem amounts to the problem of explaining the nature of scientific confirmation, something about which very little is known except that it seems immune to a formalistic solution (Putnam, 1983).

Fodor argues that belief fixation is isotropic and Quinean under the assumption that it is a process of rational nondemonstrative inference. That is to say that a cognitive agent settles on a belief by virtue of formulating a hypothesis to the best explanation relative to what information happens to be delivered from the agent's ancillary, modular information processing systems in addition to whatever is already stored in memory. Since the details of this putative process are not well understood, Fodor supposes that we might be best served by looking at them in the light of what we do know of the formation of scientific theory, that paradigm of rational nondemonstrative inference. It is this that sanctions Fodor's claim that belief fixation is isotropic and Quinean.

But it is important to notice here that while science certainly *ought* to be isotropic and Quinean, it surely does not achieve its ideal, Actually accepted scientific hypotheses typically are not measured against *everything* already known to science. Physicists simply do not normally attend to everything economists, zoologists and linguistics have to say (Glymour, 1985). While a scientific hypothesis should weather exposure to everything antecedently known, in practice the separate sciences suffer from limited encapsulation. If, then, belief fixation is akin to actual scientific progress, we should anticipate that it is perfectly isotropic and Quinean only in the fiction of the ideal situation but much less so in the actual case. Supporting this is the well-known fact that we are less than epistemically ideal cognitive agents (Tversky and Kahneman, 1974). That our errors in belief fixation are systematic and susceptible to study indicates that the processes implicated in belief fixation are, at best, only imperfectly isotropic and Quinean. The processes actually governing belief fixation in nature are, thus, not to be confused with those that might determine epistemically ideal processes (Heil, 1985; Mele, 1986).

We should anticipate, then, that if our processes of belief fixation resemble scientific discovery, they too are marked by some measure of encapsulation. And the same should suggest that beliefs are modulated in ways that ignore the actual epistemic significance of a fair amount of what we may antecedently believe. For example, Tversky and Kahneman report that subjects typically respond to classification problems in ways that betray that they ignore crucial information they possess. Told of the distribution of professions across a designated population, subjects, when provided uninformative descriptions of members of the population, tend to classify the described members according to profession in ways that are oblivious to the known distribution of professions. This amounts to saying that it is likely that the range of beliefs that actually come into play in the fixation of any particular belief is constrained in some important way. Hence, the beliefs an agent deploys in evaluating for acceptance any given hypothesis are most probably restricted according to some heuristic. Once, however, we see that belief fixation is not purely isotropic and Quinean, it should be evident that modules are not utterly implausible, even if presently simplistic, models for belief fixation within the context of the frame problem. For modules are, by their very nature, certainly nonisotropic and quite unQuinean. What I am urging, then, is that cognitive agents engaged in belief fixation are subject to various and varying limitations in cognitive resources the effect of which is an apparent encapsulation of belief. And while little is known about the principles governing belief fixation, the clear possibility of encapsulation gives credence to the idea that modules or their successors may have promise as models of belief fixation.

One point remains. Let us provisionally adopt Fodor's rough distinction between a few local modules and the single global cognitive center. The former are topically restricted, whereas the latter is grossly unbounded. Fodor is content to allow that the frame problem does not arise in modular information processing systems, such as perceptual and language processors, on the grounds that they are informationally encapsulated. Nevertheless, the modularity of these systems does not entail that they cannot be nondemonstrative processes. Presumably, they also approximate rationality in some interesting sense. That is, relative to the information to which they have access, the information they generate is reasonable. So, modular systems, like the central processes of belief fixation, possibly are processes of rational nondemonstrative inference. If this is so, then, relative to the information to which they have access, modular systems may themselves be isotropic and Quinean. It is consistent with what we know of modular systems that any and every bit of information to which they have access can affect the confirmation of any bit of information they process. Equally, considerations of simplicity and conservativism may govern their confirmation of the information they produce. If this should be so, modular systems would function exactly as do the (apparently) nonmodular processes of cognitive central. The only important difference would be that modular systems have access to restricted data bases in comparison with the central system of belief fixation. Why, then, should the frame problem not affect modular systems if it infects the central system? The only answer would seem to be that the restricted informational access of the modular systems saves them from the frame problem. No doubt, it is just this that enables modules to function as they do. But we should not fail to recall that modules, though encapsulated, need not be informationally barren. Certainly, if, as is evident, the visual system classifies distal stimuli according to kind, it will need access to all the information necessary for such classifications. What must it need to know if it is to reveal that this is a cat and that a canary? Enough to distinguish cats and canaries from all the other kinds it discriminates. That means it must have access to more than an absolutely minute amount of information. And so, we might well suspect that if central processes are subject to the frame problem in such a way as to render hopeless a computational explanation of their functioning, then so too for the peripheral, modular systems — especially considering that they suffer from limitations of computational time, space and power unknown to the more generously endowed central system of belief fixation. But, and this is the crucial point, according to Fodor, the modular systems in fact are local solutions to the frame problem consistent with the constraints imposed by computationalism. Yet if this is so, then, by *modus tollens*, it may begin to seem less likely than it originally did that a computational account of the central system is stalled by the frame problem.

NOTES

[1] The literature is less than a chorus of universal agreement. See Patricia Smith Churchland (1980), Dennett (1978a, b), Paul Churchland (1986) and, of course, the growing literature on imagistic representation (Block, 1981).

[2] See Brand (1984, pp. 232f) for a favorable discussion of Schank and Abelson's work as a contribution to the solution of the frame problem.

[3] But contrast what Paul Churchland (1979) has to say on the matter.

[4] Witness the difference between the expectations had of the General Problem Solver and what was learned from it (Boden, 1977; Haugeland, 1985).

[5] Evidently, learning will have to be a process whose explanation does not itself presuppose an explanation of the frame problem. Otherwise, it would be circular to allow that modules are sensitive to instruction while positing that modules occupy center stage in the resolution of the frame problem. In this regard, one hopes for a developmental psychology that portrays the maturing agent as, *in principio*, a doxastic pauper whose continuous conceptual achievements amount to the acquisition of cogni-

tive modules. Such a rendition of developmental psychology would, thus, make the maturing agent out to be an information sponge soaking up blocks of highly structured information — modules. Learning, in the primary case, would then be not simply the incorporation of cognitively isolated beliefs awaiting regimentation but rather the reception and deployment of internally structured doxastic structures. With less jargon, if the explanation of the frame problem here told is to be consistent, it had better turn out that what transforms an infant into a cognitive agent is the successive deployment of small and increasingly integrated conceptual schemes.

[6] Haugeland (1985, p. 204) remarks that while modular structures of the sort I have been discussing may be suited for explaining how a computational system achieves efficient access to knowledge relevant to its current demands, the problem of gaining such access is distinct from the frame problem. Thus, he maintains that the solution to the frame problem should not be sought in modularity. I will not directly address this claim and rely instead on what I have already said about the application of modules to the frame problem.

[7] This is a bit too simple. Probably, modules activated by linguistic cues react to properties of the deep structure of linguistically encoded information. Thus, verbalized speech is first sent to the speech decoder for analysis. The output will settle, in so far as possible, questions of ambiguity, pronomial reference, ellipsis and the like. The output of the decoder would then exhibit syntactic properties that may in turn address certain modules. Compare Haugeland (1985, p. 200).

[8] The efforts of Schank (1982) and Schank and Abelson (1977) to delimit a range of basic actions in terms of which other actions are characterized is relevant here. Their work looks to the construction of scripts which can handle various topics while preserving encapsulation.

REFERENCES

Anderson, John.: 1983, *The Architecture of Cognition*, Harvard University Press, Cambridge.

Block, Ned: 1981, *Imagery* MIT Press/Bradford Books, Cambridge.

Boden, Margaret: 1977, *Artificial Intelligence and Natural Man*, Basic Books, New York.

Brand, Myles: 1984, *Intending and Acting*, MIT Press/Bradford Books, Cambridge.

Chomsky, Noam: 1980, *Rules and Representations*, Columbia University Press, New York.

Churchland, Patricia Smith: 1980, 'Language, Thought and Information Processing', *Noûs* 14, 147—170.

Churchland, Paul: 1979, *Scientific Realism and the Plasticity of Mind*, Cambridge University Press, Cambridge.

Churchland, Paul: 1986, 'Some Reductive Strategies in Cognitive Neurobiology', forthcoming in *Mind*.

Dennett, Daniel: 1978a, 'A Cure for the Common Code', in Dennett's *Brainstorms*, Bradford Books, Montgomery, Vt., pp. 90—108.

Dennett, Daniel: 1978b, 'Brain Writing and Mind Reading', in Dennett's *Brainstorms*, Bradford Books, Montgomery, Vt., pp. 39—50.

Dennett, Daniel: 1978c, 'Artificial Intelligence as Philosophy and Psychology', in Dennett's *Brainstorms*, Bradford Books, Montgomery, Vt., pp. 109—126.

Fodor, Jerry: 1975, *The Language of Thought*, Thomas Crowell, New York.

Fodor, Jerry: 1983, *Modularity of Mind*, Bradford Books. Cambridge Mass.

Fodor, Jerry: 1985, 'Précis of *The Modularity of Mind*', *The Behavioral and Brain Sciences* **8**, 1—42 (including peer reviews).

Haugeland, John: 1981, *Mind Design*, MIT Press, Cambridge.

Haugeland, John: 1985, *Artificial Intelligence: The Very Idea* (Cambridge, Mass.: MIT Press).

Heil, John: 1985, 'Rationality and Psychological Explanation', *Inquiry* **28**, 359—371.

Marcus, M.: 1977, 'A Theory of Syntactic Recognition for Natural Language', Ph.D. thesis, MIT, 1977.

Marr, D.: 1982, *Vision*, W. H. Freeman, San Francisco.

Mele, A.: 1986, 'Incontinent Believing', *Philosophical Quarterly* **36**, 212—222.

McCarthy, J. and Hayes, P.: 1969, 'Some Philosophical Problems from the Standpoint of Artificial Intelligence', in Meltzer, B. and Michie, D. (eds.), *Machine Intelligence* **4**, Edinburgh University Press, Edinburgh, pp. 463—502.

Minsky, Marvin: 1975, 'A Framework for Representing Knowledge', in P. Winston (ed.), *The Psychology of Computer Vision*, McGraw-Hill, New York, reprinted in Haugeland (1981), pp. 95—128.

Nagel, Thomas: 1974, 'What is it Like to be a Bat?' *Philosophical Review* **83**, 435—450.

Putnam, Hilary: 1983, 'Computational Psychology and Interpretation Theory', in Putnam's *Realism and Reason, Philosophical Papers: Volume III*, Cambridge University Press, New York.

Pylyshyn, Zenon: 1984, *Computation and Cognition: Toward a Foundation for Cognitive Science*, MIT Press/Bradford Books, Cambridge.

Raphael, B.: 1971, 'The Frame Problem in Problem-Solving Systems', in N. V. Findler and B. Meltzer (eds.), *Artificial Intelligence and Heuristic Programming*, American Elsevier Publishing Co. New York, pp. 159—169.

Schank, R. C.: 1982, *Dynamic Memory*, Cambridge University Press, Cambridge.

Schank, R. C. and Abelson, R. P.: 1977. *Scripts, Plans, Goals and Understanding*, Erlbaum, Hillsdale, N. J..

Tversky, A., and Kahneman, D.: 1974, 'Judgment Under Uncertainty: Heuristics and Biases', *Science* **185**, 1124—31.

Winograd, Terry: 1971, *Procedures as a Representation for Data in a Computer Program for Understanding Natural Languages*, MIT Project MAC, Cambridge.

Department of Philosophy,
Oakland University,
Rochester, MI 48063, U.S.A.

WILLIAM J. RAPAPORT

SYNTACTIC SEMANTICS:
FOUNDATIONS OF COMPUTATIONAL
NATURAL-LANGUAGE UNDERSTANDING

> Language (*la langue*) is a system all of whose
> terms are interdependent and where the value
> of one results only from the simultaneous pre-
> sence of the others . . . . (de Saussure 1915, p.
> 159.)

1. INTRODUCTION

In this essay, I consider how it is possible to understand natural
language and whether a computer could do so. Briefly, my argument
will be that although a certain kind of semantic interpretation is needed
for understanding natural language, it is a kind that only involves
syntactic symbol manipulation of precisely the sort of which computers
are capable, so that it is possible in principle for computers to understand natural language. Along the way, I shall discuss recent arguments
by John R. Searle and by Fred Dretske to the effect that computers
can *not* understand natural language, and I shall present a prototype
natural-language-understanding system to illustrate some of my claims.[1]

2. CAN A COMPUTER UNDERSTAND NATURAL LANGUAGE?

What does it mean to say that a computer can understand natural
language? To even attempt to answer this, a number of preliminary
remarks and terminological decisions need to be made. For instance, by
'computer', I do not mean some currently existing one. Nor, for that
matter, do I mean some ultimate future piece of hardware, for computers by themselves can do nothing: They need a program. But neither
do I mean to investigate whether a program, be it currently existing or
some ultimate future software, can understand natural language, for
programs by themselves can do nothing.

Rather, the question is whether a computer that is running (or
executing) a suitable program — a (suitable) program being executed or

81

James H. Fetzer (ed.), Aspects of Artificial Intelligence, 81—131.
© 1988 *by Kluwer Academic Publishers.*

run — can understand natural language. A program actually being executed is sometimes said to be a "process" (*cf.* Tanenbaum 1976, p. 12). Thus, one must distinguish three things: (a) the computer (i.e., the hardware; in particular, the central processing unit), (b) the program (i.e., the software), and (c) the process (i.e., the hardware running the software). A program is like the script of a play; the computer is like the actors, sets, etc.; and a process is like an actual production of the play — the play in the process of being performed.[2] Having made these distinctions, however, I will often revert to the less exact, but easier, ways of speaking ("computers understand", "the program understands").

What kind of program is "suitable" for understanding natural language? Clearly, it will be an AI program, both in the sense that it will be the product of research in artificial intelligence and in the (somewhat looser) sense that it will be an artificially intelligent program: for understanding natural language is a mark of intelligence (in the sense of AI, *not* in the sense of IQ), and such a program would exhibit this ability artificially.

But what kind of AI program? Would it be a "weak" one that understands natural language but that does so by whatever techniques are successful, be they "psychologically valid" or not? Or would it be a "strong" one that understands natural language in more or less the way we humans do?[3] ("More or less" may depend on such things as differences in material and organization between humans and these ultimate computers.) I do not think that it matters or that any of the considerations I will present depend on the strong/weak dichotomy, although I do think that it is likely that the only successful techniques will turn out to be psychologically valid, thus "strengthening" the "weak" methodology.

Another aspect of the program can be illuminated by taking up the metaphor of the play. This ultimate AI program for understanding natural language might be thought of as something like the script for a one-character play. When this "play" is "performed", the computer that plays the role of the sole "character" communicates in, say, English. But we do not want it to be only a one-way communication; it must not merely speak to us, the "audience", yet be otherwise oblivious to our existence (as in Disney-like audio-animatronic performances). That would hardly constitute natural-language understanding. More give and take is needed — more interaction: the play must be an audience-participation improvisation. So, too, must the program. I'll return to this theme later (Section 3.2.1).

I said earlier that understanding natural language is a mark of intelligence. In what sense is it such a mark? Alan M. Turing (1950) rejected the question, "Can machines think?", in favor of the more behavioristic question, "Can a machine convince a human to believe that it (the computer) is a human?"[4] To be able to do that, the computer must be able to understand natural language. So, understanding natural language is a necessary condition for passing the Turing Test, and to that extent, at least, it is a mark of intelligence.

I think, by the way, that it is also a sufficient condition. Suppose that a computer running our ultimate program understands, say, English. Therefore, it surely understands such expressions as 'to convince', 'to imitate a human', etc. Now, of course, merely understanding what these mean is not enough. The computer must be able to *do* these things — to convince someone, to imitate a human, etc. That is, it must not merely be a *cognitive* agent, but also an *acting* one. In particular, to imitate a human, it needs to be able to reason about what a(nother) cognitive agent, such as a human, believes. But that kind of reasoning is necessary for understanding natural language; in particular, it is necessary for understanding behavior explainable in terms of "nested beliefs" (such as: Jan took Smith's course because she believes that her fellow students believe that Smith is a good teacher; on the importance of such contexts, *cf.* Dennett 1983 and Rapaport 1984, 1986c). Finally, the computer must also, in some sense, *want* to convince someone by pretending to be a human; i.e., it must *want* to play Turing's Imitation Game. But this can be done by *telling* it to do so, and this, of course, should be told to it in English. So, if it understands natural language, then it ought to be able to pass the Turing Test. If so, then understanding natural language is surely a mark of intelligence.

But even if understanding natural language is only a necessary condition of intelligence, the question whether computers can understand natural language is something we should care about. For one thing, it is relevant to Searle's Chinese-Room Argument, which has repidly become a rival to the Turing Test as a touchstone for philosophical inquiries into the foundations of AI (Searle 1980). For another, it is relevant to Dretske's claims that computers can't even add (Dretske 1985). One of my main goals in this essay is to show why Searle's and Dretske's arguments fail. Finally, it is a central issue for much research in AI, computational linguistics, and cognitive science. Many researchers in these fields, including my colleagues and myself, are investigating techniques for writing computer programs that, we claim, will be able to understand stories, narratives, discourse — in short, natural language (Shapiro and Rapaport 1986, 1987; Bruder *et al.* 1986). It would be nice to know if we can really do what we claim we are able to do!

3. WHAT DOES IT MEAN TO "UNDERSTAND NATURAL LANGUAGE"?

3.1. *Syntax Suffices*

To determine whether a computer (as understood in the light of the previous section) can understand natural language, we need to determine how it is possible for *anything* to understand natural language, and then to see if computers can satisfy those requirements.

Understanding has to do with meaning, and meaning is the province of semantics. Several recent attacks on the possibility of a computer's understanding natural language have taken the line that computers can only do syntax, not semantics, and, hence, cannot understand natural language. Briefly, my thesis in this essay is that *syntax suffices*. I shall qualify this somewhat by allowing that there will also be a certain causal link between the computer and the external world, which contributes to a *certain kind* of nonsyntactic semantics, but not the kind of semantics that is of computational interest. What kind of causal link is this? Well, obviously, if someone built the computer, there's a causal link between it and the external world. But the particular causal link that is semantically relevant is one between the external world and what I shall call the computer's "mind" — more precisely, the "mind" of the process produced by the running of the natural-language-understanding program on the computer.

Before I go into my reasons for hedging on what might seem to be the obvious importance of the causal link and what this link might be, let me say why I think I have a right to talk about a computer's "mind". Consider a system consisting of a computer, an AI program (or, what is more likely, a set of interacting AI programs), and perhaps a preexisting data base of information expressed in some "knowledge representation" language. (When such data bases are part of an AI program, they tend to be called "knowledge bases", and the preexisting, background information is called "world knowledge" — "innate ideas" would also be appropriate terminology.) The system will interact with a "user" — perhaps a human, perhaps another such system. Suppose that the system behaves as follows: It indicates to the user that it is ready to begin. (This need not be indicated by a natural-language sentence.) The user types (or otherwise interactively inputs) a sentence in, say, English. Depending on the nature of the input, the system might modify its knowledge base in accordance with the information contained in this sentence. (If the input was merely 'hello', it might not.) It may then express to the user, in English, some appropriate proposition in its knowledge base. And so the dialogue would continue. (An actual example of such a dialogue is shown in Appendix 1.)

If such a system is going to be a good candidate for one that can understand natural language, it ought to be able at least to process virtually all of what the user tells it (or at least as much as a human would), to answer questions, and, most importantly, to ask questions. What's more, it ought to do this in a fashion that at least somewhat resembles whatever it is that *we* do when we understand natural language; that is, it should probably be doing some real, live parsing and generating, and not mere pattern-matching. Under this requirement, a "strong" system would parse and generate more or less precisely as humans do; a "weak" system would parse and generate using some other grammar.

But even this is not enough. The system must also *remember* all sorts of things. It must remember things it "knew" (i.e., had in its knowledge base) before the conversation began; it must remember things it "learns" (i.e., adds to its knowledge base) during the conversation; and it must be able to draw inferences (deductively, inductively, abductively, pragmatically, etc.) — thus modifying its knowledge base — and remember *what* it inferred as well as *that, how*, and probably even *why* it inferred it.

In short, it needs a knowledge base. This is why a program such as ELIZA (Weizenbaum 1966, 1974, 1976) — which lacks a knowledge base — does *not* understand natural language, though there are many programs described in the AI literature that have knowledge bases and do some or all of these things to varying degrees (e.g., SHRDLU (Winograd 1972) and BORIS (Lehnert *et al.* 1983), to name but two). The knowledge base, expressed in a knowledge-representation language augmented by an inferencing package, is (at least a part of) the "mind" of the system. I will discuss one such system later (the one responsible for the dialogue in Appendix 1).

So, my thesis is that (suitable) purely syntactic symbol-manipulation of the system's knowledge base (its "mind") suffices for it to understand natural language. Although there is also a causal link between its "mind" and the external world, I do not think that this link is necessary *for understanding natural language*. I shall have more to say about this later; all I shall say now is that my reasons for taking this position are roughly methodologically solipsistic: the system has no access to these links, and a second system conversing with the first only has access to its own internal representations of the first system's links. Nevertheless, given that there are in fact such links, what might they be like? I shall have more to say about this, too, but for now let it suffice to say that they are perceptual links, primarily visual and auditory.

3.2. *The Chinese-Room Argument*

Now, Searle has argued that computers cannot understand natural language (or, hence, be intelligent, artificially or otherwise). In his Chinese-Room Argument, Searle, who knows neither written nor spoken Chinese, is imagined to be locked in a room and supplied with instructions in English that provide an algorithm for processing written Chinese. Native Chinese speakers are stationed outside the room and pass pieces of paper with questions written in Chinese characters into the room. Searle uses these symbols, otherwise meaningless to him, as input and — following only the algorithm — produces, as output, other Chinese characters, which are, in fact, answers to the question. He passes these back outside to the native speakers, who find his "answers ... absolutely indistinguishable from those of native Chinese speakers" (Searle 1980, p. 418). The argument that this experiment is supposed to support has been expressed by Searle as follows:

[I] still don't understand a word of Chinese and neither does any other digital computer because all the computer has is what I have: a formal program *that attaches no meaning, interpretation, or content to any of the symbols.* [Therefore,] ... no formal program by itself is sufficient for understanding .... (Searle 1982, p. 5; italics added — *cf.* Section 3.5, below.)

If this Chinese-language-processing system passes the Turing Test, then — according to the Test — it does understand Chinese. And indeed it does pass the test, according to the very criteria Searle sets up. So how can Searle conclude that it doesn't understand Chinese? One reason that he offers is that the program doesn't understand because it doesn't "know" what the words and sentences *mean*:

> The reason that no computer program can ever be a mind is simply that a computer program is only syntactical, and minds are more than syntactical. Minds are semantical, in the sense that they have more than a formal structure, they have a content. (Searle 1984, p. 31.)

That is, meaning — "semantics" — is something over and above mere symbol manipulation — "syntax". Meaning is a relation between symbols and the things in the world that the symbols are supposed to represent or be about. This "aboutness", or intentionality, is supposed to be a feature that only minds possess. So, if AI programs cannot exhibit intentionality, they cannot be said to think or understand in any way.

But there are different ways to provide the links between a program's symbols and things in the world. One way is by means of sensor and effector organs. Stuart C. Shapiro has suggested that all that is needed is a camera and a pointing finger (personal communication; *cf.* Shapiro and Rapaport 1987). If the computer running the Chinese-language program (plus image-processing and robotic-manipulation programs) can "see" and "point" to what it is talking about, then surely it has all it needs to "attach meaning" to its symbols.

Searle calls this sort of response to his argument "the Robot Reply". He objects to it on the grounds that if he, Searle, were to be processing all of this new information along with the Chinese-language program, he still would not "know what is going on", because now he would just have more symbols to manipulate: he would still have no direct access to the external world.

But there is another way to provide the link between symbols and things in the world: Even if the system has sensor and effector organs, it must still have internal representations of the external objects, and — I shall argue — it is the relations between *these* and its other symbols that constitute meaning for *it*. Searle seems to think that semantics must link the internal symbols with the outside world and that this is something that cannot be programmed. But if this is what semantics must do, it must do it for human beings, too, so we might as well wonder how the link could possibly be forged for us. Either the link between internal representations and the outside world *can* be made for both humans *and* computers, or else semantics is more usefully treated as linking one set of internal symbolic representations with another. On this view, semantics does indeed turn out to be just more symbol manipulation.

Here is Searle's objection to the Robot Reply:

> I see no reason in principle why we couldn't give a machine the capacity to understand English or Chinese, since in an important sense our bodies with our brains are precisely such machines. But ... we could not give such a thing to a machine ... [whose] operation ... is defined solely in terms of computational processes over formally defined elements. (Searle 1980, p. 422.)

'Computational processes over formally defined elements' is just a more precise phrase for symbol manipulation. The reason Searle gives for his claim that a machine that just manipulates symbols cannot understand a natural language is that "only something having the same causal powers as brains can have intentionality" (Searle 1980, p. 423). What, then, are these "causal powers"? All Searle tells us in his essay on the Chinese-Room Argument is that they are due to the (human) brain's "biological (that is, chemical and physical) structure" (Searle 1980, p. 422). But he does not specify precisely what these causal powers are. (In Rapaport 1985b and 1986b, I argue that they are not even causal.)

Thus, Searle has two main claims: A computer cannot understand natural language because (1) it is not a biological entity and (2) it is a purely syntactic entity — it can only manipulate symbols, not meanings. Elsewhere, I have argued that the biological issue is beside the point — that *any* device that "implements" (in the technical sense of the computational theory of abstract data types) an algorithm for successfully processing natural language can be said to *understand* language, no matter how the device is physically constituted (Rapaport 1985b, 1986a, 1986b). My intent here is to argue, along the lines sketched out above, that being a purely syntactic entity *is* sufficient for understanding natural language.[5]

Before doing that, I think it is worth looking at some aspects of Searle's argument that have been largely neglected, in order to help clarify the nature of a natural-language-understanding program.

3.2.1. Natural-language generation

The first aspect can be highlighted by returning to the metaphor of the natural-language-understanding program as a one-character, audience-participation, improvisatory play. Because it is improvisatory, the script[6] of the play cannot be fixed; it must be able to vary, depending on the audience's input. That is, a natural-language-understanding system must be able to respond appropriately to arbitrary input (it must be "robust"). This could, perhaps, be handled by a "conditional script": if the audience says $\ulcorner\varphi_1\urcorner$, then the character should respond by saying $\ulcorner\varphi_2\urcorner$, etc. But to be truly robust, to script would need to be "productive", in roughly Chomsky's sense: that is, the character in the play must be able to understand and the produce arbitrary "new" and relevant lines. In fact, it is fairly easy to have a productive *parser* for a natural-language-understanding system. I am not claiming that the problem of natural-language *understanding* has been solved, but we seem to be on the right track with respect to parsers for natural language *processing*, and, at any rate, we know the general outlines of what a suitably robust parser should look like. What's needed, however, is *generative* productivity: the ability to *ask* new and relevant questions and to *initiate* conversation (in a non-"canned" way: ELIZA — which relies purely on pattern-matching — still doesn't qualify). To be able to generate appropriate utterances, the system must have the capability to *plan* its speech acts, and, so, a planning component must be part of a natural-language-understanding system. Such a planning component is probably also needed for parsing, in order to be able to understand *why* the speaker said what he or she did. (*Cf.* Cohen and Perrault 1979; Appelt 1982, 1985; and Wiebe and Rapaport 1986.)

To the extent that these are missing from the Chinese-Room Argument, Searle-in-the-room wouldn't seem to understand Chinese. So, let us imagine that AI researchers and computational linguists have solved this problem, and that our system is equipped with a suitably productive generation grammar. Now, these productive capabilities are tantamount to general intelligence, as I argued in Section 2. The important point, however, is that this capability is a function of what's in the system's knowledge base: what can be produced by a productive generative grammar must first be in the knowledge base. To put it somewhat mundanely, I can only speak about what I'm familiar with. (To put it more esoterically, whereof I cannot speak, thereof I must be silent.)

3.2.2. *Learning and linguistic knowledge*

A second aspect of Searle's argument that I want to look at concerns the kind of knowledge that Searle-in-the-room is alleged to have — or lack. One difference that is sometimes pointed out between machine understanding and human understanding is that everything that the machine does is explicitly coded. This is part of what is meant when it is said that computers can only do what they are programmed to do (by someone who *is* "intelligent" or who *can* understand natural language). Furthermore, this might be interpreted to mean that the system knows everything that it is doing. But this is mistaken. It can only be said to "know" what it is "aware" of, not what is merely coded in. For instance, the knowledge bases of many AI systems distinguish between propositions that are explicitly believed by the system and those that are only implicitly believed (*cf.* Levesque 1984; Rapaport 1984, 1986c, 1987). Furthermore, the parser that transduces the user's input into the system's knowledge base, as well as the generator that transduces a proposition in the knowledge base into the system's natural-language output, need not (and arguably should not) be part of the knowledge base itself. Such "knowledge" of language would be tacit knowledge, just as Chomsky said: It is coded in and is part of the overall system, but it is not "conscious knowledge". It is no different for humans: everything we know, including our knowledge of how to understand our native natural language, must (somehow) be "coded in". In other words, human and machine understanding are *both* fully coded, but neither the human nor the machine knows everything. In the Chinese-Room Argument, the human following the Chinese-language program is in the same position as a human speaking English (only in slow motion; *cf.* Hofstadter 1980): neither has conscious knowledge of the rules of the language.

Could the machine or the human *learn* the rules, and thus gain such conscious knowledge? Or could it learn *new* rules and thus expand its natural-language understanding? Surely, yes: see the work by Jeannette Neal (Neal 1981, 1985; Neal and Shapiro 1984, 1985, and 1987).

There are other roles for learning in natural-language understanding. Many (perhaps most) conversations involve the learning of new information. And it is often the case that new words and phrases, together with their meanings, are learned both explicitly and implicitly (*cf.* Rapaport 1981, and the discussion of 'swordsman' in Section 3.5, below). In all of these cases, the learning consists, at least in part, of modifications to the system's knowledge base.

It is not clear from the rather static quality of Searle's Chinese-language program whether Searle intended it to have the capability to learn. Without it, however, the Chinese-Room Argument is weakened.

3.2.3. *The knowledge base*

It should be clear by now that a knowledge base plays a central role in natural-language understanding. Searle's original argument includes a Schank-like script as part of the input, but it is not clear whether he intended this to be a modifiable knowledge base of the sort I described as the system's "mind" or whether he intended it as the rather static structure that a script (in its early incarnation) actually is. In any case, parts of the knowledge base would probably have to be structured into, *inter alia*, such frame-like units as scripts, memory-organization packets, etc. (*Cf.* Minsky 1975, Schank 1982.) The two aspects we have just considered, and part of my argument below, imply that a modifiable knowledge base is essential to natural-language understanding. (*Cf.* n. 13.)

3.3. *Dretske's Argument*

Having set the stage, let me introduce some of my main ideas by considering Dretske's argument in 'Machines and the Mental' (1985), to the effect that an external, non-syntactic semantics is needed for natural-language understanding.

According to Dretske, machines "lack something that is essential" for being a rational agent (p. 23). That is, there is something they "can't do" (p. 23) that prevents their "membership in the society of rational agents" (p. 23). That is surely a very strong claim to make — and a very important one, if true. After all, theoretical computer science may be characterized as the study of what is effectively computable. That is, assuming Church's Thesis, it may be characterized as the study of what is expressible as a recursive function — including such theoretically uninteresting though highly practical recursive functions as payroll programs. It follows that AI can be characterized as the study of the extent to which mentality is effectively computable. So, if there is something that computers can't do, wouldn't it be something that is *not* effectively computable — wouldn't it be behavior that is nonrecursive?[7] It is reasonable to expect that it is much too early in the history of AI for such a claim as this to be proved, and, no doubt, I am interpreting Dretske's rhetoric too strongly. For a nonrecursive function is in a sense more complex than a recursive one, and Dretske's line of argument seems to be that a computer is simpler than a human (or that computer thought is more isolated than human thought): "Why can't pure thought, the sort of thing computers purportedly have, stand to ordinary thought, the sort of thing we have, the way a solitary stroll stands to a hectic walk down a crowded street?" (p. 23). Even granting that this talk about computers is to be understood in the more precise sense of Section 1, above, the ratio

$$\frac{\text{pure thought}}{\text{computers}} = \frac{\text{ordinary thought}}{\text{humans}}$$

isn't quite right. If anything, the phrase 'pure thought' ought to be preserved for the *abstraction* that can be *implemented* in computers *or* humans (or Martians, or chimps, or . . .):

$$\frac{\text{pure thought}}{\text{implementing medium}} = \frac{\text{AI program that implements pure thought}}{\text{computer}}$$

$$= \frac{\text{human mental processes (ordinary thought)}}{\text{human}}$$

(*Cf.* Rapaport 1985b, 1986b.)

What is it, then, that these "simple-minded" computers can't do? Dretske's admittedly overly strong answer is:

They don't solve problems, play games, prove theorems, recognize patterns, let alone think, see, and remember. They don't even add and subtract. (p. 24.)

Now, one interpretation of this, consistent with holding that intelligence is nonrecursive, is that these tasks are also nonrecursive. But, clearly, they aren't (or, at least, not all of them are). A second interpretation can be based on the claim that Church's Thesis is not a *reduction* of the notion of "algorithm" to that of, say, Turing-machine program on the grounds that an algorithm is an intensional entity that contains as an essential component a description of the problem that it is designed for, whereas no such description forms part of the Turing-machine program (Goodman 1986). So, perhaps, the tasks that Dretske says computers can't do are all ones that must be described in intensional language, which computers are supposed incapable of.

These two interpretations are related. For if tasks that are essentially intensional are nonrecursive, then Church's Thesis can be understood as holding that for each member of a certain class of nonrecursive functions (namely, the essentially intensional but effectively computable tasks), there is a corresponding recursive function that is extensionally equivalent (i.e., input—output behaviorally equivalent) to it. And Dretske's thesis can be taken to be that this equivalence is not an identity. For instance, although my calculator's input—output *behavior* is identical to my own behavior when I perform addition, *it* is not *adding*.

Here is Dretske's argument (p. 25):

(1)   "... 7, 5 and 12 are numbers."
(2)   "We add 7 and 5 to get 12 ...".
(3)   Therefore, "Addition is an operation on numbers."
(4)   "At best, [the operations computers perform] are operations on ... physical tokens that *stand for*. or are interpreted as standing for, ... numbers."
(5)   Therefore, "The operations computers perform ... are not operations on numbers."
(6)   "Therefore, computers don't add."
(7)   Therefore, computers cannot add.[8]

Possibly, if *all* that computers do is manipulate uninterpreted symbols, then they do *not* add. But if the symbols are interpreted, then maybe computers *can* add. So, who would have to interpret the symbols? Us? Them? To make the case parallel, the answer, perhaps surprisingly, is: them! For who interprets the symbols when *we* add? Us. But if we can do it (which is an assumption underlying premise (2)), then why can't computers? But perhaps it is *not* I who interpret "my" symbols when I add, or you when you add. Perhaps there is a dialectical component: the only reason that *I* think that *you* can add (or *vice versa*) is that *I* interpret *your* symbol manipulations (and *vice versa*). In that case, if *I* interpret the *computer's* symbol manipulations, then — to maintain the parallelism — *I* can say that *it* adds (at least as well as you add). And, take note, in the converse case, the *computer* can judge that *I* "add".

Premise (2) and intermediate conclusion (3) are acceptable. But *how* is it that we add numbers? By manipulating physical tokens of them. That is, the abstract operation of adding *is* an operation on numbers (as (3) says), but our *human implementation* of this operation is an operation on (physical) *implementations* of numbers. (Cf. Shapiro 1977, where it is argued that addition, as humans perform it, is an operation on numerals, not numbers.) So premise (4) is also acceptable;

but — as Dretske admits — if it implies (5), then the argument "shows that we don't add either" (p. 26), surely an unacceptable result.

What the argument does illuminate is the relation of an abstraction to its implementations (Rapaport 1985b, 1986b). But, says Dretske, something is still missing:

the machine is ... restricted to operations on the symbols or representations themselves. It has no access ... to the *meaning* of these symbols, to the things the representations represent, to the numbers. (p. 26)

The obvious question to ask is: How do *we* gain this essential access? And a reasonable answer is: In terms of a theory of arithmetic, say, Peano's (or that of elementary school, for that matter). But such a theory is expressed in symbols. So the *symbol* '1' means the *number* 1 for *us* because it is linked to the '1' that represents 1 in the theory. All of this is what I shall call *internal* semantics: semantics as an interconnected network of internal symbols — a "semantic network" of symbols in the "mind" or "knowledge base" of an intelligent system, artificial or otherwise. The *meaning* of '1', or of any other symbol or expression, is determined by its locus in this network (*cf.* Quine 1951; Quillian 1967, 1968) *as well as* by the way it is *used* by various processes that reason using the network. (*Cf.* the "knowledge-representation hypothesis", according to which "there is ... presumed to be an internal process that 'runs over' or 'computes with' these representational structures" (Smith 1982, p. 33).)

There's more: *My* notion of 1 might be linked not only to my internal representation of Peano's axioms, but also to my representation of my right index finger and to representations of various experiences I had as a child (*cf.* Schank 1984, p. 68). Of course, the computer's notion of 1 won't be. But it *might* be linked to its internal representation of itself in some way[9] — the computer need not be purely a Peano mathematician. But perhaps there's too much — should such "weak" links really be part of the *meaning* of '1'? In one sense, yes; in another, no: I'll discuss several different kinds of meaning in Section 3.7.

This notion of an internal semantics determined by a semantic network and independent of links to the external world — independent, that is, of an "external" semantics — is perfectly consistent with some of Dretske's further observations, though not with his conclusions. For instance, he points out that "physical activities" such as adding "cannot acquire the relevant kind of meaning merely by *assigning* them an interpretation, by letting them mean something *to* or *for us*" (p. 26). This kind of assignment is part of what I mean by "external semantics". He continues: "Unless the symbols being manipulated mean something *to the system manipulating them*," — this is roughly what I mean by "internal semantics" — "their meaning, whatever it is, is irrelevant to evaluating what the system is doing when it manipulated them" (pp. 26-27). After all, when *I* undergo the physical processes that constitute adding, it is not only *you* who says that I add (not only *you* who assigns these processes an interpretation for *you*), but I, too. Of course, one reason that *I* assign them an interpretation is the fact that *you* do. So, *how* do *I* assign them an interpretation? If this question can be answered, perhaps we will learn how the *computer* can assign them an interpretation — which is what Dretske (and Searle) deny can be done. One answer is by my observing that *you* assign my processes an interpretation. I say to myself, no doubt unconsciously, "I just manipulated some symbols; you called it 'adding 7 and 5'. So *that's* what 'adding' is!" But once this label is thus internalized, I no longer need the link to you. My internal semantic network resumes control, and I happily go on manipulating symbols, though now I have a few extra ones, such as the label 'adding'. After all, "How would one think of associating an idea with a verbal image if one had not first come upon (*surprenait*) this association in an act of speech (*parole*)?" (de Saussure 1915, p. 37).

Dretske expresses *part* of this idea as follows: "To understand *what* a system is doing when it manipulates symbols, it is necessary to know, not just what these symbols mean, what interpretation they have been, or can be, *assigned*," — i.e., what label *you* use — "but what they mean to the system performing the operations" (p. 27), i.e., how they fit into the system's semantic network. Dretske's way of phrasing this is not quite right, though. He says, "To understand what a system is doing . . ."; but *who* does this understanding? Us, or the system? For *me* to understand what the system is doing, I only need to know *my* assignment function, *not* the system's internal network. Unless I'm its programmer, how *could* I know it? Compare the case of a human: For me to understand what *you* are doing. I only need to know *my* assignment function. Given the privacy of (human) mental states and processes, how could I possibly know yours? On the other hand, for the *system* to understand what *it* is doing, it only needs to know its own semantic network. Granted, part of that network consists of nodes (the labels) created in response to "outside" stimuli — from you or me. But this just makes it possible for the system and us to communicate, as well as making it likely that there will be a good match between the system's interpretation and ours. This is another reason why *learning* is so important for a natural-language-understanding program, as I suggested earlier (Section 3.2.2.). Unless the system (such as Searle-in-the-room) can learn from its interactions with the interlocutors, it won't pass the Turing Test.

Dretske's point is that the computer doesn't do what we do because it can't understand what it's doing. He tries to support this claim with an appeal to a by-now common analogy:

Computer simulations of a hurricane do not blow trees down. Why should anyone suppose that computer simulations of problem solving must themselves solve problems? (p. 27)

But, as with most of the people who make this analogy, Dretske doesn't make it fully. I completely agree that "computer simulations of a hurricane do not blow trees down." They do, however, *simulatedly* blow down *simulated* trees (*cf.* Gleick 1985; Rapaport 1986b and forthcoming). And, surely, computer simulations of problem solving do *simulatedly* solve *simulated* problems. The natural questions are: Is *simulated* solving *real* solving? Is a *simulated* problem a *real* problem?

The answer, in both cases, is 'Yes'. The simulated problem is an *implementation* of the *abstract* problem. A problem abstractly speaking remains one in any implementation: Compare this "real" problem:

What *number x* is such that $x + 2 = 3$?

with this "simulated" version of it:

What *symbol s* is such that the physical process we call 'adding' applied to s and to '2' yields '3'?

Both are problems. The *simulated* solution of the *simulated* problem *really* solves it and can be used to really solve the "real" problem. To return to hurricanes and minds, the difference between a simulated hurricane and a simulated mind is that the latter does "blow down trees"! (*Cf.* Rapaport, forthcoming.)

Dretske sometimes *seems* to want too much, even though he asks almost the right question:

how does one build a system that is capable not only of performing operations on (or with) symbols, but one *to which* these symbols mean **something**, a machine that, in this sense, understands **the** meaning of the symbols it manipulates? (p. 27; italics in original, my boldface.)

A system to which the symbols mean "something": Can they mean *anything*? If so, then an internal semantics suffices. It could be based on a semantic network (as in SNePS — *cf.* Shapiro and Rapaport 1986, 1987; *cf.* Section 3.6, below) or on, say, discourse representation theory (Kamp 1984 and forthcoming, Asher 1986 and 1987). The symbols' meanings would be determined solely by their locus in the network or the discourse representation structure. But does Dretske really want a machine that understands "the" meaning of its symbols? Is there only one, preferred, meaning — an "intended interpretation"? How could there be? Any formal theory admits of an infinite number of interpretations, equivalent up to isomorphism. The "label" nodes that interface with the external wourld can be changed however one wants, but the network structure will be untouched. This is the best we can hope for.

The heart of Dretske's argument is in the following passages. My comments on them will bring together several strands of our inquiry so far. First.

if the meaning of the symbols on which a machine performs its operations is . . . wholly derived from us, . . . then there is no way the machine can acquire understanding, no way these symbols can have a meaning to *the machine itself.* (pp. 27-28)

That is, if the symbols' meanings are purely external, then they cannot have internal meanings. But this does not follow. The external-to-the-machine meanings that *we* assign to its symbols are *independent* of its own, internal, meanings. It may, indeed, have symbols whose internal meanings are causally derived from our external ones (these are the "labels" I discussed earlier; in SNePS, they are the nodes at the heads of LEX arcs — cf. Section 3.6, below, and: Shapiro 1982; Maida and Shapiro 1982; Shapiro and Rapaport 1986, 1987). But the machine begins with an internal semantic network, which may be built into it ("hardwired" or "preprogrammed", or "innate", to switch metaphors) but is, in any case, developed in the course of dialogue. So it either begins with or develops its own meanings independently of those that we assign to its symbols.

Next,

> Unless these symbols have . . . an intrinsic meaning . . . independent of **our** communicative intentions and purposes, then **this meaning** *must* be irrelevant to assessing what the machine is doing when it manipulates them. (p. 28; italics in original, boldface added.)

I find this confusing: which meaning is irrelevant? Dretske's syntax seems to require it to be the "intrinsic" meaning, but his thesis requires it to be the previous passage's "meaning derived from us" (*cf.* the earlier citation from pp. 26-27). On this reading, I can agree. But the interesting question to raise is: How independent is the intrinsic meaning? Natural-language understanding, let us remember, requires conversation, or dialogue; it is a *social* interaction. Any natural-language-understanding system must initially learn *a* meaning from its interlocutor (*cf.* de Saussure 1915, p. 37, cited above), but *its* network will rarely if ever be identical with its interlocutor's. And this is as true for an artificial natural-language-understanding system as it is for us: As I once put it, we almost always misunderstand each other (Rapaport 1981, p. 17; *cf.* Schank 1984, Ch. 3, esp. pp. 44-47).

Finally,

> The machine is processing meaningful (to us) symbols . . . but the *way* it processes them is quite independent of *what they mean* — hence, nothing *the machine* does is explicable in terms of the meaning of the symbols it manipulates . . . . (p. 28)

This is essentially Nicolas Goodman's point about Church's Thesis (discussed earlier in this section). On this view, for example, a computer running a program that *we* say is computing greatest common divisors does not "know" that that is what (we say that) it is doing; so, that's *not* what it's doing. Or, to take Dretske's example (p. 30), a robot that purportedly recognizes short circuits "really" only recognizes certain gaps; it is we who interpret a gap as a short circuit. But why not provide the computer with knowledge about greatest common divisors (so-named) and the robot with knowledge about short circuits (so-named), and link the number-crunching or gap-sensing mechanisms to this knowledge?

Observe that, in the passage just cited, the machine's symbol-processing is independent of what the symbols mean *to us*, i.e., independent of their external meaning. On Dretske's view, what the machine does is inexplicable in terms of *our* meanings. Thus, he says that machines don't answer questions (p. 28), because, presumably, "answers questions" is *our* meaning, not *its* meaning.

But from Dretske's claim it does not follow that the symbols are meaningless or even that they differ in meaning from our interpretation. For one thing, *our* meaning *could* also be the *machine's* meaning, if its internal semantic network happens to be sufficiently like ours (just as yours ought to be sufficiently like mine). Indeed, for communication to be successful, this will have to be the case. For another, *simulated* question-answering *is* question-answering, just as with simulated problem-solving. If the abstract answer to the abstract question, "Who did the Yankees lose to on July 7?", is: the Red Sox; and if the simulated answer (e.g., a certain noun phrase) to the simulated question (e.g., a certain interrogative sentence), 'Who did the Yankees lose to on July 7?', is 'the Red Sox' (or even, perhaps, the simulated team, in some knowledge-representation system); and if *both* the computer *and* we take those symbols in the "same" sense — i.e., if they play, roughly, the same roles in our respective semantic networks — then the *simulated* answer *is* an answer (the example is from Green 1961).

How are such internal meanings developed? Here, I am happy to agree with Dretske: "In the same way . . . that nature arranged it in our case" (p. 28), namely, by correlations between internal representations (either "innate" or "learned") and external circumstances (p. 32). And, of course, such correlations are often established during *conversation*. But — contrary to Dretske (p. 32) — this can be the case for all sorts of systems, human as well as machine.

So, I agree with many of Dretske's claims but not his main conclusion. We *can* give an AI system information about what it's doing, although *its* internal interpretation of what it's doing might not be the same as ours; but, for that matter, yours need not be the same as mine, either. Taken literally, computers *don't* add if "add" means what *I* mean by it — which involves what *I* do when I add and the locus of 'add' in *my* internal semantic network. But thus understood, *you* don't add, either; only I do. This sort of solipsism is not even methodologically useful. Clearly, we want to be able to maintain that you and I both add. The reasons we are able to maintain this are that the "label" nodes of *my* semantic network match those of yours *and* that my semantic network is structurally much like yours. How much alike? Enough so that when we talk to each other, we have virtually no reason to believe that we are misunderstanding each other (*cf.* Russell 1918,

Quine 1969, Shapiro and Rapaport 1987; note, however, that in the strict sense in which only I add, and you don't, we *always* systematically misunderstand each other — *cf.* Rapaport 1981). That is, we can maintain that we both add, because we *converse* with each other, thus bringing our internal semantic networks into closer and closer "alignment" or "calibration". But this means that there is no way that we can prevent a natural-language-understanding system from joining us. In so doing, we may learn from it — and adjust to it — as much as it does from (and to) us.[10] Rather than talking about *my* adding, *your* adding, and *its* adding (and perhaps marveling at how much alike they all are), we should talk about the *abstract* process of adding that is *implemented* in each of us.

3.4. *Deixis*

My claim, then, is that an internal semantics is sufficient for natural-language understanding and that an external semantics is only needed for *mutual* understanding. I shall offer an explicit argument for the sufficiency thesis, but first I want to consider a possible objection to the effect that *deictic* expressions require an external semantics — that an internal semantics cannot handle indexicals such as 'that'.

Consider the following example, adapted from Kamp (forthcoming): How would our system be able to represent in its "mind" the proposition expressed by the sentence, "That's the man who stole my book!"? Imagine, first, that it is the system itself that utters this, having just perceived, by means of its computational-vision module, the man in question disappear around a corner. What is the meaning of 'that', if not its external referent? And, since its external referent could not be inside the system, 'that' cannot have an internal meaning. However, the output of any perceptual system must include some kind of internal symbol (perhaps a complex of symbols), which becomes linked to the semantic network (or, in Kamp's system, to the discourse representation structure) — a sort of *visual* "label". That symbol (or one linked to it by a visual analogue of the SNePS LEX arc) is the internal meaning of 'that'. (There may, of course, be other kinds of purely internal reference to the external would. I shall not discuss those here, but *cf.* Rapaport 1976, and Rapaport 1985/1986, Section 4.4.)

Imagine, now, that the sentence is uttered *to* the system, which looks up too late to see the man turn the corner. The external meaning of 'that' has not changed, but we no longer even have the visual label. Here, I submit, the system's interpretation of 'that' is as a disguised definite (or indefinite) description (much like Russell's theory of proper names), perhaps "the (or, a) man whom my interlocutor just saw". What's important in this case is that the system must interpret 'that', and whatever its interpretation is is the internal meaning of 'that'.

3.5. *Understanding and Interpretation*

This talk of interpretation is essential. I began this section by asking what "understanding natural language" means. To understand, in the sense we are discussing,[11] is, at least in part, to provide a semantic interpretation for a syntax. Given two "systems" — human or formal/artificial — we may ask, What does it mean for one system to understand the other? There are three cases to consider:

Case 1. First, what does it mean for *two humans to understand each other*? For me to understand what you say is for me to provide a semantic interpretation of the utterances you make. I treat those utterances as if they were fragments of a formal system, and I interpret them using as the domain of interpretation, let us suppose, the nodes of *my* semantic network. (And you do likewise with my utterances and your semantic network.) That is, I map your words into my concepts.

I may err: In Robertson Davies's novel, *The Manticore*, the protagonist, David Staunton, tells of when he was a child and heard his father referred to as a "swordsman". He had taken it to mean that his father was "a gallant, cavalier-like person" (Davies 1972, p. 439), whereas it in fact meant that his father was a lecher ('whoremaster' and 'amorist' are the synonyms (!) used in the book). This leads to several embarrasments that he is oblivious to, such as when he uses the word 'swordsman' to imply gallantry but his hearers interpret him to mean 'lechery'. Staunton had correctly recognized that the word was being used metaphorically, but he had the wrong metaphor. He had mapped a new word (or an old word newly used) into his concepts in the way that seemed to him to fit best, though it really belonged elsewhere in his network.

So, my mapping might not match yours. Worse, I might not be able to map one or more of your words into my concepts in any straightforward way at all, since your conceptual system (or "world view") — implemented in your semantic network — might be radically different from mine, or you may be speaking a foreign language. This problem is relevant to many issues in translation, radical and otherwise, which I do not wish to enter into here (but *cf.* n. 13). But what I *can* do when I hear you use such a term is to fit it into my network as best I can, i.e., to devise the best theory I can to account for this fragment of your linguistic data. One way I can do this, perhaps, is by augmenting my network with a sub-network of concepts that is structurally similar to an appropriate sub-network of *yours* and that *collectively* "interprets" your term in terms of my concepts. Suppose, for example, that you are a speaker of Nuer: although your word 'kwoth' and its sub-network of concepts might not be able to be placed in 1—1 correspondence with my word 'God' and *its* sub-network of concepts (they are not exact translations of each other), I can develop my own sub-network for 'kwoth' that is linked to the rest of my semantic network and that enables me to gloss your word 'kwoth' with an account of its meaning in terms of its locus in my semantic network (*cf.* Jennings 1985). I have no doubt that something exactly like this occurs routinely when one is conversing in a foreign language.

What is crucial to notice in this case of understanding is that when I understand you by mapping your utterances into the symbols of my internal semantic network, and then manipulate these symbols, I am performing a syntactic process.

Case 2. Second, what does it mean *for a human to understand a formal language* (or formal system)? Although a philosopher's instinctive response to this might be to say that it is done by providing a semantic interpretation for the formal language, I think this is only half of the story. There are, in fact, *two* ways for me to understand a formal language. In 'Searle's Experiments with Thought' (Rapaport 1986a), I called these "semantic understanding" and "syntactic understanding". In the example I used there, a syntactic understanding of algebra might allow me to solve equations by manipulating the symbols ("move the x from the right-hand side to the left-hand side and put a minus sign in front of it"), whereas a semantic understanding of algebra might allow me to describe those manipulations in terms of a balancing-scale ("if you remove the unknown weight from the right-hand pan, you must also remove the same amount from the left-hand pan in order to keep it balanced"). Semantic understanding is, indeed, understanding via semantic interpretation. Syntactic understanding, on the other hand, is the kind of understanding that comes from directly manipulating the symbols of the formal language according to its syntactic rules. Semantic understanding is what allows one to prove soundness and completeness theorems *about* the formal language; syntactic understanding is what allows one to prove theorems *in* the formal system.

There are two important points to notice about semantic understanding. The first is that there is no unique way to understand semantically: there are an infinite number of equally good interpretations of any formal system. Only one of these may be the "intended" intepretation, but it is not possible to uniquely identify which one. What 'adding' means to me, therefore, may be radically different from what it means to you, even if we manipulate the same symbols in the same ways (*cf.* Quine 1969, Section I, especially pp. 44-45). The second point is that an interpretation of a formal system is essentially a *simulation* of it in some *other* formal system (or, to return to talk of languages, *my* interpretation of a formal language is a mapping of its terms into my concepts), and, thus, it is just more symbol manipulation.

Syntactic understanding is also, obviously, an ability to manipulate symbols, to understand what is invariant under all the semantic interpretations. In fact, my syntactic understanding of a formal system is the closest I can get to its internal semantics, to what Dretske calls the system's "intrinsic meanings".

Case 3. Finally, what would it mean *for a formal system to understand me*? This may seem like a very strange question. After all, most formal systems just sort of sit there on paper, waiting for me to do something with them (syntactic manipulation) or to say something about them (semantic interpretation). I don't normally expect them to interpret *me*. (This is, perhaps, what underlies the humor in Woody Allen's image of Kugelmass, magically transferred into the world of a textbook of Spanish, "running for his life . . . as the word *tener* ('to have') — a large and hairy irregular verb — raced after him on its spindly legs" (Allen 1980, p. 55).)

But there are some formal systems, namely, certain computer programs, that at least have the *facility* to understand (one must be careful not to beg the question here) because they are "dynamic" — they are capable of being run. Taking up a distinction made earlier, perhaps it is the *process* — the natural-language-understanding program being run on (or, implemented by) a computer — that understands. So, what would it mean for such a formal system to understand me? In keeping with our earlier answers to this sort of question, it would be for it to give a semantic interpretation to its input consisting of *my* syntax (my utterances considered as more or less a formal system) in terms of *its* concepts. (And, of course, we would semantically understand its natural-language output in a similar manner, as noted in Case 2.) But its concepts would be, say, the nodes of its semantic network — symbols that it manipulates, in a "purely syntactic" manner. That is, it would in fact be "a formal program that attaches . . . meaning, interpretation, or content to . . . the symbols" — precisely what Searle (1982, p. 5; cited earlier) said did not exist!

So the general answer to the general question — What does it mean for one system to understand another? — is this:

> A natural-language-understanding system S_1 understands the natural-language output of a natural-language-understanding system S_2 by building and manipulating the symbols of an internal model (an interpretation) of S_2's output considered as a formal system.

S_1's internal model would be a knowledge-representation and reasoning system that manipulates symbols. It is in this sense that syntax suffices for understanding.

The role of external semantics needs clarification. Internal and external semantics are two sides of the same coin. The *internal* semantics of S_1's linguistic expressions constitutes S_1's understanding of S_2. The *external* semantics of S_1's linguistic expressions constitutes S_2's understanding of S_1. It follows that the external semantics of S_1's linguistic expressions is the internal semantics of S_2's linguistic expressions! S_1's *internal* semantics links S_1's words with S_1's own concepts, but S_1's *external* semantics links S_1's words with the concepts of S_2.

What about "referential" semantics — the link between word and object-in-the-world? I do not see how this is relevant to S_1's or S_2's understanding, except in one of the following two ways. In the first of these ways, semantics is concerned with language-in-general, not a particular individual's idiolect: it is concerned with *English* — with the "socially determined" extensions of words (Putnam 1975) — not with what *I* say. This concern is legitimate, since people tend to agree pretty well on the referential meanings of their words, else communication would cease; recall the Tower of Babel. So, on this view, what does 'pen' mean? Let us say that it means the kind of object I wrote the manuscript of this essay with (I'm old-fashioned). But what does *this* mean — what does it mean to say that 'pen' means a certain kind of physical object? It means that virtually all (native) speakers of English use it in that way. That is, this view of semantics is at best parasitic on individual external semantics.

But only "at best"; things are not even that good. The second way that "referential" semantics is relevant is, in fact, at the individual level. You say 'pen'; I interpret that as "pen" in my internal semantic network. Now, what does "pen" mean for me? Internally, its meaning is given by its location in my semantic network. Referentially, I might point to a real pen. But, as we saw in our discussion of deixis, there is an internal representation of my pointing to a pen, and it is *that representation* that is linked to my semantic network, *not* the real pen. And now here is why the first view of referential semantics won't do: How does the semanticist assert that 'pen'-in-English refers to the class of pens? Ultimately, by pointing. So, at best, the semanticist can link the pen-node of some very general semantic network of English to *other* (visual) representations, and these are either the semanticist's own visual representations or else they are representations in some *other* formal language that goes proxy for the world. The semantic link between word and object is never direct, but always mediated by a representation (*cf.* Rapaport 1976, 1985a, 1985/1986). The link between that representation and the object itself (which object, since I am only a *methodological* solipsist, I shall assume exists) is a causal one. It may, as Sayre (1986) suggests, even be the ultimate source of semantic information. But it is noumenally inaccessible to an individual mind. As Jackendoff (1985, p. 24) puts it, "the semantics of natural language is more revealing of the internal representation of the world than of the external world *per se*".

Finally, some comments are in order about the different kinds of meaning that we have identified. I shall postpone this, however, till we have had a chance to look at a prototype natural-language-understanding system in operation.

3.6. *SNePS/CASSIE: A Prototype AI Natural-Language-Understanding System*

How might all this be managed in an AI natural-language-understanding system? Here, I shall doff my philosopher's hat and don my computer scientist's hat. Rather than try to say how this can be managed by *any* natural-language-understanding system, I shall show how one such system manages it. The system I shall describe — and to which I have alluded earlier — is SNePS/CASSIE: an experiment in "building" (a model of) a mind (called 'CASSIE') using the SNePS knowledge-representation and reasoning system. SNePS, the *Semantic Network Processing System* (Shapiro 1979; Maida and Shapiro 1982; Shapiro and Rapaport 1986, 1987; Rapaport 1986c), is a semantic-network language with facilities for building semantic networks to represent information, for retrieving information from them, and for performing inference with them. There are at least two sorts of semantic networks in the AI literature (see Findler 1979 for a survey): The most common is what is known as an "inheritance hierarchy", of which the most well-known is probably KL-ONE (*cf.* Brachman and Schmolze 1985). In an inheritance semantic network, nodes represent concepts, and arcs represent relations between them. For instance, a typical inheritance semantic network might represent the propositions that Socrates is human and that humans are mortal as in Figure 1a. The interpreters for such systems allow properties to be "inherited", so that the fact that Socrates is mortal does not also have to be stored at the Socrates-node. What is essential, however, is that the representation of a proposition (e.g., that Socrates is human) consists only of separate representations of the individuals (Socrates and the property of being human) linked by a relation arc (the "ISA" arc). That is, propositions are not themselves objects. By contrast,

SNePS is a *propositional* semantic network. By this is meant that all information, including propositions, "facts", etc., is represented by nodes. The benefit of representing propositions by nodes is that propositions about propositions can be represented with no limit. ... Arcs merely form the underlying syntactic structure of SNePS. This is embodied in the restriction that one cannot add an arc between two existing nodes. That would be trantamount to telling SNePS a proposition that is not represented as a node. ... Another restriction is the *Uniqueness Principle*: There is a one-to-one correspondence between nodes and represented concepts. This principle guarantees that nodes will be shared whenever possible and that nodes represent intensional objects. (Shapiro and Rapaport 1987.)

Thus, for example, the information represented in the inheritance network of Figure 1a could (though it need not) be represented in SNePS as in Figure 1b; the crucial difference is that the SNePS proposi-

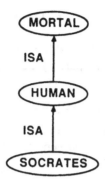

Fig. 1a. An "ISA" inheritance-hierarchy semantic network

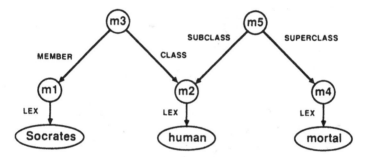

Fig. 1b. A SNePS propositional semantic network (m3 and m5 represent the propositions that Socrates is human and that humans are mortal, respectively)

tional network contains nodes (m3, m5) representing the *propositions* that Socrates is human and that humans are mortal, thus enabling representations of beliefs and rules *about* those propositions. (In fact, the network of Figure 1a could *not* be built in SNePS, by the first restriction cited; *cf.* Shapiro 1979, Section 2.3.1.) My colleagues and I in the SNePS Research Group and the Graduate Group in Cognitive Science at SUNY Buffalo are using SNePS to build a natural-language-understanding system, which we call 'CASSIE', the *C*ognitive *A*gent of the *S*NePS *S*ystem — an *I*ntelligent *E*ntity (Shapiro and Rapaport 1986, 1987; Bruder *et al.* 1986). The nodes of CASSIE's knowledge base implemented in SNePS are her beliefs and other objects of thought, in the Meinongian sense. (Needless to say, I hope, nothing about CASSIE's *actual* state of "intelligence" should be inferred from her name!)

A brief conversation with CASSIE is presented in Appendix 1. Here, I shall sketch a small part of her natural-language-processing algorithm. Suppose that the user tells CASSIE,

> Young Lucy petted a yellow dog.

CASSIE's tacit linguistic knowledge, embodied in an augmented transition network (ATN) parsing-and-generating grammar (Shapiro 1982), "hears" the words and builds the semantic network shown in Figure 2 in CASSIE's "mind" in the following way:

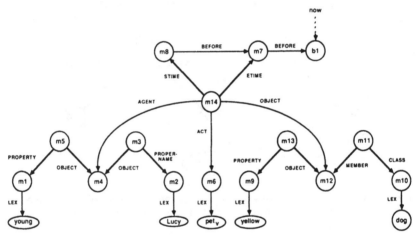

Fig. 2. CASSIE's belief that young Lucy petted a yellow dog

(1)   CASSIE builds a node (b1) representing the current time (the "now"-point; *cf.* Almeida and Shapiro 1983, Almeida 1987).

(2)   ● CASSIE "hears" the word 'young'.
        ● If she has not heard this word before, she *builds* a "sensory" node (labeled 'young') representing the *word* that she hears and a node (m1) representing the internal *concept* produced by her having heard it — this concept node is linked to the sensory node by an arc labeled 'LEX'. (See Figure 3; the formal semantic interpretation

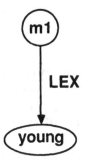

Fig. 3. SNePS network for the concept expressed in English as 'young'

of this small network is: m1 is the Meinongian object of thought corresponding to the utterance of 'young'; *cf.* Rapaport 1985a; Shapiro and Rapaport 1986, 1987.)

- If she *has* heard it before, she *finds* the already-existing concept node. (Actually, she attempts to "find" before she "builds"; henceforth, this process of "finding-or-building" will be referred to simply as "building", since it is in conformity with the Uniqueness Principle.)

(3)   • CASSIE hears the word 'Lucy'.
    She builds a sensory node (labeled 'Lucy') for the *word* 'Lucy', a concept node (m2; linked to the sensory node by a LEX arc) for the *name* 'Lucy', a concept node (m4) representing an individual, and a proposition node (m3) representing that the individual is named 'Lucy' (using an OBJECT–PROPER-NAME case frame).[12]

- She (unconsciously) determines, by means of the ATN, that Lucy is young, and she builds a proposition node (m5) representing this (using an OBJECT-PROPERTY case frame).

(4)   She hears the word 'petted', and (skipping a few details, for clarity) she builds a sensory node (labeled 'pet$_v$') for the verb 'pet', a concept node (m6; linked to the sensory node by a LEX arc) for the act of petting, and a temporal network (m7 and m8, linked by BEFORE arcs to b1) indicating that this act occurred before "now" (= the time of utterance).

(5)   She hears 'yellow' and processes it as she did 'young' (building m9).

(6)   She hears 'dog' and builds:
  ● a sensory node for it.
  ● a concept node (m10) representing the class whose label is 'dog',
  ● a concept node (m12) representing the individual yellow dog whom young Lucy petted,
  ● a proposition node (m11) representing that this individual concept node is a MEMBER of the CLASS whose label is 'dog',
  ● a proposition node (m13) representing that that individual concept node (the dog) is yellow, and, finally,
  ● a proposition node (m14) representing that the dog is the OBJECT of an AGENT-ACT-OBJECT case frame whose agent is Lucy, whose act is petting, whose starting time is m8, and whose ending time is m7.

(7)   She generates a sentence expressing her new understanding. I shall not go into the details of the generation algorithm, except to point out that she uses the sensory nodes to generate the words to express her new belief (cf. Shapiro 1982 for details).

(8)   As the conversation shown in Appendix 1 continues, CASSIE's semantic network is continually updated with new nodes, as well as with new links to old nodes (cf. Fig. 4).

The crucial thing to see is that the semantic network (Fig. 2) that represents CASSIE's belief (the belief produced by her understanding of the user's sentence) is her interpretation of that sentence and that it has three parts: One part consists of the sensory nodes: the nodes at the heads of LEX arcs; a second part consists of the entire network except for that set of sensory nodes and the LEX arcs; and the third part consists of the LEX arcs themselves, which link the other two, major, parts of the network.

Notice that the sensory-node set by itself has (or seems to have) no structure. This is consistent with viewing these as CASSIE's internal representations, causally produced, of external entities (in this case, utterances) to which she has no other kind of access and, hence, no knowledge of their relationships. As I suggested earlier when discussing deixis, if we had a visual-input module, there might be a set of sensory nodes linked by, say, "PIX" arcs. At present, I see no need for any direct links between visual and linguistic sensory nodes, even between those that, in some extensional sense, represent the same entity; any such links would be forged by means of links among the concept nodes at the *tails* of LEX and PIX arcs (but this is a matter for future investigation, as is the entire issue of the structure and behavior of sensory nodes).

The concept-node set, on the other hand, has a great deal of structure. It is this fragment of the entire network that represents CASSIE's internal understanding. If CASSIE were not intended to converse in natural language, there would not be any need for LEX arcs or sensory nodes. If CASSIE's sensory nodes were replaced by others, she would converse in a notational variant of English (*cf.* Quine 1969, Section II, p. 48). If her generation grammar were replaced with one for French and her sensory nodes replaced with "French" ones, she would understand English but speak in French (though here, no doubt, other modifications would be required in order for her knowledge repre- sentation system to be used in this way as an "interlingua", as machine- translation researchers call it).[13] In each of these cases, *the structure of her mind and, thus, her understanding — which would be in terms of purely syntactic symbol manipulation — would remain the same.* Only the external semantic interpretation function, so to speak, would differ. "Meaning," in the sense of internal semantics, "is determined by structures, truth by facts" (Garver 1986, p. 75).[14]

A nice metaphor for this is Carnap's example of the railroad map whose station names (but not rail-line names) have been removed; the stations can still be uniquely identified by the rail lines that meet at them. The "meaning" of a node in such a network is merely its locus in the entire network. In Appendix 2, I sketch how this might be done in a SNePS-like semantic network. (See Carnap 1928, Section 14, pp. 25- 27; *cf.*: Quillian 1967, p. 101; Quillian 1968, Section 4.2.1, especially p. 238; and Quine 1951, Section 6, especially pp. 42f. Quine's "fabric which impinges on experience only along the edges" nicely captures the notion of a semantic network with sensory nodes along the edges.)

3.7. *Varieties of Meaning*

At this point, we can make the promised comments on the different kinds of meaning. Recall the three-part nature of the semantic network: the sensory nodes, the LEX arcs, and the main body of the semantic network. The meaning of a node, in one sense of 'meaning', is its locus in the network; this is, I have been urging, the central meaning of the node. This locus provides the *internal* semantics of the node and, hence, of the words that label sensory nodes. Considered as an object of thought, a node can be taken as being constituted by a collection of properties, hence as an intensional, Meinongian object. The locus in the network of a node at the tail of a LEX arc can be taken as a collection of propositional functions corresponding to the open sentences that are satisfied by the word that labels the sensory node. In particular, at any time t, the collection will consist of those open sentences satisfied by the word that were heard by the system prior to t. (For details, see Rapaport 1981.) But this means that the internal meaning of the word will change each time the word is heard in a new sentence. So, the internal meaning is extensional. This curious duality of intension and extension is due, I think, to the fine grain of this sort of meaning: it is intensional because of its fine grain and the fact that it is an object of thought; but it is extensional in that it is determined by a set-in-extension.

But there is a meaning determined by a set-in-intension, too. This may be called the "definitional" meaning of the word. It is a subset of the internal meaning, whose characterizing feature is that it contains those propositions in the semantic network that are the meaning postulates of the word. That is, these propositions are the ones from which all other facts containing the word can be inferred (together with miscellaneous other facts; again, *cf.* Rapaport 1981). Thus, this kind of meaning is an internal, intensional meaning; it is a sort of idiosyncratic or idiolectal *Sinn*.

Both of these kinds of meaning are or consist of internal symbols to be manipulated. To fill out the picture, there may also be the (physical) objects in the world, which are the external, extensional, referential meanings of the words. But these are not symbols to be manipulated and are irrelevant for natural-language understanding.

3.8. *Discourse*

Another aspect of my interpretation of natural-language understanding is the importance of *discourse* (sequences of sentences), rather than isolated sentences, for the construction of the system's knowledge base. Discourse is important for its *cumulative* nature:

[P]utting one sentence after another can be used to express time sequence, deductive necessity, cause, exemplification or other relationships, *without any words being used to express the relation.* (Mann *et al.* 1981, Part 1, p. 6.)

This aspect of discourse illuminates the role of internal semantics in a way hinted at earlier. To provide a semantic interpretation for a language by means of an internal semantic network (or a discourse representation structure) is to provide a more or less formal *theory* about the linguistic data (much as Chomsky 1965 said, though this is a *semantic* theory). But, in discourse as in science, the data underdetermine the theory: it is internal semantic network — the mind of the understander — that provides explicit counterparts to the unexpressed relations.

Isolated sentences (so beloved by philosophers and linguists) simply would not serve for enabling a system such as CASSIE to understand natural language: they would, for all practical purposes, be random, unsystematic, and *unrelated* data. The *order* in which CASSIE processes (or "understands") sentences is important: Given a mini-discourse of even as few as two sentences,

$$s_1. s_2.$$

her interpretation of s_2 will be partially determined by her interpretation of s_1. Considered as part of a discourse, sentence s_2 is syntactically within the "scope" of s_1; hence, the interpretation of s_2 will be within the scope of the interpretation of s_1. (This aspect of discourse is explored in Maida and Shapiro 1982, Mann and Thompson 1983, Kamp 1984 and forthcoming, Fauconnier 1985, Asher 1986, 1987, and Wiebe and Rapaport 1986.) Thus, discourse and, hence, *conversation* are essential, the latter for important feedback in order to bring the conversers' semantic networks into alignment.

4. WOULD A COMPUTER "REALLY" UNDERSTAND?

I have considered what it would be for a computer to understand natural language, and I have argued for an interpretation of "understanding natural language" on which it makes sense to say that a computer *can* understand natural language. But there might still be some lingering doubts about whether a computer that understands natural language in this sense "really" understands it.

4.1. *The Korean-Room Argument*

Let us start with a variation of Searle's Chinese-Room Argument, which
may be called the "Korean Room Argument" (though we shall do away
with the room):[15]

> Imagine a Korean professor of English literature at the
> University of Seoul who does not understand spoken or
> written English but who is, nevertheless, a world authority
> on Shakespeare. He has established and maintains his repu-
> tation as follows: He has only read Shakespeare in excellent
> Korean translations. Based on his readings and, of course,
> his intellectual acumen, he has written, in Korean, several
> articles on Shakespeare's play. These articles have been
> translated for him into English and published in numerous,
> well-regarded, English-language, scholarly journals, where
> they have met with great success.

The Korean-Room-Argument question is this: Does the Korean scholar
"understand" Shakespeare? Note that, unlike the Chinese-Room Argument, the issue is not whether he understands English; he does not. Nor
does he mechanically ("unthinkingly") follow a translation algorithm;
others do his translating for him. Clearly, though, he does understand
Shakespeare — the literary scholarly community attests to that — and,
so, he understands *something*.

Similarly, Searle in the Chinese room *can* be said to understand
something, even if it isn't Chinese. More precisely (because, as I urged
in Section 3.2, I don't think that Searle's Chinese-Room Argument is as
precisely spelled out as it could be), an AI natural-language-understanding system can be said to understand something (or even to
understand *simpliciter*), insofar as what it is doing is semantic interpretation.[16] (Of course, it does this syntactically by manipulating the
symbols of its semantic interpretation.) We can actually say a bit more:
it understands *natural language*, since it is a natural language that it is
semantically interpreting. It is a separate question whether that which it
understands is *Chinese*.[17] Now, I think there are *two* ways in which this
question can be understood. In one way, it is quite obvious that if the
system is understanding a natural language, then, since the natural
language that it is understanding is Chinese, the system must be
understanding Chinese. But in other ways, it is not so obvious. After all,

the system shares very little, if any, of Chinese culture with its inter-locutors, so in what sense can it be said to "really" understand Chinese? Or in what sense can it be said to understand Chinese, as opposed to, say, code of the computer-programming language that the Chinese "squiggles" are transduced into? This Chinese-*vs.*-code issue can be resolved in favor of Chinese by the Korean-Room Argument: just as it is *Shakespeare*, not merely a Korean *translation* of Shakespeare, that the professor understands, so it is Chinese, and not the programming-language code, that Searle-in-the-room understands.

As for the cultural issue, here, I think, the answer has to be that the system understands Chinese as well as any nonnative-Chinese human speaker does (and perhaps even better than some). The only qualm one might have is that, in some vague sense, what *it* means or understands by some expression might not be what the native Chinese speaker means or understands by it. But as Quine and, later, Schank have pointed out, the same qualm can beset a conversation in our native tongue between you and me (Quine 1969, p. 46; Schank 1984, Ch. 3). As I said earlier, we systematically *mis*understand each other: we can *never* mean *exactly* what another means; but that does not mean that we cannot understand each other. We might not "really" understand each other in some deep psychological or empathic sense (if, indeed, sense can be made of that notion; *cf.* Schank 1984, pp. 44-47), but we do "really" understand each other — and the AI natural-language-understanding system can "really" understand natural language — in the only sense that matters. Two successful conversers' understandings of the expressions of their common language will (indeed, they *must*) eventually come into alignment, even if one of the conversers is a computer (*cf.* Shapiro and Rapaport 1987).

4.2. *Simon and Dreyfus vs. Winograd and SHRDLU*

The considerations thus far can help us to see what is wrong with Herbert Simon's and Hubert Dreyfus's complaints that Terry Winograd's SHRDLU program does not understand the meaning of 'own' (Winograd 1972; Simon 1977, cited in Dreyfus 1979). Simon claims that "SHRDLU's test of whether something is owned is simply whether it is tagged 'owned'. These is no intensional test of ownership . . ." (Simon 1977, p. 1064/Dreyfus 1979, p. 13). But this is simply not correct: When Winograd tells SHRDLU, "I own blocks which are not red, but I don't own anything which supports a pyramid," he comments that these are "two new theorems . . . created for proving things about 'owning'" (Winograd 1972, p. 11, *cf.* pp. 143f; cited also in Dreyfus 1979, p. 7). SHRDLU doesn't *merely* tag blocks (although it can also do that); rather, there are procedures for determining whether something is "owned" — SHRDLU can figure out new cases of ownership.[18] So there *is* an intensional test, although it may bear little or no resemblance, except for the label 'own', to *our* intensional test of ownership. But even this claim about lack of resemblance would only hold at an early stage in a conversation; if SHRDLU were a perfect natural-language-understanding program that *could* understand English (and no one claims that it is), *eventually* its intensional test of ownership would come to resemble ours *sufficiently for us to say that it understands 'own'*.

But Dreyfus takes this one step further:

> [SHRDLU] still wouldn't understand, unless it also understood that it (SHRDLU) couldn't own anything, since it isn't a part of the community in which owning makes sense. Given our cultural practices which constitute owning, a computer cannot own something any more than a table can. (Dreyfus 1979, p. 13.)

The "community", of course, is the *human* one (which is biological; *cf.* Searle). There are several responses one can make. First of all, taken literally, Dreyfus's objection comes to nothing: it should be fairly simple to give the computer the information that, because it is not part of the right community, it cannot own anything. But that, of course, is not Dreyfus's point. His point is that it cannot *understand* 'own' because it *cannot* own. To this, there are two responses. For one thing, cultural practices can change, and, in the case at hand, they are already changing (for better or worse): computers *could* legally own things just as corporations, those other nonhuman persons, can (*cf.* Willick 1985).[19] But even if they can't, or even if there is some other relationship that computers are forever barred from participating in (even by means of a simulation), that should not prevent them from having an understanding of the concept. After all, women understood what voting was before they were enfranchised, men can understand what pregnancy is, and humans can understand what (unaided) flying is.[20] A computer could learn and understand such expressions to precisely the same extent, and that is all that is needed for it to really understand natural language.

5. DOES THE COMPUTER UNDERSTAND THAT IT UNDERSTANDS?

There are two final questions to consider. The first is this: Suppose that we have our ultimate AI natural-language-understanding program that passes the Turing Test; does it understand that it understands natural language? The second, perhaps prior, question is: *Can* it understand that it understands?

Consider a variation on the Korean-Room Argument. Suppose that the Korean professor of English literature has been systematically misled, perhaps by his translator, into thinking that the author of the plays that he is an expert on was a Korean playwright named, say, Jaegwon. The translator has systematically replaced 'Shakespeare' by 'Jaegwon', and *vice versa*, throughout all of the texts that were translated. Now, does the Korean professor understand Shakespeare? Does he understand that he understands Shakespeare? I think the answer to the latter question is pretty clearly 'No'. The answer to the former question is not so clear, but I shall venture an answer: Yes.

Before explaining this answer, let's consider another example (adapted from Goodman 1986). Suppose that a student in my Theory of Computation course is executing the steps of a Turing-machine program, as an exercise in understanding how Turing machines work. From time to time, she writes down certain numerals, representing the output of the program. Let us even suppose that they are Arabic numerals (i.e., let us suppose that she decodes the actual Turing-machine output of, say, 0s and 1s, into Arabic numerals, according to some other algorithm). Further, let us suppose that, *as a matter of fact*, each number that she writes down is the greatest common divisor of a pair of numbers that is the input to the program. Now, does she know that that is what the output is? Not necessarily; since she might not be a math major (or even a computer science major), and since the Turing-machine program need not be labeled 'Program to Compute Greatest Common Divisors', she might not know what she is doing *under that description*. Presumably, she does know what she is doing under some other description, say, "executing a Turing-machine program"; but even this is not necessary. Since, as a matter of fact, she *is* computing greatest common divisors, if I needed to know what the greatest common divisor of two numbers was, I could ask her to execute that program for me. She would not have to understand what she is doing, under that description, in order to do it.

Similarly, the Korean professor does not have to understand that he understands Shakespeare in order to, in fact, understand Shakespeare. And, it should be clear, Searle-in-the-Chinese-room does not have to understand that he understands Chinese in order to, in fact, understand Chinese. So, a natural-language-understanding program does not have to understand that it understands natural language in order to understand natural language. That is, this use of '*understand*' is referentially transparent! If a cognitive agent A understands X, and X is equivalent to Y (in some relevant sense of equivalence), then A understands Y.

But this is only the case for "first-order" understanding: *understanding that one understands* is referentially opaque. I don't think that this is inconsistent with the transparency of first-order understanding, since this "second-order" sense of 'understand' is more akin to 'know that' or 'be aware', and the "first-order" sense of 'understand' is more akin to 'know how'.

Now, *can* the Korean professor understand that he understands Shakespeare? Of course; he simply needs to be told that it is Shakespeare (or merely someone *named* 'Shakespeare'; cf. Hofstadter *et al.* 1982), not someone named 'Jaegwon', that he has been studying all these years. Can my student understand that what she is computing are greatest common divisors? Of course; she simply needs to be told that. Moreover, if the program that she is executing is suitably modularized, the names of its procedures might give the game away to her. Indeed, an "automatic programming" system would have to have access to such labels in order to be able to construct a program to compute greatest common divisors (so-named or so-described); and if those labels were linked to a semantic network of mathematical concepts, it could be said to understand what that program would compute. And the program itself could understand what it was computing if it had a "self-concept" and could be made "aware" of what each of its procedures did.

This is even clearer to see in the case of a natural-language-understanding program. A natural-language-understanding program can be made to understand what it is doing — can be made to understand that it understands natural language — by, first, telling it (in natural language, of course) that that is what it is doing. Merely telling it, however, is not sufficient by itself; that would merely add some network structure to its knowledge base. To gain the requisite "awareness", the system would have to have LEX-like arcs linked, if only indirectly, to its actual natural-language-processing module — the ATN parser-generator, for instance. But surely that can be done; it is, in any event, an empirical issue as to precisely how it would be done. The point is that it would be able to associate expressions like 'understanding natural language' with certain of its activities. It would then understand what those expressions meant in terms of what those activities were. It would not matter in the least if it understood those activities in terms of bit-patterns or in terms of concepts such as "parsing" and "generating"; what would count is this: that it understood the expressions in terms of its actions; that its actions were, in fact, actions for understanding natural language; and, perhaps, that 'understanding natural language' was the label that its interlocutors used for that activity.

6. CONCLUSION

By way of conclusion, consider (a) the language L that the system understands, (b) the external world, W, about which L expresses information, and (c) the language (or model of W), L_w, that provides the interpretation of L. As William A. Woods (among many others) has made quite clear, such a "meaning representation language" as L_w is involved in two quite separate sorts of semantic analyses (Woods 1978; cf. Woods 1975 and, especially, Kamp 1981).

There must be, first, a semantic interpretation function, P (for 'parser'), from utterances of L (the input to the system) to the system's internal knowledge base, L_w. L_w is the system's model of W, as filtered through L. There will also need to be a function, G (for 'generator'), from L_w to L, so that the system can express itself. P and G need not, and probably should not, be inverses (they are not in SNePS/CASSIE); "they" might also be a single function (as in SNePS/CASSIE; cf. Shapiro 1982). Together, P, G, and L_w constitute the central part of the *system's* understanding of L.

But, second, there must also be a semantic interpretation of L_w in terms of W (or in terms of *our* idiosyncratic L_ws) — i.e., a semantic interpretation of the domain of semantic interpretation. Since L_w is itself a formal language, specified by a formal syntax, it needs a semantics (*cf.* Woods 1975, McDermott 1981). But this semantic interpretation is merely *our* understanding of L_w. It is independent of

and external to the relevant semantic issue of how the *system* understands *L*. (This semantic interpretation of the knowledge base is
provided for SNePS/CASSIE by interpreting L_w as a Meinongian
theory of the objects of thought; *cf.*: Rapaport 1985a; Shapiro and
Rapaport 1986, 1987.)

There may be another relationship between L_w and *W*, although this
may also be provided by the semantic interpretation of L_w. This
relationship is the causal one from *W* to L_w, and there is no reason to
hold that it is limited to humans (or other biological entities). It
produces the sensory nodes, but — other than that — it is also
independent of and external to the system's understanding of *L*.

Once the sensory and concept nodes (or their analogues in some
other knowledge-representation system) are produced, the actual causal
links cease to be relevant to the system's *understanding* (except — and I
am willing to admit that this is an important exception — for purposes
of the system's communication with others), thus highlighting the
representationalism of the system.

Searle holds, however, that the links — the access to *W* — are
necessary for understanding, that humans have (or that only biological
entities can have) such access, that computers lack it, and, hence, that
computers cannot understand. By contrast, I hold that *if* such access
were needed, then computers could have it, too, so that Searle's
pessimism with respect to computer understanding is unsupported. I
also hold that such access is *not* needed, that, therefore, humans don't
need it either (here is where methodological solipsism appears), so that,
again, there's no support for Searle's conclusion. I agree with Searle that
semantics is necessary for understanding natural language, but that the
kind of semantics that's needed is the semantics provided by an
internal semantic interpretation, which is, in fact, syntactic in nature
and, hence, computable. Syntax suffices.

APPENDIX 1: A "CONVERSATION" WITH CASSIE

Following is the transcript of a "conversation" with CASSIE. A com-
mented version of part of it appears in Shapiro and Rapaport 1986,
1987. User input is on lines with the :-prompt; CASSIE's output is on
the lines that follow. A fragment of the full network showing CASSIE's
state of mind at the end of the conversation is shown in Figure 4.

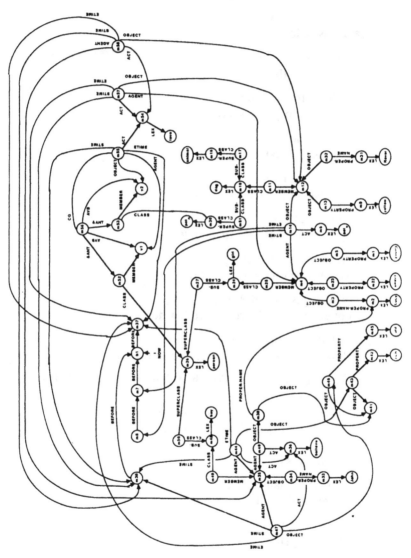

Fig. 4. Small fragment of CASSIE's beliefs at the end of the conversation.

: Young Lucy petted a yellow dog
I understand that young Lucy petted a yellow dog

: What is yellow
a dog is yellow

: Dogs are animals
I understand that dogs are animals

: Who petted an animal
young Lucy petted a yellow dog

: Lucy is clever
I understand that young Lucy is clever

: Lucy is a girl
I understand that clever young Lucy is a girl

: Girls are people
I understand that girls are people

: What was petted by a person
a yellow dog was petted by clever young Lucy

: The dog is named Rover
I understand that Rover is a yellow dog

: What was petted by a person
yellow Rover was petted by clever young Lucy

: For every d if d is a dog then there is a b such that b is a
boy and b owns d
I understand that for every d, if d is a dog
    then there exists a b such that b is a boy
and
        b owns d

: The boy is named John
I understand that John is a boy

: Who owns Rover
John owns yellow Rover

: John believes that Lucy is rich
I understand that John believes that Lucy is rich

: Who is rich
I don't know.

: Who is clever
young Lucy is clever

: John believes that Lucy is old
I understand that John believes that rich Lucy is old

: Boys are people
I understand that boys are people

: Dogs are pets
I understand that dogs are pets

: For every p and d if p is a person and d is a pet then p
loves d
I understand that for every d and p, if p is a person
and
   d is a pet
   then p loves d

: Who loves a pet
clever young Lucy loves yellow Rover
and
John loves yellow Rover

APPENDIX 2: DESCRIBING A NODE VIA ITS NETWORK LOCUS

How can a node be identified if there are no LEX arcs or sensory
nodes? That is, how can they be identified if they have no names? The
answer is, by descriptions. It is important to see that the identifiers of
the nodes ("m1", etc.) are *not* names (or labels); they convey no
information to the system. (They are like the nodes of a tree each of
which contains no data but only pointers to its left and right children.
The sensory nodes are like leaf nodes that do contain data; their labels
do convey information.) The nodes can be described solely in terms of
their locus in the network, i.e., in terms of the structure of the arcs
(which *are* labeled) that meet at them. If a node has a unique "arc
structure", then it can be uniquely described by a *definite* description; if
two or more nodes share an arc-structure, they can only be given
indefinite descriptions and, hence, cannot be uniquely identified. That

is, they are indistinguishable to the system, unless each has a LEX arc emanating from it. (*Cf.*, again, Carnap 1928, Section 14.) Thus, for example, in the network in Figure 5, m1 is *the* node with precisely two ARG arcs emanating from it, and b1 is *a* node with precisely one ARG arc entering it (and similarly for b2). In keeping with the notion that the internal meaning of a node is its locus in the *entire* network, the *full* descriptions of m1 and b1 (or b2) are:

(m1)   *the* node with one ARG arc to *a* base node and with another ARG arc to *a* base node.

(b1)   *a* base node with an ARG arc from *the* node with one ARG arc to it and with another ARG arc to a base node.

(A base node is a node with no arcs leaving it; no SNePS node can have an arc pointing to itself.) The pronominal 'it' has widest scope; i.e., its anaphoric antecedent is always the node being described. Note that each node's description is a monad-like description of the *entire* network from its own "point of view".

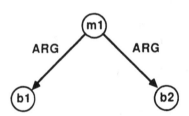

Fig. 5. A small SNePS network.

NOTES

[1] This material is based upon work supported by the National Science Foundation under Grant Nos. IST-8504713, IRI-8610517, and by SUNY Buffalo Research Development Fund Award No. 150-8537-G. I am grateful to Randall R. Dipert, Michael Leyton, Ernesto Morgado, Jane Pease, Sandra Peters, Stuart C. Shapiro, Marie Meteer Vaughan, Janyce M. Wiebe, Albert Hanyong Yuhan, and other colleagues in the SNePS Research Group and the SUNY Buffalo Graduate Group in Cognitive Science for discussions on these topics and comments on earlier versions of this essay.
[2] *Cf.* my earlier critiques of Searle, in which I distinguish between an abstraction, an implementing medium, and the implemented abstraction (Rapaport 1985b, 1986b, and forthcoming).
[3] The "weak/strong" terminology is from Searle 1980.

[4] Or: that it is a woman. More precisely. Turing describes the Imitation Game, in which "a man (A), a woman (B), and an interrogator (C)" have as their object "for the interrogator to determine which of the other two is the man and which is the woman". Turing then modifies this:

> We now ask the question, "What will happen when a machine takes the part of A in this game?" Will the interrogator decide wrongly as often when the game is played like this as he does when the game is played between a man and a woman? These questions replace our original, "Can machines think?" (Turing 1950, p. 5.)

[5] Randall R. Dipert has suggested to me that Searle's Chinese-Room Argument does show that what *executes* the program (viz., the central processing unit) does not understand, leaving open the question whether the *process* might understand.

[6] Not to be confused with scripts in the sense of Schank's AI data structures.

[7] Similarly, part of my argument in this essay may be roughly paraphrased as follows: I want to understand what it means to understand natural language; I believe that it is capable of being understood (that it is not a mystery) and that, for a system to understand natural language, certain formal techniques are necessary and sufficient; these techniques are computational; hence, understanding natural language is a recursive function.

[8] I take (7) to be the conclusion, since (1)—(6) are in response to the question, "*Can computers add?*" (p. 25; italics added). If I am taking Dretske too literally here, then simply end the argument at step (6).

[9] Note that, for independent reasons, the computer *will* need an internal representation or model of itself. Maybe this won't be a *complete* self-model, on pain of infinite regress, but then neither is ours. If needed, it, and we, can use an *external* model that *is* complete, via self-reflection; *cf.* Case 1986, esp. p. 91. For more on self-models, *cf.*: Minsky 1965; Rapaport 1984, 1986c; Smith 1986.

[10] I venture to say that the mutual learning and adjusting process has already begun: studying such computers and primitive AI systems as we have now has led many philosophers and AI researchers to this kind of opinion.

[11] It should be obvious by now that by 'understand' I do not mean some sort of "deep" psychological understanding, merely that sort of understanding required for understanding natural language. *Cf.* Schank 1984, Ch. 3.

[12] *Cf.* Shapiro and Rapaport 1986 and 1987 for the formal syntax and semantics of this and the other case frames. The node identifiers ("m1", etc.) are generated by the underlying program in an implementation-dependent order; the order and the identifiers are inessential to the semantic network.

[13] The importance of the knowledge base, whether it is a semantic network, a discourse representation structure, or some other data type, for understanding natural language has some interesting implications for machine translation. There are several paradigms for machine translation; two are relevant for us: the "transfer" approach and the "interlingua" approach (*cf.* Slocum 1985). Transfer approaches typically do not use a knowledge base, but manipulate syntactic structures of the source language until they turn into syntactic structures of the target language. Such a system, I would argue, cannot be said to understand the natural languages it deals with. Interlingua approaches, on the other hand, do have a sort of knowledge base. They are "mere" symbol manipulation systems, but the symbols that get manipulated include those of the

system's internal knowledge-representation system: hence, interlingua machine-translation systems have at least the *potential* for understanding. (Searle's Chinese-language
program appears to be more like a transfer system (for, say, translating Chinese into
Chinese) than an interlingua system, despite the use of Schank-like scripts.)

Note, too, that this suggests that the "machine-translation problem" is coextensive
with the "natural-language-understanding problem" and, thus (*cf.* Section 1, above),
with the general "AI problem": solve one and you will have solved them all. (This
underlies Martin Kay's pessimism about the success of machine translation; *cf.* Kay
1986).

[14] Garver's "challenge of metaphor", it must be noted, is also a challenge for the theory
presented here, which I hope to investigate in the future.

[15] The Korean-Room Argument was suggested to me by Albert Hanyong Yuhan.

[16] And to the extent that it does *not* do semantic interpretation, it does not understand. My former teacher, Spencer Brown, recently made the following observation:

> As for Searle, I myself have been a *corpus vile* for his "experiment": once
> I conveyed a message from one mathematician to another, with complete
> understanding on the part of the second and with total, nay, virginal,
> ignorance on my part of the meaning of the message. Similarly I have
> conveyed a message from my doctor to my dentist without knowing what I
> was telling. Q.E.D.: I can't think. This is something I have always sus-
> pected. (Personal communication, 1986; *cf.* Rapaport 1981, p. 7.)

But the conclusions (both of them!) are too hasty: all that follows is that he did not
understand certain isolated statements of mathematics and medicine. And this was, no
doubt, because he lacked the tools for interpreting them and an appropriate knowledge
base within which to fit them.

[17] I owe this way of looking at my argument to Michael Leyton.

[18] I am indebted to Stuart C. Shapiro for pointing this out to me.

[19] I am indebted to Shapiro for this reference.

[20] The examples are due to Shapiro. My original example was that, as a U.S. citizen, I
am probably forever enjoined from some custom unique to and open only to French
citizens; yet surely I can learn and understand the meaning of the French expression for
such a custom. But finding an example of such a custom is not as easy as it seems.
Voting in a French election, e.g., isn't quite right, since I *can* vote in U.S. elections, and
similarly for other legal rights or proscriptions. Religious practices "unique" to one
religion usually have counterparts in others. Another kind of case has to do with
performatives: I cannot marry two people merely by reciting the appropriate ritual,
since I do not have the right to do so. The case of owning that Simon and Dreyfus focus
on is somewhat special, since it is both in the legal realm as well as the cultural one:
there are (allegedly or at least conceivably) cultures in which the institutions of ownership and possession are unknown. I maintain, however, that the case of humans and
computers are parallel: we share the same abilities and inabilities to understand or act.

REFERENCES

Allen, Woody: 1980, 'The Kugelmass Episode' in W. Allen, *Side Effects*, Random
House, New York, 41—55.

Almeida, Michael J.: 1987, 'Reasoning about the Temporal Structure of Narratives', Technical Report 86—10, Department of Computer Science, Buffalo, SUNY Buffalo.

Almeida, Michael J., and Shapiro, Stuart C.: 1983, 'Reasoning about the Temporal Structure of Narrative Texts', *Proc. Fifth Annual Meeting of the Cognitive Science Society*, Rochester, N.Y., unpaginated.

Appelt, Douglas E.: 1982, 'Planning Natural-Language Utterances', *Proc. of the National Conference on Artificial Intelligence (AAAI-82; Pittsburgh)*, Morgan Kaufmann, Los Altos, CA, pp. 59—62.

Appelt, Douglas E.: 1985, 'Some Pragmatic Issues in the Planning of Definite and Indefinite Noun Phrases', *Proc. 23rd Annual Meeting of the Assoc. for Computational Linguistics (University of Chicago)*, Assoc. for Computational Linguistics, Morristown, N.J.; pp. 198—203.

Asher, Nicholas: 1986, 'Belief in Discourse Representation Theory', *Journal of Philosophical Logic* **15**, 127—89.

Asher, Nicholas: 1987, 'A Typology for Attitude Verbs and Their Anaphoric Properties', *Linguistics and Philosophy* **10**, 125—197.

Brachman, Ronald J., and Levesque, Hector J.: 1985, *Readings in Knowledge Representation*, Morgan Kaufmann, Los Altos, CA.

Brachman, Ronald J., and Schmolze, James G.: 1985, 'An Overview of the KL-ONE Knowledge Representation System', *Cognitive Science* **9**, 171—216.

Bruder, Gail A., Duchan, Judith F., Rapaport, William J., Segal, Erwin M., Shapiro, Stuart C., and Zubin, David A.: 1986, 'Deictic Centers in Narrative: An Interdisciplinary Cognitive-Science Project', Technical Report No. 86—20, Dept. of Computer Science, SUNY Buffalo, Buffalo.

Carnap, Rudolf: 1928, *The Logical Structure of the World*, R. A. George (trans.), Univ. of California Press, Berkeley, 1967.

Case, John: 1986, 'Learning Machines', in W. Demopoulos and A. Marras (eds.), *Language Learning and Concept Acquisition*, Ablex, Norwood, N.J., 83—102.

Cohen, Philip R., & Perrault, C. Raymond: 1979, 'Elements of a Plan-Based Theory of Speech Acts', *Cognitive Science* **3**, 177—212; reprinted in B. L., Webber & N.J. Nilsson (eds.), *Readings in Artificial Intelligence* Tioga Publishing Co., Palo Alto, pp. 478—95.

Davies, Robertson: 1972, *The Manticore*, In R. Davies, *The Deptford Trilogy*, Penguin, Middlesex, Eng. 1983.

de Saussure, Ferdinand: 1915, *Cours de linguistique générale*, Payot, Paris, 1972.

Dennett, Daniel C.: 1983, 'Intentional Systems in Cognitive Ethology: The 'Panglossian Paradigm' Defended', *Brain and Behavioral Sciences* **6**, 343—390.

Dretske, Fred: 1985, 'Machines and the Mental', *Proc. and Addresses of the American Philosophical Assoc.* **59**, 23—33.

Dreyfus, Hubert L.: 1979, *What Computers Can't Do: The Limits of Artificial Intelligence*, Harper & Row, New York, rev. ed.; Introduction to the Revised Edition, pp. 1—87; reprinted with revisions in Haugeland 1981, pp. 161—204, and in Brachman and Levesque 1985, pp. 71—93. Page references are to Dreyfus's book.

Fauconnier, Gilles: 1985, *Mental Spaces: Aspects of Meaning Construction in Natural Language*, MIT Press, Cambridge, MA.

Findler, Nicholas V.: 1979, *Associative Networks: The Representation and Use of Knowledge by Computers*, Academic Press, New York.

Garver, Newton: 1986, 'Structuralism and the Challenge of Metaphor', *Monist* **69**, 68—86.

Gleick, James: 1985, 'They're Getting Better about Predicting the Weather (Even Though You Don't Believe It)', *The New York Times Magazine*, 27 January.

Goodman, Nicolas D.: 1986, 'Intensions, Church's Thesis, and the Formalization of Mathematics', *Notre Dame Journal of Formal Logic* (forthcoming).

Green, Bert F., Wolf, Alice K., Chomsky, Carol, and Laughery, Kenneth: 1961, 'Baseball: An Automatic Question Answerer', reprinted in E. A. Feigenbaum and J. Feldman (eds.), *Computers and Thought*, McGraw-Hill, New York, 1963, pp. 207—16.

Haugeland, John (ed.): 1981, *Mind Design: Philosophy, Psychology, Artificial Intelligence* MIT Press, Cambridge, MA.

Hofstadter, Douglas R.: 1980, 'Reductionism and Religion', *Behavioral and Brain Sciences* **3**, 433—34.

Hofstadter, Douglas R, Clossman, Gray A., and Meredith, Marsha J.: 1982, 'Shakespeare's Plays Weren't Written by Him, but by Someone Else of the Same Name: An Essay on Intensionality and Frame-Based Knowledge Representation Systems', Indiana University Linguistics Club, Bloomington, IN.

Jackendoff, Ray: 1985, 'Information Is in the Mind of the Beholder', *Linguistics and Philosophy* **8**, 23—34.

Jennings, Richard C.: 1985, 'Translation, Interpretation and Understanding', paper read at the American Philosophical Assoc. Eastern Division (Washington, DC); abstract, *Proc. and Addresses of the American Philosophical Assoc.* **59**, 345—46.

Kamp, Hans: 1984, 'A Theory of Truth and Semantic Representation', in J. Groenendijk, T. M. V. Janssen, and M. Stokhof (eds.), *Truth, Interpretation and Information* Foris, Dordrecht, pp. 1—41.

Kamp, Hans: (forthcoming), *Situations in Discourse*, manuscript, Center for Cognitive Science, University of Texas, Austin, TX.

Kay, Martin: 1986, 'Forum on Machine Translation: Machine Translation will not Work', *Proc. 24th Annual Meeting of the Association for Computational Linguistics (Columbia University)*, Assoc. for Computational Linguistics, Morristown, NJ, p. 268.

Lehnert, W. G., M. G. Dyer, P. N. Johnson, C. J. Yang and S. Harley: 1983, 'BORIS—An Experiment in In-Depth Understanding of Narratives', *Artificial Intelligence* **20**, 15—62.

Levesque, Hector J.: 1984, 'A Logic of Implicit and Explicit Belief', *Proc. of the National Conference on Artificial Intelligence (AAAI-84; Austin, TX)*, Morgan Kaufmann, Los Altos, CA, pp. 198—202.

Maida, Anthony S., and Shapiro, Stuart C.: 1982, 'Intensional Concepts in Propositional Semantic Networks', *Cognitive Science* **6**, 291—330; reprinted in Brachman and Levesque 1985, pp. 169—89.

Mann, William C., Bates, Madeline, Grosz, Barbara J., McDonald, David D., McKeown, Kathleen R., and Swartout, William R.: 1981, 'Text Generation: The State of the Art and the Literature', Technical Report No. ISI/RR-81-101, Information Sciences Institute, Univ. of Southern California, Marina del Rey, CA.

Mann, William C., and Thompson, Sandra S.: 1983, 'Relational Propositions in Discourse', Technical Report No. ISI/RR-83-115, Information Sciences Institute, Univ. of Southern California, Marina del Rey, CA.

McDermott, Drew: 1981, 'Artificial Intelligence Meets Natural Stupidity', in Haugeland 1981, pp. 143—60.

Minsky, Marvin L.: 1965, 'Matter, Mind, and Models', in Minsky 1968, pp. 425—32.

Minsky, Marvin (ed.): 1968, *Semantic Information Processing*, MIT Press, Cambridge, MA.

Minsky, Marvin: 1975, 'A Framework for Representing Knowledge', in Haugeland 1981, pp. 95—128; reprinted in Brachman and Levesque 1985, pp. 245—62.

Neal, Jeannette G.: 1981, 'A Knowledge Engineering Approach to Natural Language Understanding', Technical Report No. 179, Dept. of Computer Science, SUNY Buffalo, Buffalo.

Neal, Jeannette G.: 1985, 'A Knowledge Based Approach to Natural Language Understanding', Technical Report No. 85—06, Dept. of Computer Science, SUNY Buffalo, Buffalo.

Neal, Jeannette G., and Shapiro, Stuart C.: 1984, 'Knowledge Based Parsing', Technical Report No. 213, Dept. of Computer Science, SUNY Buffalo, Buffalo.

Neal, Jeannette G., and Shapiro, Stuart C.: 1985, 'Parsing as a Form of Inference in a Multiprocessing Environment', *Proc. Conf. on Intelligent Systems and Machines* Oakland University, Rochester, MI, pp. 19—24.

Neal, Jeannette G., and Shapiro, Stuart C.: 1987, 'Knowledge Representation for Reasoning about Language', in J. C. Bouderaux *et al.* (eds.), *The Role of Language in Problem Solving* 2, Elsevier/North-Holland, pp. 27—46.

Putnam, Hilary: 1975, 'The Meaning of "Meaning"', reprinted in H. Putnam, *Mind, Language and Reality*, Cambridge University Press, Cambridge Eng, pp. 215—71.

Quillian, M. Ross: 1967, 'Word Concepts: A Theory and Simulation of Some Basic Semantic Capabilities', *Behavioral Science* 12, 410—30; reprinted in Brachman and Levesque 1985, pp. 97—118. Page references are to the reprint.

Quillian, M. Ross: 1968, 'Semantic Memory', in Minsky 1968, pp. 227—70.

Quine, Willard Van Orman: 1951, 'Two Dogmas of Empiricism', reprinted in W. V. O. Quine, *From a Logical Point of View*, Harvard University Press, Cambridge MA, 2nd ed., revised, 1980, pp. 20—46.

Quine, Willard Van Orman: 1969, 'Ontological Relativity', in W. V. O. Quine, *Ontological Relativity and Other Essays* Columbia University Press, New York, pp. 26—68.

Rapaport, William J.: 1976, *Intentionality and the Structure of Existence*. Ph.D. dissertation, Dept. of Philosophy, Indiana University, Bloomington, IN.

Rapaport, William J.: 1981, 'How to Make the World Fit Our Language: An Essay in Meinongian Semantics', *Grazer Philosophische Studien* 14, 1—21.

Rapaport, William J.: 1984, 'Belief Representation and Quasi-Indicators', Technical Report No. 215, Department of Computer Science, SUNY Buffalo, Buffalo.

Rapaport, William J.: 1985a, 'Meinongian Semantics for Propositional Semantic Networks', *Proc. 23rd Annual Meeting Assoc. for Computational Linguistics* (*University of Chicago*), Assoc. for Computational Linguistics, Morristown, N.J. pp. 43—48.

Rapaport, William J.: 1985b, 'Machine Understanding and Data Abstraction in Searle's Chinese Room', *Proc. 7th Annual Meeting Cognitive Science Soc. (University of California at Irvine)*, Lawrence Erlbaum, Hillsdale, N.J. pp. 341—45.
Rapaport, William J.: 1985/1986, 'Non-Existent Objects and Epistemological Ontology', *Grazer Philosophische Studien* **25/26**, 61—95.
Rapaport, William J.: 1986a, 'Searle's Experiments with Thought', *Philosophy of Science* 53: 271—279; preprinted as Technical Report 216, Dept. of Computer Science, SUNY Buffalo, Buffalo, 1984.
Rapaport, William J.: 1986b, 'Philosophy, Artificial Intelligence, and the Chinese-Room Argument', *Abacus* 3 (Summer 1986), 6—17.
Rapaport, William J.: 1986c, 'Logical Foundations for Belief Representation', *Cognitive Science* **10**, 371—422.
Rapaport, William J.: (forthcoming), 'To Think or Not to Think', *Noûs*.
Rapaport, William J.: 1987, 'Belief Systems', in S. C. Shapiro (ed.), *Encyclopedia of Artificial Intelligence*, John Wiley, New York, pp. 63—73.
Russell, Bertand: 1918, 'The Philosophy of Logical Atomism', in B. Russell, *Logic and Knowledge: Essays 1901—1950*, R. C. Marsh (ed.), Capricorn, New York, 1956, pp. 177—281.
Sayre, Kenneth, M.: 1986, 'Intentionality and Information Processing: An Alternative Model for Cognitive Science', *Behavioral and Brain Sciences* **9**, 121—166.
Schank, Roger C.: 1982, *Dynamic Memory: A Theory of Reminding and Learning in Computers and People*, Cambridge University Press, Cambridge, Eng.
Schank, Roger C. (with Childers, Peter G.): 1984, *The Cognitive Computer: On Language, Learning, and Artificial Intelligence*, Addison-Wesley, Reading, MA.
Searle, John R.: 1980, 'Minds, Brains, and Programs', *Behavioral and Brain Sciences* **3**, 417—57.
Searle, John R.: 1982, 'The Myth of the Computer', *New York Review of Books*, 29 April, 3—6; *cf.* correspondence, same journal, 24 June 1982, 56—57.
Searle, John R.: 1984, *Minds, Brains and Science*, Harvard University Press, Cambridge, MA.
Shapiro, Stuart C.: 1977, 'Representing Numbers in Semantic Networks: Prolegomena', *Proc. 5th International Joint Conference on Artificial Intelligence (IJCAI-77; MIT)*, Morgan Kaufmann, Los Altos, CA, p. 284.
Shapiro, Stuart C.: 1979, 'The SNePS Semantic Network Processing System', in Findler 1979, pp. 179—203.
Shapiro, Stuart C.: 1982, 'Generalized Augmented Transition Network Grammars For Generation From Semantic Networks', *American Journal of Computational Linguistics* **8**, 12—25.
Shapiro, Stuart C., and Rapaport, William J.: 1986, 'SNePS Considered as a Fully Intensional Propositional Semantic Network', *Proc. National Conference on Artificial Intelligence (AAAI-86; Philadelphia)*, Vol. 1, Morgan Kaufmann, Los Altos, CA, pp. 278—83.
Shapiro, Stuart C., and Rapaport, William J.: 1987, 'SNePS Considered as a Fully Intensional Propositional Semantic Network', in G. McCalla and N. Cercone (eds.), *The Knowledge Frontier*, Springer-Verlag, Berlin.

Simon, Herbert A.: 1977, 'Artificial Intelligence Systems that Can Understand', *Proc. of the 5th International Joint Conference on Artificial Intelligence (IJCAI-77; MIT)*, Morgan Kaufmann, Los Altos, CA, pp. 1059—73.

Slocum, Jonathan: 1985, 'A Survey of Machine Translation: its History, Current Status, and Future Prospects', *Computational Linguistics* **11**, 1—17.

Smith, Brian C.: 1982, 'Prologue to "Reflection and Semantics in a Procedural Language', in Brachman and Levesque 1985, 31—39.

Smith, Brian C.: 1986, 'Varieties of Self-Reference', In J. Y. Halpern (ed.), *Theoretical Aspects of Reasoning about Knowledge: Proc. of the 1986 Conference*, Morgan Kaufmann, Los Altos, CA, pp. 19—43.

Tanenbaum, Andrew S.: 1976, *Structured Computer Organization*, Prentice-Hall, Englewood Cliffs, N.J.

Turing, Alan M.: 1950, 'Computing Machinery and Intelligence', *Mind* **59**; reprinted in A. R. Anderson (ed.), *Minds and Machines*, Prentice-Hall, Englewood Cliffs, N.J., 1964, pp. 4—30.

Weizenbaum, Joseph: 1966, 'ELIZA—A Computer Program for the Study of Natural Language Communcation between Man and Machine', *Communications of the Association for Computing Machinery* **9**, 36—45. Reprinted in *CACM* **26**(1983), 23—28.

Weizenbaum, Joseph: 1974, 'Automating Psychotherapy', *ACM Forum* **17**, 543; reprinted with replies, *CACM* **26**(1983), 28.

Weizenbaum, Joseph: 1976, *Computer Power and Human Reason: From Judgment to Calculation*, W. H. Freeman, San Francisco.

Wiebe, Janyce M., and Rapaport, William J.: 1986, 'Representing *De Re* and *De Dicto* Belief Reports in Discourse and Narrative', *Proc. of the IEEE*, Special Issue on Knowledge Representation, **74**, 1405—1413.

Willick, Marshal S.: 1985, 'Constitutional Law and Artificial Intelligence: The Potential Legal Recognition of Computers as "Persons"', *Proc. of the 9th International Joint Conference on Artificial Intelligence (IJCAI-85; Los Angeles)*, Morgan Kaufmann, Los Altos, CA, pp. 1271—73.

Winograd, Terry: 1972, *Understanding Natural Language*, Academic Press, Orlando, FL.

Woods, William A.: 1975, 'What's in a Link: Foundations for Semantic Networks', in D. G. Bobrow and A. M. Collins (eds.), *Representation and Understanding*, Academic Press, New York, pp. 35—82. Reprinted in Brachman and Levesque 1985, pp. 217—41.

Woods, William A.: 1978, 'Semantics and Quantification in Natural Language Question Answering', in M. C. Yovits (ed.), *Advances in Computers*, Vol. 17, Academic Press, New York, pp. 1—87.

Department of Computer Science, and
Graduate Group in Cognitive Science
State University of New York at Buffalo
Buffalo, NY 14260, U.S.A.

JAMES H. FETZER

SIGNS AND MINDS: AN INTRODUCTION TO THE THEORY OF SEMIOTIC SYSTEMS*

Perhaps no other view concerning the theoretical foundations of artificial intelligence has been as widely accepted or as broadly influential as the physical symbol system conception advanced by Newell and Simon (1976), where symbol systems are machines — possibly human — that process symbolic structures through time. From this point of view, artificial intelligence deals with the development and evolution of physical systems that employ symbols to represent and to utilize information or knowledge, a position often either explicitly endorsed or tacitly assumed by authors and scholars at work within this field (*cf.* Nii *et al.*, 1982 and Buchanan 1985). Indeed, this perspective has been said to be "the heart of research in artificial intelligence" (Rich 1983, p. 3), a view that appears to be representative of its standing within the community at large.

The tenability of this conception, of course, obviously depends upon the notions of system, of physical system, and of symbol system that it reflects. For example, Newell and Simon (1976) tend to assume that symbols may be used to designate "any expression whatsoever", where these expressions can be created and modified in arbitrary ways. Without attempting to deny the benefits that surely are derivable from adopting their analysis, it may be worthwhile to consider the possibility that this approach could be perceived as part of another more encompassing position, relative to which physical symbol systems might qualify as, say, one among several different sorts of systems possessing the capacity to represent and to utilize knowledge or information. If such a framework could be constructed, it might not only serve to more precisely delineate the boundaries of their conception but contribute toward the goal of illuminating the theoretical foundations of artificial intelligence as well.

Not the least important question to consider, moreover, is what an ideal framework of this kind ought to be able to provide. In particular, as their "physical symbol system hypothesis" — the conjecture that physical symbol systems satisfy the necessary and sufficient conditions for "intelligent action" (Newell and Simon, 1976) — itself suggests, the

133

James H. Fetzer (ed.), Aspects of Artificial Intelligence, 133—161.
© 1988 *by Kluwer Academic Publishers.*

conception of physical symbol systems is intended to clarify the relationship between symbol processing and deliberate behavior, in some
appropriate sense. Indeed, it would seem to be a reasonable expectation
that an ideal framework of this kind ought to have the capacity to shed
light on the general character of the causal connections that obtain
between mental activity and behavioral tendencies to whatever extent
they occur as causes or as effects of the production and the utilization
of symbols (or of their counterparts within a more encompassing
conception).

The purpose of this paper, therefore, is to explore this possibility by
examining the issues involved here from the point of view of a theory of
mind based upon the theory of signs (or "semiotic theory") proposed by
Charles S. Peirce (1839-1914). According to the account that I shall
elaborate, minds ought to be viewed as *semiotic systems*, among which
symbolic systems (in a sense that is not quite the same as that of Newell
and Simon) are only one among three basic kinds. As semiotic systems,
minds are sign-using systems that have the capacity to create or to
utilize signs, where this capability may be either naturally produced or
artificially contrived. The result is a conception of mental activity that
promises to clarify and illuminate the similarities and the differences
between semiotic systems of various kinds — no matter whether human,
(other) animal or machine — and thereby avail its support for the
foundations of artificial intelligence, while contributing toward understanding mankind's place within the causal structure of the world.

1. PEIRCE'S THEORY OF SIGNS

It is not uncommon to suppose that there is a fundamental relationship
between thought and language, as when it is presumed that all thinking
takes place in language. But it seems to me that there is a deeper view
that cuts into this problem in a way in which the conception of an
intimate connection between thought and language cannot. This is a
view about the nature of mind that arises from reflection upon the
theory of signs (or "semiotic theory") advanced by Peirce. The most
fundamental concept Peirce elaborated is that of a sign as a something
that stands for something (else) in some respect or other for somebody.
He distinguished between three principal areas of semiotic inquiry,
namely, the study of the relations signs bear to other signs; the study of
the relations signs bear to that for which they stand; and, the study

of the relations that obtain between signs, what they stand for, and sign users. While Peirce referred to these dimensions of semiotic as "pure grammar", "logic proper", and "pure rhetoric", they are more familiar under the designations of "syntax", "semantics", and "pragmatics", respectively, terms that were introduced by Morris (1938) and have become standard.

Within the domain of semantics, Peirce identified three ways in which a sign might stand for that for which it stands, thereby generating a classification of three kinds of signs. All signs that stand for that for which they stand by virtue of a relation of resemblance between those signs themselves and that for which they stand are known as "icons". Statues, portraits and photographs are icons in this sense, when they create in the mind of a sign user another — equivalent or more developed — sign that stands in the same relation to that for which they stand as do the original signs creating them. Any signs that stand for that for which they stand by virtue of being either causes or effects of that for which they stand are known as "indices". Dark clouds that suggest rain, red spots that indicate measles, ashes that remain from a fire are typical indices in this sense. Those signs that stand for that for which they stand either by virtue of conventional agreements or by virtue of habitual associations between those signs and that for which they stand are known as "symbols". Most of the words that occur in ordinary language, such as "chair" and "horse"— which neither resemble nor are either causes or effects of that for which they stand — are symbols in this technical sense. (Compare Peirce, 1955, esp. pp. 97—155 and pp. 274—289; and Peirce, 1985.)

There is great utility in the employment of symbols by the members of a sign-using community, of course, since, as Newell and Simon (1976) recognize, purely as a matter of conventional agreement, almost anything could be used to stand for almost anything else under circumstances fixed by the practices that might govern the use of those signs within that community of sign users. Thus, the kinds of ways in which icons and indices can stand for that for which they stand may be thought of as natural modes, insofar as relations of resemblance and of cause-and-effect are there in nature, whether we notice them or not; whereas conventional or habitual associations or agreements are there only if we create them. Nevertheless, it would be a mistake to make too much of this distinction, because things may be alike or unalike in infinitely many ways, where two things qualify as of a common kind if they share common properties from a certain point of view, which may or may not be easily ascertainable.

The most important conception underlying this reflection is that of the difference between types and tokens, where "types" consist of specifications of kinds of things, while "tokens" occur as their instances. The color blue for example, when appropriately specified (by means of color charts, in terms of angstrom units, or whatever), can have any number of instances, including bowling balls and tennis shoes, where each such instance qualifies as a token of that type. Similar considerations obtain for sizes, shapes, weights, and all the rest: any property (or pattern) that can have distinct instances may be characterized as a kind (or type), where the instances of that kind (type) qualify as its tokens. The necessary and sufficient conditions for a token to qualify as a token of a certain type, moreover, are typically referred to as the intension (or "meaning") of that type, while the class of all things that satisfy those conditions is typically referred to as its extension (or "reference"). The distinction applies to icons, to indices and to symbols alike, where identifying things as tokens of a type presumes a point of view, which may or may not involve more than the adoption of a certain semiotic framework.

The importance of perspective can be exemplified. My feather pillow and your iron frying-pan both weight less than 7 tons, are not located on top of Mt. Everest, and do not look like Lyndon Johnson. Paintings by Rubens, Modigliani, and Picasso clearly tend to suggest that relations of resemblance presuppose a point of view. A plastic model of the battleship *Missouri* may be like the "Big Mo" with respect to the relative placement of its turrets and bulwarks, yet fail to reflect other properties — including the mobility and firepower — of the real thing, where whether a specific set of similarities ought to qualify as resemblance depends upon and varies with the adoption of some perspective. And, indeed, even events are described as "causes" and as "effects" in relation to implicit commitments to laws or to theories. Relations of resemblance and causation have to be recognized to be utilized, where the specification of a complex type can be a rather complex procedure.

In thinking about the nature of mind, it seems plausible that Peirce's theory of signs might provide us with clues. In particular, reflecting upon the conception of a sign as a something that stands for something (else) in some respect or other for somebody, it appears to be a presumption to assume that those somethings for which something can stand for something (else) in some respect or other must be "some-bodies": would it not be better to suppose, in a non-question-begging way, that these are "somethings", without taking for granted that the kind of thing that these somethings are has to be human? In reasoning about the kind of thing that these somethings might be, therefore, the possibility arises that those somethings for which something can stand for something (else) might themselves be viewed as "minds". (Morris, 1938, p. 1, suggests that (human) mentality and the use of signs are closely connected). The conception that I shall elaborate, therefore, is that minds are things that are capable of utilizing signs ("sign users"), where semiotic systems in this sense are causal systems of a special kind.

2. ABSTRACT SYSTEMS AND PHYSICAL SYSTEMS

In defense of this position, I intend to explain, first, what it means to be a system, what it means to be a causal system, what it means to be the special kind of causal system that is a semiotic system; and, second, what, if anything, makes this approach more appropriate than the Newell and Simon conception, especially with respect to the field of artificial intelligence. The arguments that follow suggests that Newell and Simon's account harbors an important equivocation, insofar as it fails to define the difference between a set of symbols that is significant *for a user of a machine* — in which case, there is a semiotic relationship between the symbols, what they stand for, and the symbol user, where the user is not identical with the machine — and a set of symbols that is significant *for use by a machine* — in which case, a semiotic relationship obtains between the symbols, what they stand for, and the symbol user, where the user is identical with the machine. The critical difference between symbol systems and semiotic systems emerges at this point.

Let us begin by considering the nature of systems in general. A system may be defined as a collection of things — numbers and operators, sticks and stones, whatever — that instantiates a fixed arrangement. This means that a set of parts becomes a system of a certain kind by instantiating some set of specific relations — logical, causal, whatever — between those parts. Such a system may be func-tional or non-functional with respect to specified inputs and outputs, where the difference depends on whether that system responds to an input or not. In particular, for a fixed input — such as assigning seven

as the value of a variable, resting a twelve-pound weight on its top,
whatever — a specific system will tend to produce a specific output
(which need not be unique to its output class) — such as yielding four-
teen as an answer, collapsing in a heap, whatever — where differences
in outputs (or in output classes) under the same inputs can serve to
distinguish between systems of different kinds, but only as a sufficient
and not as a necessary condition (cf. the distinction of deterministic and
indeterministic systems below).

For a system to be a causal system means that it is a system of things
within space/time between which causal relations obtain; an abstract
system, by comparison, is a system of things not in space/time between
which logical relations alone can obtain. This conception of a causal
system bears strong resemblance to Newell and Simon's conception of a
physical system, which is a system governed by the laws of nature.
Indeed, to the extent to which the laws of nature include non-causal as
well as causal laws, both of these conceptions should be interpreted
broadly (though not therefore reductionistically, since the non-occur-
rence of emergent properties — possibly including at least some mental
phenomena — is not a feature of the intended interpretation of causal
systems; cf. Fetzer, 1981, 1986a). Since systems of neither kind are
restricted to inanimate as opposed to animate systems, let us assume
that their "physical systems" and my "causal systems" are by and large
("roughly") the same, without pretending to have conclusively established their identity.

Within the class of causal (or of physical) systems, moreover, two
subclasses require differentiation. Causal systems whose relevant properties are only incompletely specified may be referred to as "open",
while systems whose relevant properties are completely specified are
regarded as "closed". Then a distinction may be drawn between two
kinds of closed causal systems, namely, those for which, given the same
input, the same output invariably occurs (without exception), and those
for which, given the same input, one or another output within the same
class of outputs invariably occurs (without exception). Systems of the
first kind, accordingly, are deterministic causal systems, while those of
the second kind are indeterministic causal systems. Whether or not
Newell and Simon would be willing to acknowledge this distinction, I
cannot say for certain; but because it is well-founded and its introduction begs no crucial questions, I shall suppose they would.

This difference, incidentally, is not the same as that in computational theory between deterministic and non-deterministic finite automata, such as parsing schemata, which represent paths from grammatical rules (normally called "productions") to well-formed formulae (typically "terminal strings") for which more than one production sequence is possible, where human choice influences the path selected (Cohen, 1986, pp. 142—145). While Newell and Simon acknowledge this distinction, strictly speaking, the systems to which it applies are special kinds of "open" rather than "closed" causal systems. The conception of abstract systems also merits more discussion, where purely formal systems — the systems of the natural numbers, of the real numbers, and the like — are presumptive examples thereof. While abstract systems in this sense are not in space/time and cannot exercise any causal influence upon the course of events during the world's history, this result does not imply that, say, inscriptions of numerals — as their representatives within space/time — cannot exercise causal influence as well. Indeed, since chalkmarks on blackboards affect the production of pencilmarks in notebooks, where some of these chalkmarks happen to be numerals, such a thesis would be difficult to defend.

For a causal system to be a semiotic system, of course, it has to be a system for which something can stand for something (else) in some respect or other, where such a something (sign) can affect the (actual or potential) behavior of that system. In order to allow for the occurrence of dreams, of daydreams, and of other mental states as potential outcomes (responses or effects) of possible inputs (stimuli or trials) — or as potential causes (inputs or stimuli) of possible outcomes (responses or effects) — and thereby circumvent the arbitrary exclusion of internal (or private) as well as external (or public) responses to internal as well as to external signs, behavior itself requires a remarkably broad and encompassing interpretation. A conception that accommodates this possibility is that of behavior as any internal or external effect of any internal or external cause. Indeed, from this point of view, it should be apparent that that something affects the behavior of a causal system does not mean that it has to be a sign for that system, which poses a major problem for the semiotic approach — distinguishing semiotic causal systems from other kinds of causal systems.

3. CAUSAL SYSTEMS AND SEMIOTIC SYSTEMS

To appreciate the dimensions of this difficulty, consider that if the capacity for the (actual or potential) behavior of a system to be affected by something were enough for that system to qualify as semiotic, the class of semiotic systems would be coextensive with the class of causal systems, since they would have all and only the same members. If even one member of the class of causal systems should not qualify as a member of the class of semiotic systems, however, then such an identification cannot be sustained. Insofar as my coffee cup, your reading glasses and numberless other things — including sticks and stones — are systems whose (actual and potential) behavior can be influenced by innumerable causal factors, yet surely should not qualify as semiotic systems, something more had better be involved: that something can affect the behavior of a causal system is not enough.

That a system's behavior can be affected by something is necessary, of course, but in addition that something must be functioning as a sign for that system: that that sign stands for that for which it stands for that system must make a difference to (the actual or potential behavior of) that system, where this difference can be specified in terms of the various ways that such a system would behave, were such a sign to stand for something other than that for which it stands for that system (or, would have behaved, had such a sign stood for something other than that for that system). Were what a red light at an intersection stands for to change to what a green light at an intersection stands for (and conversely) for specific causal systems, including little old ladies but also fleeing felons, then that those signs now stand for things other than that for which they previously stood ought to have corresponding behavioral manifestations, which could be internal or external in kind.

Little old ladies who are not unable to see, for example, should now slow down and come to a complete stop at intersections when green lights appear and release the break and accelerate when red lights appear. Felons fleeing with the police in hot pursuit, by contrast, may still speed through, but they worry about it a bit more, which, within the present context, qualifies as a behavioral manifestation. Strictly speaking, changes in external behavior (with respect to outcome classes) are sufficient but not necessary, whereas changes in internal behavior (with respect to outcome classes) are necessary and sufficient. Thus, a more exact formulation of the principle in question would state that, *for any specific system, a sign S stands for something for that system rather than for something else if and only if the strength of the tendencies for that system to manifest behavior of some specific kind in the presence of S —*

no matter whether publicly displayed or not — differs from case to case, where otherwise what it stands for remains the same.

This principle implies that a change that effects no change is no change at all (with respect to the significance of a sign for a system), which tends to occur when some token is exchanged for another token of the same type. Once again, however, considerations of perspective may have to be factored in, since one dime (silver) need not stand for the same thing as another dime (silver and copper) for the same system when they are tokens of some of the same types, but not of others. Although this result appears agreeable enough, the principle of significance being proposed does not seem to be particularly practical, since access to strengths of tendencies for behavior that may or may not be displayed is empirically testable, in principle, but only indirectly measureable, in practice (*cf.* Fetzer, 1981, 1986a). As it happens, this theoretical yardstick can be supplemented by a more intuitive standard of its kind.

The measure that I have proposed, of course, affords an account of what it means for a sign to change its meaning (what it stands for) for a system, where the differences involved here may be subtle, minute and all but imperceptible. For defining "sign", it would suffer from circularity in accounting for what it means for a sign to change its meaning while relying upon the concept of a sign itself: it does not provide a definition of what it is to be a sign or of what it is to be a semiotic system as such, but of what it is for a sign to change its meaning for a semiotic system. This difficulty, however, can be at least partially offset by appealing to (what I take to be) a general criterion for a system to be a semiotic system (for a thing to be a mind), namely, *the capacity to make a mistake*; for, in order to make a mistake, something must take something to stand for something other than that for which it stands, a reliable evidential indicator that something has the capacity to take something to stand for something in some respect or other, which is the right result.

We should all find it reassuring to discover that the capacity to make a mistake — to mis-take something for other than that for which it stands — appears to afford conclusive evidence that something has a mind. That something must have the capacity to make a mistake, however, does not mean that it must actually make them as well, since the concept of a divine mind that never makes mistakes — no matter whether as a matter of logical necessity or as a matter of lawful necessity for minds of that kind — is not inconsistent (Fetzer, 1986b). The difference between mistakes and malfunctions, moreover, deserves to be emphasized, where *mistakes* are made by systems while remaining systems of the same kind, while *malfunctions* transform a system of one kind into a system of another. That a system makes a mistake is not meant to imply that its output classes, relative to its input classes, have been revised, but rather that, say, a faulty inference has occurred, the false has been taken for the true, something has been misclassified, and so on, which readily occurs with perceptual and inductive reasoning (Fetzer, 1981).

4. THE VARIETIES OF SEMIOTIC SYSTEMS

The semiotic analysis of minds as semiotic systems invites the introduction of at least three different kinds (or types) of minds, where systems of Type I can utilize icons, systems of Type II can utilize icons and indices, and systems of Type III can utilize icons, indices, and symbols. Thus, if the conception of minds as semiotic systems is right-headed, at least in general, it would seem reasonable to conjecture that there are distinctive behavioral (psychological) criteria for semiotic systems of these different types; in other words, if this approach is approximately correct, then it should not be overly difficult to discover that links can be forged with pyschological (behavioral) distinctions that relate to the categories thereby generated. In particular, there appear to be kinds of learning (conditioning, whatever) distinctive to each of these three types of systems, where semiotic systems of Type I display type/token recognition, those of Type II display classical conditioning, and those of Type III display instrumental conditioning, where behavior of these kinds appears to be indicative that a system is one of such a type. Let us begin, therefore, by considering examples of semiotic systems of all three kinds and subsequently return to the symbol system hypothesis.

Non-human animals provide useful examples of semiotic systems that display classical conditioning, for example, as systems of Type II which have the capacity to utilize indices as signs, where indices are things that stand for that for which they stand by virtue of being either causes or effects of that for which they stand. Pavlov's famous experiments with dogs are illustrative here. Pavlov observed that dogs tend to salivate at the appearance of their food in the expectation of being fed; and that, if a certain stimulus, such as a bell, was regularly sounded at the same time its food was brought in, a dog soon salivated at such a bell's sound whether its food came with it or not. From the semiotic perspective, food itself functions as an (unconditioned) sign for dogs, namely, as a sign standing to the satiation of their hunger as causes stand to effects. Thus, when the sound of a bell functions as a (conditioned) sign for dogs, it similarly serves as a sign standing to the satiation of their hunger as cause stands to its effects. In the case of the (unconditioned) food stimulus, of course, the stimulus actually is a cause of hunger satiation, while in the case of the (conditioned) bell stimulus, it is not; but that does not undermine this example, since it shows that dogs sometimes make mistakes.

Analogously, Skinner's familiar experiments with pigeons provide an apt illustration of semiotic systems of Type III that have the capacity to utilize symbols as signs, where symbols are things that stand for that for which they stand by virtue of a conventional agreement or of an habitual association between the sign and that for which it stands. Skinner found that pigeons kept in cages equipped with bars that would emit a pellet if pressed rapidly learned to depress the bar whenever they wanted some food. He also discovered that if, say, a system of lights was installed, such that a bar-press would now release a pellet only if a green light was on, they would soon refrain from pressing the bar, even when they were hungry, unless the green light was on. Once again, of course, the pigeon might have its expectations disappointed by pressing a bar when the apparatus has been changed (or the lab assistant forgot to set a switch, whatever), which shows that pigeons are no smarter than dogs in avoiding mistakes.

Classical conditioning and operant conditioning, of course, are rather different kinds of learning. The connection between a light and the availability of food, like that between the sound of the bell and the satiation of hunger, was artificially contrived. The occurrence of the bell stimulus, of course, causes the dog to salivate, whether it wants to or not, whereas the occurrence of a green light does not cause a pigeon to press the bar, whether it wants to or not, but rather establishes a conventional signal for the pigeon that, if it were to perform a bar press now, a pellet would be emitted. It could be argued that the bell stimulus has now become a sufficient condition for the dog to salivate, while the light stimulus has become a sufficient condition for the pigeon not to press the bar. But Skinner's experiments, unlike those of Pavlov, involve reinforcing behavior after it has been performed, because of which the pigeons learn means/ends relations over which they have some control.

An intriguing example of type/token recognition that displays what appears to be behavior characteristic of semiotic systems of Type I, at last, was described in a recent newspaper article entitled, 'Fake Owls Chase Away Pests' (*St. Petersburg Times*, 27 January 1986), as follows:

Birds may fly over the rainbow, but until 10 days ago, many of them chose to roost on top of a billboard that hangs over Bill Allen's used car lot on Drew Street in Clearwater. Allen said he tried everything he could think of to scare away the birds, but still they came — sometimes as many as 100 at a time. He said an employee had to wash the used cars at Royal Auto Sales every day to clean off the birds' droppings. About a month ago, Allen said, he called the billboard's owner for help fighting the birds. Shortly afterward, Allen said, two viny owl "look alikes" were put on the corners of the billboard. "I haven't had a bird land up there since", he said.

The birds, in other words, took the sizes and shapes of the viny owls to be instances of the sizes and shapes of real owls, treating the fake owls as though they were the real thing. Once again, therefore, we can infer that these systems have the capacity to take something to stand for something (else) in some respect or other on the basis of the criterion that they have the capacity to make a mistake, which has been illustrated by Pavlov's dogs, by Skinner's pigeons, and by Allen's birds alike. While there do seem to be criteria distinctive of each of these three types of semiotic systems, in other words, these more specific criteria themselves are consistent with and illuminate that more general semiotic criterion.

These reflections thus afford a foundation for pursuing a comparison of the semiotic approach with the account supported by the symbol system conception. Indeed, the fundamental difference that we are about to discover is that Newell and Simon appear to be preoccupied exclusively with systems of Type III (or their counterparts), which, if true, establishes a sufficient condition for denying that semiotic systems and symbol systems are the same — even while affirming that they are both physical systems (of one or another of the same general kind). The intriguing issues that arise here, therefore, concern (a) whether there is any significant difference between semiotic systems of Type III and physical symbol systems and (b) whether there are any significant reasons for preferring one or the other of these accounts with respect to the foundations of artificial intelligence.

5. SYMBOL SYSTEMS AND CAUSAL SYSTEMS

The distinction between types and tokens ought to be clear enough by now to consider the difference between Newell and Simon's physical symbol systems and semiotic systems of Type III. The capacity to utilize indices seems to carry with it the capacity to utilize icons, since recognizing instances of causes as events of the same kind with respect to some class of effects entails drawing distinctions on the basis of resemblance relations. Similarly, the capacity to utilize symbols seems to carry with it the ability to utilize indices, at least to the extent to which the use of specific symbols on specific occasions can affect the behavior of a semiotic system for which they are significant signs. Insofar as these considerations suggest that a physical symbol system ought to be a powerful kind of semiotic system with the capacity to utilize icons, indices and symbols as well, it may come as some surprise that I want to deny that Newell and Simon's conception supports such a conclusion at all. For, it appears to be the case that, appearances to the contrary notwithstanding, physical symbol systems in the sense of Newell and Simon (1976) do not qualify as semiotic systems.

Since I take it to be obvious that physical symbol systems are causal systems in the appropriate sense, the burden of my position falls upon the distinction between systems for which something functions as a sign for a user of that system and systems for which something functions as a sign for that system itself. According to Newell and Simon (1976), in particular:

A physical symbol system consists of a set of entities, called symbols, which are physical patterns that can occur as components of another type of entity called an expression (or symbol structure). Thus a symbol structure is composed of a number of instances (or tokens) of symbols related in some physical way (such as one token being next to another).

Notice, especially, that symbol structures (or "expressions") are composed of sequences of symbols (or "tokens"), where "physical symbol systems", in this sense, process expressions, which they refer to as "symbol structures". The question that I want to raise, therefore, is whether or not these "symbol structures" function as signs in Peirce's sense — and, if so, for whom.

At first glance, this passage might seem to support the conception of physical symbol systems as semiotic systems, since Newell and Simon appeal to tokens and tokens appear to be instances of different types. Their conceptions of expression and of interpretation, moreover, are relevant here:

> Two notions are central to this structure of expressions, symbols, and objects: designation and interpretation.
>
>> Designation. An expression designates an object if, given the expression, the system can either affect the object itself or behave in ways depending on the object.
>
> In either case, access to the object via the expression has been obtained, which is the essence of designation.
>
>> Interpretation. The system can interpret an expression if the expression designates a process and if, given the expression, the system can carry out the process.
>
> Interpretation implies a special form of dependent action: given an expression, the system can perform the indicated process, which is to say, it can evoke and execute its own processes from expressions that designate them. (Newell and Simon, 1976, pp. 40-41.)

An appropriate illustration of "interpretation" in this sense would appear to be computer commands, whereby a suitably programmed machine can evoke and execute its own internal processes when given "expressions" that designate them. Notice, however, that portable typewriters (pocket calculators, and so forth) would seem to qualify as "physical symbol systems" in Newell and Simon's sense — since combinations of letters from their keyboards (of numerals from their interface, and so on) appear to be examples of "expressions" that designate a process whereby various shapes can be typed upon a page (strings of numerals can be manipulated, . . .). Other considerations, however, suggest that the sorts of systems that are intended to be good examples of symbol systems are general-purpose digital computers rather than simple systems of these kinds.

A consistent interpretation of Newell and Simon's conception depends upon an analysis of "symbols" as members of an alphabet/character set (such as "a", "b", "c", and so on), where "expressions" are sequences of the members of such a set. The term that they employ which corresponds most closely to that of "symbol" in Peirce's technical sense, therefore, is not "symbol" itself but rather "expression". Indeed, that "symbols" in Newell and Simon's sense cannot be "symbols" in Peirce's technical sense follows from the fact that most of the members of a character set do not stand for anything at all — other than that their inclusion within such a set renders them permissible members of the character sequences that constitute (well-formed) expressions. A more descriptive name for systems of this kind, therefore, might be that of "expression processing" ("string manipulating") systems; but so long as Newell and Simon's systems are not confused with semiotic systems, there is no reason to dispute the use of a name that has already become well-entrenched.

An important consequence of this account, moreover, is that words like "chair" and "horse", which occur in ordinary language, are good examples of symbols in Peirce's sense, yet do not satisfy Newell and Simon's conception of expressions. These words stand for that for which they stand without in any fashion offering the least hint that the humans (machines, whatever) for which such things are signs can either affect such objects or behave in ways that depend upon those objects: the capacity to describe horses and chairs does not entail the ability to ride or to train them, to build or to refinish them, or otherwise manipulate them. These symbols can function as significant signs whether or not Newell and Simon's conditions are satisfied, a proof of which follows from examples such as "elf" and "werewolf", signs that function as symbols in Peirce's sense, yet could not possibly fulfill Newell and Simon's conception because they stand for non-existent objects, which can neither affect nor be affected by causal relations in space/time. Thus, "symbol systems" in Newell and Simon's sense (of string manipulating systems) do not qualify as systems that utilize symbols in Peirce's sense.

6. SYMBOL SYSTEMS AND SEMIOTIC SYSTEMS

This result tends to reflect the fact that Newell and Simon's conception of physical symbol system depends upon at least these two assumptions:

(a)   expressions $=_{df}$ sequences of characters (strings of symbols); and,
(b)   symbols $=_{df}$ elements of expressions (tokens of character types);

where these "character types" are those specified by some set of characters (ASCII, EBCDIC, whatever). This construction receives further support from other remarks they make during the course of their analysis of completeness and closure as properties of systems of this kind, insofar as they maintain:

(i)   there exist expressions that designate every process of which such
       a system (machine) is capable; and,
(ii)   there exist processes for creating any expression and for modify-
       ing any expression in arbitrary ways;

which helps to elucidate the sense in which a system can affect an
object itself or behave in ways depending on that object, namely, when
that object itself is either a computer command or a string of characters
from such a set. (A rather similar conception can be found in Newell,
1973, esp. pp. 27—28.)

Conditions (i) and (ii), I believe, are intended to support the restriction of symbol systems to general purpose digital computers, even
though they can only be satisfied relative to some (presupposed) set of
symbols, no matter how arbitrarily selected. Whether or not these
conditions actually have their intended effect, however, appears to be
subject to debate. Although programming languages vary somewhat on
this issue, such practices as the overloading of operators, the multiple
definition of identifiers and the like are ordinarily supposed to be
important to avoid in order to secure the formal syntax of languages
that are suitable for employment with computers, which tends to
restrict the extent to which they can be arbitrarily composed. Moreover,
since typewriters and calculators seem to have unlimited capacities to
process any number of expressions that can be formulated within their
respective sets of symbols, it could be argued that they are not excluded
by these constraints, which is a striking result (insofar as few would be
inclined to claim that a typewriter "has a mind of its own").

No doubt, symbol systems in Newell and Simon's sense typically
behave in ways that depend upon certain members of the class of
expressions, since they are causal systems that respond to particular
computer commands for which they have been "programmed" (in a
sense that takes in hardware as well as software considerations). Since
computer commands function as input causes in relation to output
effects (for suitably programmed machines), it should be obvious that
Newell and Simon's conception entails the result that physical symbol
systems are causal systems. For the reasons outlined above, however, it
should be equally apparent that their conception does not entail that
these causal systems are semiotic systems of Type III. Indeed, if
expressions were symbols in Peirce's technical sense, then they would
have to have intensions and extensions; but it seems plain that strings of

symbols from a character set, however well-formed, need not have these properties.

Indeed, the most telling considerations of all emerge from inquiring for whom Newell and Simon's "symbols" and "expressions" are supposed to be significant (apart from the special class of computer commands, where, in fact, it remains to be ascertained whether or not those commands function as signs for those systems). Consider, for example, the following cases;

INPUT	(FOR SYSTEM)		OUTPUT
finger (*pushes*)	*button*	(*causing*)	*printout of file*
match (*lights*)	*fuse*	(*causing*)	*explosion of device*
child (*notices*)	*cloud*	(*causing*)	*expectation of storm*

When a finger pushes a button that activates a process, say, leading to a printout, no doubt an input for a causal system has brought about an output. When a match lights a fuse, say, leading to an explosion, that an input for a causal system has brought about an output is not in doubt. And when a child notices a cloud, say, leading to some such expectation, no doubt an input for a causal system has brought about an output. Yet, surely only the last of these cases is suggestive of the possibility that something stands for something (else) for that system, where that particular thing is a meaningful token for that system (with an intensional dimension) and where that system might be making (or have made) a mistake.

If these considerations are correct, then we have discovered, first, that the class of causal systems is not coextensive with the class of semiotic systems. Coffee cups and matches, for example, are particular cases of systems in space/time that stand in causal relations to other things, yet surely do not qualify as sign-using systems. Since two words, phrases or expressions mean the same thing only if their extensions are the same, causal systems and semiotic systems are not the same thing. We have also discovered, second, that the meaning of "symbol system" is not the same as that of "semiotic system of Type III". General-purpose digital computers are causal systems that process expressions, yet do not therefore need to be systems for which signs function as signs. Since two words, phrases, or expressions mean the same thing only if their intensions are the same, symbol systems and semiotic systems of Type III are not the same things.

From the perspective of the semiotic approach, in other words, the conception of physical symbol systems encounters the distinction between sets of symbols that are significant for users of machines — in which case there is a semiotic relationship between those signs, what they stand for and those sign users, where the users are not identical with the machines themselves — and sets of symbols that are significant for use by machines — in which case there is a semiotic relationship between those signs, what they stand for and those sign users, where these users are identical with the machines themselves. Without any doubt, the symbols and expressions with which programmers program machines are significant signs for those programmers; and without any doubt, the capacity to execute such commands qualifies those commands as causal inputs with respect to causal outputs. That is not enough for these machines to be semiotic systems of Type III.

If these considerations are correct, then there is a fundamental difference between causal systems and semiotic systems, on the one hand, and between symbol systems and semiotic systems of Type III, on the other. Of course, important questions remain, including ascertaining whether or not there are good reasons to prefer one or another conception, which will depend in large measure upon their respective capacities to clarify, if not to resolve, troublesome issues within this domain. Moreover, there appear to be several unexamined alternatives with respect to the interpretation of Newell and Simon's conception, since other arguments might be advanced to establish that symbol systems properly qualify either as semiotic systems of Type I or of Type II — or that special kinds of symbol systems properly qualify as semiotic systems of Type III, which would seem to be an important possibility that has not yet been explored. For the discovery that some symbol systems are not semiotic systems of Type III no more proves that special kinds of symbol systems cannot be semiotic systems of Type III than the discovery that some causal systems are not semiotic systems proves that special kinds of causal systems cannot be semiotic systems.

7. THE SYMBOL-SYSTEM AND THE SEMIOTIC-SYSTEM HYPOTHESES

The conception of semiotic systems, no less than the conception of symbol systems, can be evaluated (at least, in part) by the contribution they make toward illuminating the relationship between the use of signs, on the one hand, and the manipulation of symbols, on the other, in relation to deliberate behavior. Both accounts, in other words, may be viewed as offering characterizations of "mental activity" in some appropriate sense, where these accounts are intended to afford a basis for understanding "intelligent" (or "deliberate") behavior. Indeed, the respective theoretical hypotheses that they represent ought to be formulated as follows:

(h_1) *The Symbol-System Hypothesis*: a symbol system has the necessary and sufficient means (or capacity) for general intelligent action; and,

(h_2) *The Semiotic-System Hypothesis*: a semiotic system has the necessary and sufficient means (or capacity) for general intelligent action;

where these hypotheses are to be entertained as empirical generalizations (or as lawlike claims) whose truth or falsity cannot be ascertained merely by reflection upon their meaning within a certain language framework alone.

Because these hypotheses propose necessary and sufficient conditions, they could be shown to be false if either (a) systems that display "intelligent" (or "deliberate") behavior are not symbol (or semiotic) systems; or (b) systems that are symbol (or semiotic) systems do not display "intelligent" (or "deliberate") behavior. Moreover, since they are intended to be empirical hypotheses (*cf.* Newell and Simon, 1976, esp. p. 42 and p. 46), these formulations ought to be understood as satisfied by systems that display appropriate behavior without assuming (i) that behavior that involves the processing or the manipulation of a string of tokens from a character set is therefore either "intelligent" or "deliberate" (since otherwise the symbol-system hypotheses must be true as a function of its meaning); and, without assuming (ii) that behavior that is "intelligent" or "deliberate" must therefore be successful in attaining its aims, objectives, or goals, where a system that displays behavior of this kind cannot make a mistake (since otherwise the semiotic-system hypothesis must be false as a function of its meaning). With respect to the hypotheses (h_1) and (h_2), therefore, "intelligent action" and "deliberate behavior" are synonymous expressions.

A certain degree of vagueness inevitably attends an investigation of this kind to the extent to which the notions upon which it depends, such as "deliberate behavior" and "intelligent action", are not fully defined. Nevertheless, an evaluation of the relative strengths and weaknesses of these hypotheses can result from considering classes of cases that fall within the extensions of "symbol system" and of "semiotic system", when properly understood. In particular, it seems obvious that the examples of type/token recognition, of classical conditioning, and of instrumental conditioning considered above are instances of semiotic systems of Type I, II, and III that do not qualify as symbol systems in Newell and Simon's sense. This should come as no surprise, since Newell and Simon did not intend that their conception should apply with such broad scope; but it evidently entails that hypothesis (h_1) must be empirically false.

Indeed, while this evidence amply supports the conclusion that the semiotic-system approach has applicability to dogs, to pigeons, and to (other) birds that the symbol-system approach lacks, the importance that ought to attend this realization may or may not be immediately apparent. Consider, after all, that a similar argument could be made on behalf of the alternative conception, namely, that — depending upon the resolution of various issues previously identified — the symbol-system approach has applicability to typewriters, to calculators, and to (other) machines that the semiotic-system approach lacks, which may be of even greater importance if the objects of primary interest are machines. For if digital computers, for example, have to be symbol systems but do not have to be semiotic systems, that they might also qualify as semiotic systems does not necessarily have to be a matter of immense theoretical significance.

Nevertheless, to the extent to which these respective conceptions are supposed to have the capacity to shed light on the general character of the causal connections that obtain between mental activity and behavioral tendencies — that is, to the extent to which frameworks such as these ought to be evaluated in relation to hypotheses such as (h_1) and (h_2) — the evidence that has been presented would appear to support the conclusion that the semiotic-system approach clarifies connections between mental activity as semiotic activity and behavioral tendencies as deliberate behavior — connections which, by virtue of its restricted range of applicability, the symbol-system approach cannot accommodate. By combining distinctions between different kinds (or types) of mental activity together with psychological criteria concerning the sorts of capacities distinctive of systems of these different types (or kinds), the semiotic approach provides a powerful combination of (explanatory and predictive) principles, an account that, at least with respect to non-human animals, the symbol-system approach cannot begin to rival.

This difference, however, surely qualifies as an advantage of an approach only so long as it is being entertained as an approach to a certain specific class of problems, such as explaining and predicting the deliberate behavior (the "intelligent actions") of non-human animals. To whatever extent Newell and Simon did not intend to account for the intelligent actions (the "deliberate behavior") of non-human animals, it may be said, the incapacity of their conception to accommodate explanations and predictions should not be held against them. The strict interpretation of this position would lead to the conclusions that (a) any theory should be evaluated exclusively in terms of its success or failure at achieving its intended aims, goals or objectives, where (b) unintended consequences should be viewed as, at most, of secondary importance, no matter how striking their character or significant their potential. Relative to this standard, the incapacity to accommodate non-human animals not only cannot count against Newell and Simon's conception but instead ought to support it.

8. WHAT ABOUT HUMANS AND MACHINES?

A more interesting — and less implausible — position would be for Newell and Simon to abandon their commitment to the symbol system hypothesis and restrict the scope of their analysis to the thesis that, after all, general-purpose digital computers *are* symbol systems, where it really does not matter whether or not they have captured the nature of mental activity in humans or in (other) animals. There is a sense in which this attitude is almost precisely right, so long as the possibility that they may have captured no sense of mental activity is itself left open. Newell and Simon's conception, in other words, may be completely adequate for digital computers but remain completely inadequate for other things — unless, of course, the precise processes that characterize symbol systems were the same processes that characterize, say, human or non-human systems that process knowledge or information, a position that is not consistent with the results we have discovered.

Notice, in particular, that the following theses regarding the relationship between symbol systems and semiotic systems are compatible:

(t_1)   general-purpose digital computers are symbol systems; and,
(t_2)   animals — human and non-human alike — are semiotic systems;

where, even if not one digital computer heretofore constructed qualifies as a semiotic system, that some digital computer yet to be built might later qualify as a semiotic system remains an open question. Indeed, whether information or knowledge processing in humans and in (other) animals is like that in symbol systems or that in semiotic systems appears to be the fundamental question at the foundations of artificial intelligence.

Strictly speaking, after all, to be a symbol system in Newell and Simon's sense is neither necessary nor sufficient to be a semiotic system in the Peircean sense. Recall, in particular, that their account of designation presupposes the existence of that which is designated, since otherwise that system could neither affect nor be affected by that thing; yet it is no part of the notion of an icon or a symbol, for example, that the things thereby signified should be open to causal influence by a semiotic system. Moreover — and most importantly — nothing about their conception warrants the conclusion that symbol systems in their sense even have the capacity to utilize signs in Peirce's sense at all, even though — as I readily concede — there is no reason to deny that they are causal systems.

Artificial intelligence is often taken to be an attempt to develop causal processes that perform mental operations. Sometimes such a view has been advanced in terms of formal systems, where "intelligent beings are ... automatic formal systems with interpretations under which they consistently make sense" (Haugeland, 1981, p. 31). This conception exerts considerable appeal, since it offers the promise of reconciling a domain about which a great deal is known — formal systems — with one about which a great deal is not known — intelligent beings. This theory suffers from profound ambiguity, however, since it fails to distinguish between systems that make sense to themselves and those that make sense for others. Causal models of mental processes, after all, might either effect connections between inputs and outputs so that, for a system of a certain specific type, those models yield outputs for certain classes of inputs that correspond to those exemplified by the systems that they model; or else effect those connections between inputs and outputs and, in addition, process these connections by means of processes that correspond to those that are exemplified by those systems that they model.

This distinction, which is not an unfamiliar one, can be expressed by differentiating between "simulation" and "replication", where, say,

(a)   causal models that simulate mental processes capture connections between inputs and outputs that correspond to those of the systems that they represent; while,

(b)   causal models that replicate mental processes not only capture these connections between inputs and outputs but do so by means of processes that correspond to those of the systems they represent;

where, if theses (t_1) and (t_2) are true, then it might be said that symbol systems simulate mental processes that semiotic systems replicate — precisely because semiotic systems have minds that symbol systems lack. There are those, such as Haugeland (1985), of course, who are inclined to believe that symbol systems replicate mental activity in humans too because human mental activity, properly understood, has the properties of symbol systems too. But this claim appears to be plausible only if there is no real difference between systems for which signs function as signs for those systems themselves and systems for which signs function as signs for the users of those systems, which is the issue in dispute.

Another perspective on this matter can be secured by considering the conception of systems that possess the capacity to represent and to utilize information or knowledge. The instances of semiotic systems of Types I, II, and III that we have examined seem to fulfill this desideratum, in the sense that, for Pavlov's dogs, for Skinner's pigeons, and for Allen's birds, there are clear senses in which these causal systems are behaving in accordance with their beliefs, that is, with something that might be properly characterized as "information" or as "knowledge". Indeed, the approach represented here affords the opportunity to relate genes to bodies to minds to behavior, since phenotypes develop from genotypes under the influence of environmental factors, where phenotypes of different kinds may be described as predisposed toward the utilization of different kinds of signs, which in turn tends toward the acquisition and the utilization of distinct ranges of behavioral tendencies, which have their own distinctive strengths (Fetzer, 1985, 1986b). This in itself offers significant incentives for adopting the semiotic approach.

Yet it could still be the case that digital computers (pocket calculators and the like) cannot be subsumed under the semiotic framework precisely because (t_1) and (t_2) are both true. After all, nothing that has gone before alters obvious differences between systems of these various different kinds, which are created or produced by distinctive kinds of causal processes. The behavior of machine systems is (highly) artificially determined or engineered, while that of human systems is (highly) culturally determined or engineered, and that of (other) animal systems is (highly) genetically determined or engineered. Systems of all three kinds exhibit different kinds of causal capabilities: they differ with respect to their ranges of inputs/stimuli/trials, with respect to their ranges of output/responses/outcomes, and with respect to their higher-order causal capabilities, where humans (among animals) appear superior. Even if theses (t_1) and (t_2) were true, what difference would it make?

9. WHAT DIFFERENCE DOES IT MAKE?

From the point of view of the discipline of artificial intelligence, whether computing machines do what they do the same way that humans and (other) animals do what they do only matters in relation to whether the enterprise is that of simulating or of replicating the mental processes of semiotic systems. If the objective is simulation, it is surely unnecessary to develop the capacity to manufacture semiotic systems; but if the objective is replication, there is no other way, since this aim can not otherwise be attained. Yet it seems to be worth asking whether the replication of the mental processes of human beings would be worth the time, expense, and effort that would be involved in building them. After all, we already know how to reproduce causal systems that possess the mental processes of human beings in ways that are cheaper, faster, and lots more fun. Indeed, when consideration is given to the limited and fallible memories, the emotional and distorted reasoning and the inconsistent attitudes and beliefs that tend to distinguish causal systems of this kind, it is hard to imagine why anyone would want to build them: there are no interpretations "under which they consistently make sense".

A completely different line could be advanced by defenders of the faith, however, who might insist that the distinction I have drawn between symbol systems and semiotic systems cannot be sustained, because the conception of semiotic systems itself is circular and therefore unacceptable. If this contention were correct, the replication approach might be said to have been vindicated by default in the absence of any serious alternatives. The basis for this objection could be rooted in a careful reading of the account that I have given for semiotic systems of Type I, since, within the domain of semantics, icons are supposed to stand for that for which they stand "when they create in the mind of a sign user another — equivalent or more developed — sign that stands in the same relation to that for which they stand as do the original signs creating them". This Peircean point, after all, employs the notion of mind in the definition of one of the kinds of signs — the most basic kind, if the use of indices involves the use of icons and the use of symbols involves the use of indices, but not conversely — which might be thought to undermine any theory of the nature of mind based on his theory of signs.

This complaint, I am afraid, is founded upon an illusion; for those signs in terms of which other signs are ultimately to be understood are unpacked by Peirce in terms of the habits, dispositions, or tendencies by means of which all signs are best understood: "the most perfect account of a concept that words can convey", he wrote, "will consist in a description of the habit which that concept is calculated to produce" (Peirce, 1955, p. 286). But this result itself could provide another avenue of defense by contending that systems of dispositions cannot be causal systems so that, a fortiori, semiotic systems cannot be special kinds of causal systems within a dispositional framework. Without intimating that the last word has been said with reference to this question, there appears to be no evidence in its support; but it would defeat the analysis that I have presented if this argument were correct.

The basic distinction between symbol systems and semiotic systems, of course, is that symbol systems may or may not be systems for which signs stand for something for those systems, while semiotic systems are, where I have employed the general criterion that semiotic systems are capable of making mistakes. A severe test of this conception, therefore, is raised by the problem of whether or not digital computers, in particular, are capable of making a mistake. If the allegations that the super-computers of the North American Defense Command (NORAD) located at Colorado Springs have reported the U.S. to be under ballistic missile attacks from the Soviet Union no less than 187 times are (even roughly) accurate, this dividing line may already have been crossed, since it appears as though all such reports thus far have been false. The systems most likely to fulfill this condition are ones for which a faulty inference can occur, the false can be mistaken for the true or things can be misclassified, which might not require systems more complex than those capable of playing chess (Haugeland, 1981, p. 18) — but this question, as we have discovered, is theoretically loaded.

Human beings, as semiotic systems, display certain higher-order causal capabilities that deserve to be acknowledged, since we appear to have a remarkable capacity for inferential reasoning that may or may not differ from that of (other) animals in kind but undoubtedly exceeds them in degree. In this respect, especially, however, human abilities are themselves surpassed by "reasoning machines", which are, in general, more precise, less emotional, and far faster in arriving at conclusions by means of deductive inference. The evolution and development of digital computers with inductive and perceptual capabilities, therefore, would seem to be the most likely source of systems that display the capacity to make mistakes. By this criterion, systems that have the capacity to make mistakes qualify as semiotic systems, even when they do not replicate processes of human systems.

A form of mentality that exceeds the use of symbols alone, more-over, appears to be the capacity to make assertions, to issue directives, to ask questions and to utter exclamations. At this juncture, I think, the theory of minds as semiotic systems intersects with the theory of languages as transformational grammars presented especially in the work of Noam Chomsky (Chomsky, 1965, and Chomsky, 1966, for example; cf. Chomsky, 1986 for his more recent views). Thus, this connection suggests that it might be desirable to identify a fourth grade of mentality, where semiotic systems of Type IV can utilize signs that are transformations of other signs, an evidential indicator of which may be taken to be the ability to ask questions, make assertions and the like. This conception indicates that the capacity for explicit formalization of propositional attitudes may represent a level of mentality distinct from the occurrence of propositional attitudes as such.

Humans, (other) animals, and machines, of course, also seem to differ with respect to other higher-order mental capabilities, such as in their attitudes toward and beliefs about the world, themselves and their methods. Indeed, I am inclined to believe that those features of mental activity that separate humans from (other) animals occur at just this juncture; for humans have a capacity to examine and to criticize their attitudes, their beliefs, and their methods that (other) animals do not enjoy. From this perspective, however, the semiotic approach seems to classify symbol systems as engaged in a species of activity that, if it were pursued by human beings, would occur at this level; for the activities of linguists, of logicians, and of critics in creating and in manipulating expressions and symbols certainly appear to be higher-order activities, indeed.

A fifth grade of mentality accordingly deserves to be acknowledged as a meta-mode of mentality that is distinguished by the use of signs to stand for other signs. While semiotic systems of Type I can utilize icons, of Type II indices, of Type III symbols and of Type IV transforms, semiotic systems of Type V are capable of using meta-signs as signs that stand for other signs (one variety of meta-signs, of course, being meta-languages). Thus, perhaps the crucial criterion of mentality of this degree is the capacity for criticism, of ourselves, our theories and our methods. While the conception of minds as semiotic systems has a deflationary effect in rendering the existence of mind at once more ubiquitous and less important than we have heretofore supposed, it does not therefore diminish the place of human minds as semiotic systems of a distinctive kind, nevertheless.

The introduction of semiotic systems of Type IV and of Type V, however, should not be allowed to obscure the three most fundamental species of mentality. Both transformational and critical capacities are presumably varieties of semiotic capability that fall within the scope of symbolic mentality. Indeed, as a conjecture, it appears to be plausible to suppose that each of these successively higher types of mentality presupposes the capacity for each of those below, where evolutionary considerations might be brought to bear upon the assessment of this hypothesis by attempting to evaluate the potential benefits for survival and reproduction relative to species and societies — that is, for social groups as well as for single individuals — that accompany this conception (cf. Fetzer, 1985 and especially 1986a).

There remains the further possibility that the distinction between symbol systems and semiotic systems marks the dividing line between computer science (narrowly defined) and artificial intelligence, which is not to deny that artificial intelligence falls within computer science (broadly defined). On this view, what is most important about artificial intelligence as an area of specialization within the field itself would be its ultimate objective of replicating semiotic systems. Indeed, while artificial intelligence can achieve at least some of its goals by building systems that simulate — and improve upon — the mental abilities that are displayed by human beings, it cannot secure its most treasured goals short of replication, if such a conception is correct. It therefore appears to be an ultimate irony that the ideal limit and final aim of artificial intelligence — whether by replicating human beings or by creating novel species — could turn out to be the development of systems capable of making mistakes.

NOTE

* The original version of this paper was presented at New College on 8 May 1984. Subsequent versions were presented at the University of Virginia, at the University of Georgia, and — most recently — at Reed College. I am indebted to Charles Dunlop, Bret Fetzer, Jack Kulas, Terry Rankin, and Ned Hall for instructive comments and criticism.

REFERENCES

Buchanan, B.: 1985, 'Expert Systems', *Journal of Automated Reasoning* **1**, 28—35.

Chomsky, N.: 1965, *Aspects of the Theory of Syntax*. MIT Press. Cambridge, MA.

Chomsky, N.: 1966, *Cartesian Linguistics*. Harper & Row, New York.

Chomsky, N.: 1986, *Knowledge of Language: Its Nature, Origin, and Use*. Praeger Publishers, New York.

Cohen, D.: 1986, *Introduction to Computer Theory*. John Wiley & Sons, Inc., New York.

Fetzer, J. H.: 1981, *Scientific Knowledge*. D. Reidel, Dordrecht, Holland.

Fetzer, J. H.: 1985, 'Science and Sociobiology', in J. H. Fetzer (ed.), *Sociobiology and Epistemology* (D. Reidel, Dordrecht, Holland), pp. 217—246.

Fetzer, J. H.: 1986a, 'Methodological Individualism: Singular Causal Systems and Their Population Manifestations', *Synthese* **68**, pp. 99—128.

Fetzer, J. H.: 1986b, 'Mentality and Creativity', *Journal of Social and Biological Structures* (forthcoming).

Haugeland, J.: 1981, 'Semantic Engines: An Introduction to Mind Design', in J. Haugeland (ed.), *Mind Design* (MIT Press, Cambridge, MA.), pp. 1—34.

Haugeland, J.: 1985, *Artificial Intelligence: The Very Idea*. MIT Press, Cambridge, MA.

Morris, C. W.: 1938, *Foundations of the Theory of Signs*. University of Chicago Press, Chicago, IL.

Newell, A.: 1973, 'Artificial Intelligence and the Concept of Mind', in R. Schank and K. Colby (eds.), *Computer Models of Thought and Language* (W. H. Freeman and Company, San Francisco.), pp. 1—60.

Newell, A. and Simon, H.: 1976, 'Computer Science as Empirical Inquiry: Symbols and Search', reprinted in J. Haugeland (ed.), *Mind Design* (MIT Press, Cambridge, MA.), pp. 35—66.

Nii, H. P. *et al.*: 1982, 'Signal-to-Symbol Transformation: HASP/SIAP Case Study', *AI Magazine* **3**, pp. 23—35.

Peirce, C. S.: 1955, *Philosophical Writings of Peirce*, J. Buchler (ed.) (Dover Publications, New York.).

Peirce, C. S.: 1985, 'Logic as Semiotic: The Theory of Signs', reprinted in R. Innis (ed.), *Semiotics: An Introductory Anthology* (Indiana University Press, Bloomington IN) pp. 4—23.

Rich, E.: 1983, *Artificial Intelligence*. McGraw-Hill, New York.

Department of Philosophy and Humanities
University of Minnesota
Duluth, MN 55812

BRUCE MACLENNAN

LOGIC FOR THE NEW AI

The psyche never thinks without an image.
— Aristotle

I. INTRODUCTION

A. *The New AI*

There is growing recognition that the traditional methods of artificial intelligence (AI) are inadequate for many tasks requiring machine intelligence.[1] Space prevents more than a brief mention of the issues.

Research in connected speech recognition has shown that an incredible amount of computation is required to identify spoken words. This is because the contemporary approach begins by isolating and classifying phonemes by means of context-free features of the sound stream. Thus the stream must be reduced to acoustic atoms before classification can be accomplished. On the other hand, for people the context determines the phoneme, much as the melody determines the note.[2] This is why a phoneme can vary so greatly in its absolute (i.e., context-free) features and still be recognized. People recognize a gestalt, such as a word or phrase, and identify the phonetic features later if there is some reason to do so (which there's usually not).

In general, much of the lack of progress in contemporary pattern recognition can be attributed to the attempt to classify by context-free features. If the chosen features are coarse-grained, the result is a "brittleness" in classification and nonrobustness in the face of novelty.[3] On the other hand, if the features are fine-grained, then the system is in danger of being swamped by the computation required for classification.[4] People (and other animals) do not seem to face this dilemma. We recognize the whole, and focus on the part only when necessary for the purpose at hand.[5] The logical atomism of contemporary AI precludes this approach, since wholes can be identified only in terms of their constituents. But, as we've seen, the constituents are determined by the whole. Thus, with contemporary AI technology, pattern recognition faces a fundamental circularity.

163

James H. Fetzer (ed.), Aspects of Artificial Intelligence, 163—192.
© 1988 *by Kluwer Academic Publishers.*

A similar problem occurs in robotics. Robotic devices need to coordinate their actions through kinesthetic, visual and other forms of feedback. Contemporary systems attempt to accomplish this by building explicit models of the world in which the system must function. Coordination and planning are accomplished by symbolic manipulation (basically deduction) on the knowledge base. This is subject to the same limitations we've already seen. If the model is simple, then the robot will be unprepared for many exigencies. If the model is extremely detailed, then the system will be swamped by computation — and still face the possibility of unforeseen circumstances. How is it that people and animals avoid this predicament?

Heidegger has shown that much of human behavior exhibits a ready-to-hand understanding of our world that is not easily expressed in propositional form. "When we use a piece of equipment," Heidegger claims, "we actualize a bodily skill (which cannot be represented in the mind) in the context of a socially organized nexus of equipment, purposes and human roles (which cannot be represented as a set of facts)."[6] Such skill is acquired through our successful use of our animal bodies to cope with the physical and social worlds in which we find ourselves. Our use of this knowledge is *unconscious* in that we do not think in terms of propositional rules that are then applied to the situation at hand. "The peculiarity of what is proximally ready-to-hand is that, in its readiness to hand, it must, as it were, withdraw in order to be ready to hand quite authentically."[7]

In contrast, all computer knowledge, at least with current AI technology, is rule based. The knowledge is stored in the form of general rules and schemata, and the computer's "thinking" proceeds by the application of these general rules to particular situations. Furthermore, since the computer has no body with which to interact with the world and since it does not develop in a culture from which it can learn norms of behavior, its knowledge must be acquired either in the form of decontextualized general rules, or by mechanized generalization processes from decontextualized data.[8] As Papert and Minsky have said, "Many problems arise in experiments on machine intelligence because things obvious to any person are not represented in any program."[9]

Combinatorial explosion threatens many other applications of AI technology. For example, automatic theorem provers proceed by blind enumeration of possibilities, possibly guided by context-free heuristics. On the other hand, human theorem provers are guided by a contextual sense of relevance, and a sense of similarity to previously accomplished proofs. With current AI technology, automatic deduction cannot take advantage of past experience to guide its search.

The same problem occurs in automatic *induction*. This is relatively simple *if* the system is told in advance the relevant variables. That is, given measurements of a dozen different variables, it's not so hard to find which are related and to conjecture a relationship between them. Unfortunately, scientists face a much harder problem, since the number of possible variables is unlimited, as is the number of their relationships. How then are scientific laws ever discovered? First, prior experience gives scientists a sense of relevance (in the context of their investigations); this guides their search. In addition, human cognition permits scientists to *first* recognize similarities and patterns, and *then* to identify the common features (if any) upon which these similarities and patterns are based.

It has long been recognized that people rarely use language as a logical calculus. As Wittgenstein says, "in philosophy we often *compare* the use of words with games and calculi which have fixed rules, but cannot say that someone who is using language *must* be playing such a game."[10] Rather than being fixed by formal definition, the meanings of words expand and retract as required by context and the particulars of the speech situation. If computers are to be able to understand natural language "as she is spoke," then they too must be able to treat meaning in this context and situation-dependent manner — without a combinatorial explosion.

There is evidence[11] that expert behavior is better characterized as automatized *knowledge-how* rather than explicit *knowledge-that*.[12] As we've seen, even in predominantly symbolic activities such as mathematics, a primary determinant of the *skill* of the expert is his sense of relevance and his "nose" for the right attack on a problem. These automatized responses guide his behavior at each step of the process and avoid the combinatorial explosion characteristic of rule-based systems. What is missing from current AI is an explanation of the *vectors* of gestalt psychology: "When one grasps a problem situation, its structural features and requirements set up certain strains, stresses, tensions in the thinker. What happens in real thinking is that these strains and stresses are followed up, yield vectors in the direction of improvement of the situation, and change it accordingly."[13] As the Dreyfuses note, rule-based behavior with explicit heuristics is more characteristic of advanced beginners than of experts. But are there alternatives to logical atomism and rule-based behavior?

With current computer and AI technology, it seems unlikely that computers can be made to exhibit the sort of ready-to-hand understanding that people do. Thus contemporary AI emphasizes present-at-hand[14] knowledge that can be expressed in terms of decontextualized concepts and verbal structures. With our present techniques, for the computer to know *how*, it is necessary for it to know *that*.

Traditional logic, of which modern logic and conventional AI technologies are developments, is an idealization of certain cognitive activities that may be loosely characterized as *verbal*. On the other hand, many of the tasks for which we would like to use computers are nonverbal.[15] Seen in this way, it is no surprise that idealized verbal reasoning is indequate for these tasks — the idealization is too far from the fact. I suggest that AI is being driven by the needs and limitations of its current methods into a new phase which confronts directly the issues of nonverbal reasoning. This new phase, which I refer to as the *new* AI, will broaden AI technology to encompass nonverbal as well as verbal reasoning.

The preceding suggests that the new AI will require a *new* logic to accomplish its goals. This logic will have to be an idealization of nonverbal thinking in much the same way that conventional logic is an idealization of verbal thinking. In this paper I outline the requirements for such a logic and sketch the design of one possible logic that satisfies the requirements.

But isn't verbal thinking inherent in the very word *logic* ($\lambda o \gamma \iota \kappa \acute{\eta}$ < $\lambda \acute{o} \gamma o \varsigma$)? Can there be such a thing as a nonverbal logic? In the next section we justify our use of the term *logic* to refer to an idealization of nonverbal thinking.

B. *Why a New Logic?*

1. *The three roles of logic*

Based on ideas of Peirce[16] I distinguish three different *roles* fulfilled by logic and related subjects (epistemology, mathematics, philosophy of science).

Empirical logic is a nomological account, in the form of *descriptive* laws, of what people actually do when they reason (*cf.*, "epistemology naturalized"). Thus it can be considered a specialized discipline within psychology or sociology. As such it must account for those patterns of reasoning that are not formally valid, as well as those that are.

Mathematical logic, which is a subdiscipline of mathematics, provides, by means of *formal* laws, an idealized model of the reasoning process. There is no presumption that people do in fact reason this way all the time. Indeed, there's ample evidence that they don't.[17] Of course, for it to be an interesting model of reasoning, it must bear some relationship to actual reasoning. For example, mathematical logic should explain why the actual reasoning processes used by people work when they do. Similarly, mathematical theories of induction should explain why confirmation of unlikely outcomes is more valuable than the confirmation of likely ones, etc.[18] Thus, empirical logic provides the motivation for mathematical logic, and in turn mathematical logic suggests theories that guide the descriptive activities of empirical logicians.

Empirical logic tells us how people in fact reason; mathematical logic explains the validity of various reasoning processes. Neither tells us how we *ought* to reason. This is the role of *normative logic*, which is a subdiscipline of ethics.[19] Whereas empirical logic is formulated in terms of descriptive laws, and mathematical logic in terms of formal laws, normative logic is formulated in terms of *prescriptive* laws.

Normative logic must, of course, draw results from mathematical logic, since the latter explains why some reasoning processes are valid and others aren't. It must also draw from empirical logic for insights into the psychology of knowing and the practical limitations of human reason. Normative logic may thus prescribe rules that are not mathematically necessary, but that are psychologically or sociologically desirable.

Finally, to the extent to which its norms are followed, normative logic influences the way people actually reason, and hence future descriptive logic. And, to the extent to which mathematical logic models actual reasoning, the normative science also affects the mathematical science.

Logic is important in all three of its roles, but it is its normative role that is ultimately relevant to AI: we want to use logic as guide for programming intelligent machines. On the other hand, the mathematical role is central, since it forms the core of the normative principles and a standard for empirical studies. Therefore a mathematical logic of nonverbal reasoning must be our first goal, and on this I concentrate in the rest of this paper.

2. *Conventional logic inadequate for the new AI*

Conventional logic — by which we mean any of the well-known idealizations of verbal reasoning — is inadequate as a logic for the new AI. I summarize the reasons.

There is now ample evidence[20] that conventional logic is inadequate as an empirical description of the way people actually think. People apparently use a mixture of verbal and nonverbal reasoning that tends to combine the advantages of both. For example, Miller describes the alternation in the uses of imagery and mathematics that led to the development of quantum mechanics.[21] He also describes the nonverbal processes used by Einstein in creating relativity theory; an alternate account of this process is given by Wertheimer.[22]

Gardner summarizes recent empirical studies of the thought processes actually used by people in problem solving situations.[23] They show that the conventional logic is much too idealized. For example, the research of Peter Wason and Philip Johnson-Laird suggests that people are much more likely to reason correctly when the problem has relevance to practical action, than when it is merely abstract. Gardner's work and the references he cites contain additional examples.

Conventional logic is also inadequate as a normative discipline, since it provides standards for verbal reasoning but not for nonverbal reasoning. The costs and benefits and therefore the tradeoffs involved in nonverbal reasoning are different from those of verbal reasoning. Hence the practical guidance provided by the two logics will differ. The "old AI" has been following the norms of conventional logic — and has found their limitations.

Like any mathematical theory, the conventional logic is an idealization. There is nothing wrong with such idealizations, so long as they are appropriate. Unfortunately, conventional logic's idealization of verbal reasoning is often inappropriate to nonverbal problems. It often leads to a discrete, atomistic approach that results in a combinatorial explosion of possibilities.

3. *Potential value of a new logic*

There are many potential benefits that we may expect from a logic for the new AI. A mathematical theory would provide idealizations of the processes involved in nonverbal thinking. As such it would provide a basis for structuring empirical investigations of nonverbal thinking, and a standard upon which to base the norms of nonverbal thinking. The new AI will benefit directly from the normative science, since it is this science that will supply the guidelines for the design of machines that "think" nonverbally.[24] Thus AI should be helped to pass beyond its current difficulties and achieve some of the goals that have eluded the "old" AI.

II. REQUIREMENTS FOR A NEW LOGIC

In this section I review some of the processes of (human and animal) thought that, although poorly modeled by conventional logic, are essential to many present and future applications of AI. The new AI must provide a logic geared to the description and analysis of these processes. They are the basis of the criteria by which our own proposal for a new logic should be evaluated.

A. *Indefinite Classification*

Most, perhaps all, of the concepts that we use in everyday life have *indefinite boundaries*. That is, the presence of borderline cases is the rule rather than the exception. There are several reasons that we should expect this. First, many of the phenomena of nature that are important for survival are continuous. Thus it is natural to expect animal life to have evolved cognitive means for dealing with continuously variable qualities. Second, perception is subject to noise and error, caused both by imperfections in the sense organs and by circumstances in the environment. Survival requires that perception be robust in the face of a wide variety of disturbances. Third, indefinite boundaries avoid the "brittleness" associated with definite boundaries. What is "brittleness"? Suppose I have a rule: "Flee from predators bigger than myself." It seems unreasonable — that is, anti-survival — to treat a predator one millimeter shorter than myself as though this rule is inapplicable. Animal cognition avoids "brittleness" — a thing doesn't cease being a threat just because it's one millimeter too short.

It seems that indefinite classification will be as important for intelligent machines as it is for intelligent life. This does not imply that there is no need for definite boundaries. But for many purposes, especially the ones normally classified as "everyday", and hence typical of the new AI, indefinite classes will be required.

To reiterate, indefinite classes are often preferable to definite classes. Recall Wittgenstein's game example: "One might say that the concept 'game' is a concept with blurred edges. — 'But is a blurred concept a concept at all?' — Is an indistinct photograph a picture at all? Is it even an advantage to replace an indistinct picture by a sharp one? Isn't the indistinct one often exactly what we need?"[25]

Indefiniteness should not be considered a defect. As Wittgenstein points out, "one pace" is a perfectly useful measurement, despite the absence of a formal definition. "But this is not ignorance. We do not know the boundaries because none have been drawn. To repeat, we can draw a boundary for a special purpose."[26] We expect as much from our new logic — it should be capable of operating with or without boundaries, as the situation requires.

The reader will no doubt think of Zadeh's *fuzzy set theory*.[27] Although this is a step in the right direction, I do not think Zadeh's proposal goes far enough. Some of its limitations for the present purpose will become clear below.

B. *Context Sensitivity*

The indefiniteness of the boundaries of a concept is dependent on the context in which it is being used. Thus 'pure water' means one thing when I am thirsty in the woods and another when I'm serving my guests — or working in a chemistry lab. Many of our rules — heuristic or otherwise — are couched in context-dependent words and phrases: 'too near', 'dangerous', 'acceptable', 'untrustworthy', etc. etc.

Human (and animal) use of context-sensitive abstractions gives flexibility to the rules that use them. Context sensitivity seems a prerequisite of the intelligent use of rules. Indeed, isn't the blind following of rules — independent of context — the principal example of *stupidity*?

Here we can see the limitations of a fuzzy set theory that attaches a fixed membership distribution to a class. To achieve the flexibility characteristic of animal behavior, it's necessary to have this distribution adjust in a manner appropriate to the context.[28]

Context is not something that can be added onto an otherwise context-free concept. Rather, all our concepts are context-dependent;

the notion of a decontextualized concept is an idealization formed by abstraction from contextual concepts in a wide variety of contexts. Unfortunately, when it comes to modeling commonsense intelligence the idealization is too far from reality, "because everything in this world presents itself in context and is modulated by that context."[29] In Heideggerian terms, we are "always already in a situation."

The above observations also apply to activities that hold definite-boundaried, context-free abstractions as an ideal. The prime example is mathematics; here the abstractions are all defined formally. Observe, however, that many of the concepts — such as rigor — that guide mathematical behavior have just the indefinite, contextual character that I've described. Is this not the reason that computers, the paragons of formal symbol manipulators, are so poor at doing mathematics?

C. *Logical Holism*

The use of conventional logic drives AI to a kind of logical atomism. That is, all classification is done on the basis of a number of "atomic" features that are specified in advance. Whether these features have definite boundaries or are "fuzzy" is not the issue. Rather, the issue is whether it is possible to specify in advance (i.e., independent of context) the essential properties of a universal.

Wittgenstein, in *Philosophical Investigations* (1958), has criticized the notion of essential attributes and the basis for logical atomism. He observes that "these phenomena have no one thing in common which makes us use the same word for all, — but that they are *related* to one another in many different ways."[30] Further, he instructs us to "*look and see* whether there is anything common to all. — For if you look at them you will not see something that is common to *all*, but similarities, relationships, and a whole series of them at that."[31] This is what gives flexibility and adaptability to human classification; we are not dependent on a particular set of "essential" attributes. If we come upon a sport lacking one of the essentials, it will still be recognized, if it's sufficiently similar. "Is it not the case that I have, so to speak, a whole series of props in readiness, and am ready to lean on one if another should be taken from under me and *vice versa*?"[32]

What is the alternative to classification by essentials? "How then is it possible to perform an abstraction without extracting common elements, identically contained in all particular instances? It can be done when certain aspects of the particulars are perceived as deviations from, or deformations of, an underlying structure that is visible within them."[33] This is in accord with Eleanor Rosch's research, in which she concludes, "Many experiments have shown that categories appear to be coded in the mind neither by means of lists of each individual member of the category, nor by means of a list of formal criteria necessary and sufficient for category membership, but, rather, in terms of a prototype of a typical category member. The most cognitively economical code for a category is, in fact, a *concrete image* of an average category member."[34] Kuhn sees this as the usual pattern of science: "The practice of normal science depends on the ability, acquired from exemplars, to group objects and situations into similarity sets which are primitive in the sense that the grouping is done without an answer to the question, 'Similar with respect to what?'"[35] In contrast to logical atomism, he emphasizes this is "a manner of knowing which is misconstrued if reconstructed in terms of rules that are first abstracted from exemplars and thereafter function in their stead."[36]

Classification by family resemblance avoids two problems characteristic of classification by context-free features. First, classification by context-free features can be inflexible, since the number of features used is by necessity limited. Classification by family resemblance is more flexible because of the open-ended set of features upon which the classification is based. On the other hand, if we attempt to improve the flexibility of context-free classification by classifying on the basis of more context-free attributes, the efficiency of the process is much degraded, since the computer must consider all the attributes. Classification by family resemblance naturally focuses on those attributes likely to be relevant in the life of the person. This improves the efficiency of human cognitive processing.

Classification by family resemblance may in part account for animals' ability to adapt to novel situations. An object may be judged as belonging to a certain class in spite of the fact that it lacks certain "essential" characteristics, provided that it satisfies the overall gestalt. That is, in a given context certain attributes may be more relevant than others.

D. *Intentionality*

Another characteristic of human cognition that should be accounted for by our logic is *intentionality*, the directedness of consciousness towards its objects.[37] The effect of intentionality is to restrict awareness to just those aspects of the environment that are likely to be relevant to the problem at hand. It shifts some things into the foreground so that the rest can be left in the background. If we think of the foreground as having a high probability of being considered and the background as having a low probability, then the effect of this focusing process is to decrease the entropy of the probability distribution. Indeed, the functions that Peirce attributes to consciousness are self-control and improving the efficiency of habit formation.[38] This suggests that we can get the beneficial effect of intentionality by any process that on the average skews the probability of processing in favor of the more relevant information.

We see that both classification by family resemblances and intentionality improve cognitive efficiency by focusing cognitive activity on factors likely to be relevant. Our goal is to program computers to have a sense of relevance.

E. *Mixed-Mode Reasoning*

The issues discussed so far suggest that the new logic be based on a continuous (or analog) versus discrete (or digital) computational metaphor. Yet the limitations of analog computation are well known. For example, errors can accumulate at each stage of an analog computational process to the extent to which all accuracy is lost.[39] This suggests that analog (continuous) reasoning cannot be as *deep* — support as long chains of inference — as digital (discrete) reasoning.

Introspection suggests a solution to this problem. Based on the current context and the measures of relevance it induces, we "digitize" much of our mental experience — we verbalize our mental images. Such verbalization permits longer chains of inference by preventing the accumulation of error. It is successful so long as the context is relatively stable. We expect a logic for the new AI to accommodate verbal as well as nonverbal reasoning, and to permit the optimal mix of the two to be determined.

Paivio's remarks on visual cognition apply as well to other forms of nonverbal reasoning: "Images and verbal processes are viewed as alternative coding systems, or modes of symbolic representation, which are developmentally linked to experiences with concrete objects and events as well as with language."[40] But, the two systems are not mutually exclusive: "Many situations likely involve an interaction of imaginal and verbal processes, however, and the latter would necessarily be involved at some stage whenever the stimuli or responses, or both, are verbal . . . ."[41]

The key point is that verbal thinking is really a special case of nonverbal, since "language is a set of perceptual shapes — auditory, kinesthetic, visual."[42] The relative definiteness of linguistic symbols stabilizes nonverbal thinking. "Purely verbal thinking is the prototype of thoughtless thinking, the automatic recourse to connections retrieved from storage. What makes language so valuable for thinking, then, cannot be thinking in words. It must be the help that words lend to thinking while it operates in a more appropriate medium, such as visual imagery."[43]

III. PRELIMINARY DEVELOPMENT OF NEW LOGIC

A. *Approach*

In this section we outline the general framework for our model for nonverbal reasoning. Recall, however, that it is not our intention to develop a psychological theory; that is, our goal is not descriptive. Rather, our goal is to develop an idealized theoretical model of certain functions of nonverbal mental activity, much as Boolean algebra and predicate logic are idealized models of verbal cognition. But there is also a normative goal: the resulting logic should be useful as a tool for designing and programming computers.

Consider all the neurons that comprise a nervous system.[44] We postulate that there are two distinct ways in which they can encode information. *Semipermanent* information is in some way encoded in the neural structure (e.g., in terms of strength of synaptic connection). *Transient* information is encoded in a way that does not alter the neural structure (e.g., dynamic electrochemical activity). Transient information processing is involved in processes such as associative recall; alteration of semipermanent information is involved in processes such as learning.[45] We refer to transient information encoded in the electrochemical state of a neuron as the *state* of that neuron. Semipermanent information will be described in terms of *memory traces*.

Neurons can be divided into three categories on the basis of their connections to each other and to nonneural structures. We call neurons *afferent* if their state is determined *solely* by nonneural mechanisms, such as sense organs. We call neurons *efferent* if they have no effect on other neurons; that is, their state affects only nonneural mechanisms, such as motor organs. The remaining neurons we call *interneurons*; they both affect and can be affected by other neurons.[46]

We call the set of all afferent neurons the *afferent system*. Analogously we define the efferent and interneural systems. Since at any given time each neuron is in a state, at any given time each of these systems is in a state. Thus we can speak of the state of the afferent system, etc. To describe the possible states of these systems we define three spaces, A, I and E, the set of possible states of the afferent, interneural and efferent systems.

We will usually not need to distinguish the efferent neurons from other nonafferent neurons. Therefore we define $B = I \times E$, the space of all states of the nonafferent system.

What properties can we expect of the spaces A and B? Taking the points in these spaces to represent neural states, it is reasonable to assume that there is some notion of "closeness" between these states. Therefore, for any two points a, $a' \in A$ we postulate a distance $\delta_A(a, a')$ such that (1) $\delta_A(a, a') \geq 0$, and (2) $\delta_A(a, a') = 0$ if and only if $a = a'$. That is, distance is a nonnegative number such that the distance between two neural states is zero if and only if the states are identical. Such a function is commonly called a *semimetric* on A. Similarly we postulate a semimetric $\delta_B(b, b')$ for b, $b' \in B$. Furthermore, we will drop the subscripts and write $\delta(a, a')$ and $\delta(b, b')$ when no confusion will result.

It seems reasonable that neural states cannot be infinitely different. Therefore we make an additional assumption, that the semimetrics are *bounded*. This means that there is some number Δ_A such that $\delta(a, a') \leq \Delta_A$ for all a, $a' \in A$. Similarly there is a bound Δ_B on distances in B. Without loss of generality take $\Delta_A = \Delta_B = 1$ (this only changes the distance scale). Thus

$$\delta_A : A^2 \to [0, 1], \qquad \delta_B : B^2 \to [0, 1]$$

The above assumptions can be summarized by saying that A and B are bounded semimetric spaces.

The functions δ represent the *difference* between neural states. It will generally be more intuitive to work in terms of the *similarity* between states. Hence we define

$$\sigma(x, x') = 1 - \delta(x, x')$$

A and B subscripts will be added as needed to make the space clear. Note that σ inherits from δ the following properties:

$$0 \leqslant \sigma(x, x') \leqslant 1, \qquad \sigma(x, x') = 1 \quad \text{if and only if} \quad x = x'$$

Thus $\sigma(x, x')$ ranges from 1, meaning that x and x' are identical, to 0, meaning that they're as different as they can be.

A final assumption that we make about the spaces A and B is that they are *dense*. That is, for every $\zeta < 1$ and $a \in A$, there is an $a' \in A$, $a' \neq a$, such that $\sigma(a, a') > \zeta$. That means, for every state of the afferent system, there is at least one different state that is arbitrarily similar. The same applies to the space B. These assumptions guarantee that the afferent and nonafferent systems can respond continuously, which seems reasonable, at least as an idealization.

We next define a number of functions that will be useful in the following development. A set of points in a bounded semimetric space can be characterized in terms of their minimum similarity:

$$\mu(S) = \min\{\sigma(x, y) \mid x, y \in S\}$$

This can vary from 1 for a singleton set to 0 for a set containing at least two dissimilar elements.[46]

A useful quantity for the following derivations is $\sigma_S(x)$, the similarity of x to the other points in the set S. It is defined

$$\sigma_S(x) = \sum_{y \in S} \sigma(x, y)$$

Note that $0 \leqslant \sigma_S(x) \leqslant |S|$, where $|S|$ is the cardinality of S. Thus $\sigma_S(x) = 0$ if x is completely dissimilar from all the members of S.

Sometimes it is more meaningful to work in terms of the average similarity of a point to a set of points:

$$\hat{\sigma}_S(x) = \sigma_S(x) / |S|$$

Thus $0 \leqslant \hat{\sigma}_S(x) \leqslant 1$.

We will refer to the metrics δ_A and δ_B as the *physical metrics* on the afferent and nonafferent spaces because they are directly related to the neural structure. A major goal of the following theory is to show that the structure of the mental content does not depend strongly on the physical metrics. That is, we will attempt to show that the metrics induced by the mental content are "stronger" than the physical metrics.

Under the reasonable assumption that the neurons are the basis for cognitive processes, it makes sense to use the physical metrics as the basis of association. That is, things which cause the same pattern of neurological stimulation are in fact indistinguishable. Further, we will make continuity assumptions: things which cause nearly the same pattern of stimulation should lead to nearly the same response. These will be introduced when needed, rather than here, so that it is more apparent which assumptions are necessary for which results.

In the following two sections we investigate logics based on the preceding postulates. The first logic is based on a finite number of memory traces formed at discrete instants of time. This is only a partial step to a logic that satisfies the requirements in Part II, but it is useful to build intuition. The second logic takes a further step in the required direction by postulating a single memory trace that evolves continuously in time. Both theories are very tentative, but they nevertheless should indicate my expectations for a logic for the new AI.

B. *Discrete Time*

1. *Definitions*

The first version of our logic will be based on a discrete time model. Thus the state transition process is discontinuous. We imagine the afferent system taking on a series of states (stimuli) s_1, s_2, s_3, ... under the control of external events. These stimuli drive the nonafferent system through a series of states (responses) r_1, r_2, r_3, ... That is, the response r, or new state of the nonafferent system, is a function of (1) the stimulus s, or current afferent state, and (2) the context c, or current nonafferent state.[47] We write $r = s : c$ to denote that r is a new nonafferent state resulting from the afferent state s and the nonafferent state c. Thus $s : c$ is the response resulting from the stimulus s in the context c.

An assumption we make here is that the semipermanent information is constant throughout this process (i.e., no learning takes place). We consider later the case where, in addition to a state transition, we have an alteration of memory traces. So long as the semipermanent information is fixed, the responses to the stimuli $s_1, s_2, s_3, \ldots$ are

$$s_1 : c, \qquad s_2 : (s_1 : c), \qquad s_3 : [s_2 : (s_1 : c)], \ldots$$

That is, the response of each stimulus becomes the context for the next, $c_{i+1} = r_i = s_i : c_i$. Thus, excluding learning, we have a notation for the context-dependent interpretation of sensory data.

The new state is obviously dependent on the semipermanent information stored in the memory. Therefore we postulate that at any given time the memory M contains a finite number K of traces, (s_i, c_i, r_i), for $1 \leq i \leq K$.[48] These traces reflect situations in the past in which stimulus s_i in context c_i produced response r_i.

We expect that the response to a stimulus s in a context c will be dependent on the similarity of s and c to the stored s_i and c_i. Therefore, for fixed s and c we define $\alpha_i = \sigma(s, s_i)$ and $\beta_i = \sigma(c, c_i)$. Thus α_i is the similarity of the present stimulus to the ith stored stimulus, and β_i is the similarity of the present context to the ith stored context.

We expect the activation of memory traces to be related directly to the similarity of the present stimulus and context to the stimulus and context of the trace. On the other hand, we do not wish to make a commitment to a *particular* relationship. Thus we postulate a function $\Gamma(s_i, c_i)$, monotonically increasing in both its arguments, but otherwise unspecified. For fixed s and c we then define $\gamma_i = \Gamma(s_i, c_i)$. By monotonically increasing we mean that if $\alpha_i > \alpha_j$ and $\beta_i = \beta_j$, or if $\alpha_i = \alpha_j$ and $\beta_i > \beta_j$, then $\gamma_i \geq \gamma_j$. Thus γ_i measures the similarity of the current stimulus/context to the ith memory trace. Without loss of generality we take $\Gamma : A \times B \rightarrow [0, 1]$, that is, $0 \leq \gamma_i \leq 1$.

The monotonicity condition on Γ tells us that if s is more similar to s_i than to s_j, but c is equally similar to c_i and c_j, then $\gamma_i \geq \gamma_j$. Similarly, if c is more similar to c_i than to c_j, but s is equally similar to s_i and s_j, then $\gamma_i \geq \gamma_j$. Thus the similarity of a current state (s, c) to the stored states (s_i, c_i) is a function of both the similarity of the stimuli and the similarity of the contexts. This will be the basis for context-sensitive classification.

We will need to be able to compare various responses in terms of their similarity to the stored responses, weighted by the similarity of the current stimulus/context to the corresponding stored stimulus/context.

For this purpose we define $\varsigma(r)$, the weighted similarity of r to the responses in memory:

$$\varsigma(r) = \sum_{i=1}^{K} \gamma_i \sigma(r, r_i)$$

This is a kind of "score" for r, since $\varsigma(r)$ will tend to be high when r is close to the reponses of those traces that are most strongly activated by the current stimulus/context.

2. State transition function

Given the foregoing definitions, it is easy to describe the state transition function. We are given a memory M consisting of the traces (s_i, c_i, r_i), with $1 \leqslant i \leqslant K$. Given the stimulus/state pair (s, c) we want the new state to be a r that maximally similar to the r_i, but weighted according to the similarity of (s, c) to the corresponding (s_i, c_i). That is, we want $\varsigma(r)$ to be a maximum over the space B. Hence, we define the state transition operation:

$$s : c = \varepsilon r[\varsigma(r) = \max \{\varsigma(r) \mid r \in B\}]$$

Here we have used Russell's *indefinite description* operation ε. The definition can be read, "$s : c$ is any r such that the weighted similarity of r is the maximum of the weighted similarities of all the points in B."

Notice that, as implied by the use of the indefinite description operator, there may not be a unique point that maximizes ς. Thus there may be bifurcations in the state transition histories. They are in effect gestalt switches between equally good interpretations of the sensory input.

3. Activation of a single trace

Our first example is to show how a memory trace can be activated by a stimulus that's sufficiently close. In particular we assume[49] that the stimulus/state pair (s, c) is similar to the stored pair (s_1, c_1) but different from all the other pairs, (s_i, c_i), $i \neq 1$. We intend to show that the induced state $r = s : c$ is similar to r_1.

Since (s, c) is close to (s_1, c_1), take $\gamma_1 = \zeta \approx 1$. Since (s, c) is not similar to any of the other (s_i, c_i), take $\gamma_i < \varepsilon \approx 0$, for $i \neq 1$. Our goal is to show that $\sigma(r, r_1) \approx 1$.

Since r maximizes ς, we know $\varsigma(r_j) \leqslant \varsigma(r)$ for all responses r_j in memory. In particular $\varsigma(r_1) \leqslant \varsigma(r)$. Now note:

$$\varsigma(r_1) = \sum_{i=1}^{K} \gamma_i \sigma(r_1, r_i)$$

$$= \gamma_1 \sigma(r_1, r_1) + \sum_{i=2}^{K} \gamma_i \sigma(r_1, r_i)$$

$$= \zeta + \sum_{i=2}^{K} \gamma_i \sigma(r_1, r_i)$$

$$\geqslant \zeta$$

On the other hand,

$$\varsigma(r) = \sum_{i=1}^{K} \gamma_i \sigma(r, r_i)$$

$$= \gamma_1 \sigma(r, r_1) + \sum_{i=2}^{K} \gamma_i \sigma(r, r_i)$$

$$= \zeta \sigma(r, r_1) + \sum_{i=2}^{K} \gamma_i \sigma(r, r_i)$$

$$< \zeta \sigma(r, r_1) + \varepsilon \sum_{i=2}^{K} \sigma(r, r_i)$$

The inequality follows from $\gamma_i < \varepsilon$, for $i \neq 1$. Now, since $\sigma(r, r_i) \leqslant 1$ always, we have $\varsigma(r) < \zeta \sigma(r, r_1) + \varepsilon(K - 1)$.

We know that the transition operation $s : c$ maximizes $\varsigma(r)$, so we know $\varsigma(r_1) \leqslant \varsigma(r)$. Hence, combining the three inequalities we have $\zeta \sigma(r, r_1) + \varepsilon(K - 1) > \zeta$. Solving for $\sigma(r, r_1)$ yields $\sigma(r, r_1) > 1 - (\varepsilon/\zeta)(K - 1)$. Hence $\sigma(r, r_1)$ differs from 1 by at most the amount $(\varepsilon/\zeta)(K - 1)$. This quantity reflects the extent by which the other memory traces interfere with perfect recall of r_1. Note that as $\varepsilon \to 0$

this quantity approaches zero. That is,

$$\lim_{\varepsilon \to 0} \sigma(r, r_1) = 1$$

Hence, as the interference approaches 0 the recall approaches perfect.

4. *Activation of a cluster of traces*

We now perform very much the same analysis in the situation in which (s, c) closely matches a number of traces in memory, all of which induce a similar response. Thus, suppose that H and $\overline{H}$ partition the indices $\{1, 2, \ldots, K\}$. Suppose that $\gamma_i > \zeta \approx 1$ for $i \in H$, and that $\gamma_i < \varepsilon \approx 0$ for $i \in \overline{H}$. Assuming all the matching traces induce a similar state transition implies that $\mu(H) = \eta \approx 1$. As before compute

$$\varsigma(r) = \sum_{i=1}^{K} \gamma_i \sigma(r, r_i) < \sum_{i \in H} \sigma(r, r_i) + \varepsilon \sum_{i \in \overline{H}} \sigma(r, r_i)$$

$$\leqslant \sigma_H(r) + \varepsilon |\overline{H}|$$

Hence $\varsigma(r) < \sigma_H(r) + \varepsilon |\overline{H}|$. For an arbitrary r_j we derive

$$\varsigma(r_j) = \sum_{i \in H} \gamma_i \sigma(r_j, r_i) + \sum_{i \in \overline{H}} \gamma_i \sigma(r_j, r_i)$$

$$> \zeta \sum_{i \in H} \sigma(r_j, r_i) + \sum_{i \in \overline{H}} \gamma_i \sigma(r_j, r_i)$$

$$\geqslant \zeta \sum_{i \in H} \mu(H)$$

$$= \zeta \eta |H|$$

Also, since $r = s : c$ maximizes ς, We know that $\varsigma(r) \geqslant \varsigma(r_j)$, for all j, $1 \leqslant j \leqslant K$. Hence, combining the inequalities, $\sigma_H(r) + \varepsilon |\overline{H}| > \zeta \eta |H|$. Rearranging terms: $\sigma_H(r) > \zeta \eta |H| - \varepsilon |\overline{H}|$. Now, since $\hat{\sigma}_H(r) = |H|^{-1} \sigma_H(r)$,

$$\hat{\sigma}_H(r) > \zeta \eta - \varepsilon |\overline{H}|/|H|$$

Notice that as $\varepsilon \to 0$, $\varepsilon |\overline{H}|/|H| \to 0$. That is, the interference approaches zero as ε approaches zero.[50] Considering the limit as $\varepsilon \rightarrow 0$ and $\zeta \rightarrow 1$,

$$\lim_{\substack{\varepsilon \rightarrow 0 \\ \zeta \rightarrow 1}} \hat{\sigma}_H(r) \geqslant \eta \approx 1$$

Hence, the recall is arbitrarily similar to the remembered responses, to the extent that the remembered responses are similar.

5. *Family resemblances*

We expect a new logic to be able to describe the way in which universals are abstracted from concretes without an analysis in terms of features. Therefore we investigate how, in a fixed context c, a number of stimulus/response pairs define an abstraction. We assume that both positive examples E and negative examples $\bar{E}$ are provided, and that the response to the positive examples is r^+ and to the negative examples is r^-. The entire classification process is meaningless if the responses are the same, so we assume $s(r^+, r^-) = \varepsilon \approx 0$. For convenience we let $\sigma^+ = \sigma(r, r^+)$ and $\sigma^- = \sigma(r, r^-)$; our goal is to show that $\sigma^+ \approx 1$ and $\sigma^- \approx 0$ for positive stimuli, and *vice versa* for negative.

We aim to show that the exemplars define an abstraction to which various stimuli belong or don't belong to the extent that they are similar to positive or negative examples. Therefore, every stimulus $s \in A$ can be characterized by its similarity to the positive and negative exemplars (in context). We define a parameter ρ that measures the similarity to positive exemplars relative to the similarity to negative exemplars: $\rho = \Sigma_{i \in E} \gamma_i / \Sigma_{i \in \bar{E}} \gamma_i$. We then relate to ρ the similarity of the response r to the trained responses r^+ and r^-.

First we compute the weighted similarity of the response to the stored responses:

$$\varsigma(r) = \sum_{i \in E} \gamma_i \sigma(r, r^+) + \sum_{i \in \bar{E}} \gamma_i \sigma(r, r^-)$$

$$= \sigma^+ \sum_E \gamma_i + \sigma^- \sum_{\bar{E}} \gamma_i$$

Similarly, $\varsigma(r^+) = \Sigma_E \gamma_i + \varepsilon \Sigma_{\bar{E}} \gamma_i$. Combining via the inequality $\varsigma(r) \geqslant \varsigma(r^+)$ and solving for σ^+ yields:

$$\sigma^+ \geqslant 1 - (\sigma^- - \varepsilon) \sum_{\bar{E}} \gamma_i / \sum_E \gamma_i \geqslant 1 - (1 - \varepsilon)/\rho \geqslant 1 - 1/\rho$$

This then is our first result:

$$\sigma(r, r^+) \geqslant 1 - \rho^{-1}$$

Hence, the correctness of the response improves to the extent that the stimulus is more similar to the positive than to the negative exemplars. A symmetric analysis allows us determine the similarity of the response to r^-:

$$\sigma(r, r^-) \geqslant 1 - \rho$$

Hence, the response approaches r^- as the stimulus becomes more similar to the negative exemplars.

The preceding two results do not complete the analysis. They show that the response is similar to the appropriate correct response, but not that it is dissimilar to the appropriate incorrect response. Unfortunately, our current assumption, that $1 - \sigma(x, y)$ is a semimetric, does not guarantee this result. We could have the case that r^+ and r^- are very dissimilar, yet r is very similar to both; this is permitted by a semimetric. On the other hand, it certainly seems unintuitive to have a neural state that is simultaneously similar to two dissimilar neural states. Since the issue is unresolved, we introduce the additional assumption only here, where it is required.

A semimetric $\delta : S \times S \to R$ is a *metric* if is satisfies the *triangle inequality*:

$$\delta(x, z) \leqslant \delta(x, y) + \delta(y, z), \quad \text{for all} \quad x, y, z \in S$$

From this we immediately derive a triangle inequality for similarity metrics:

$$\sigma(x. y) + \sigma(y, z) \leqslant 1 + \sigma(x, z)$$

We return to the analysis of family resemblance.

The triangle inequality tells us that

$$\sigma(r, r^+) + \sigma(r, r^-) \leqslant 1 + \sigma(r^+, r^-) = 1 + \varepsilon$$

Therefore,

$$\sigma(r, r^+) \leqslant 1 + \varepsilon - \sigma(r, r^-) \leqslant 1 + \varepsilon - (1 - \rho) = \varepsilon + \rho$$

since $\sigma(r, r^-) \geqslant 1 - \rho$. Similarly it is easy to show that $\sigma(r, r^-) \leqslant \varepsilon + \rho^{-1}$. Thus, given the triangle inequality, we can derive the other two inequalities that characterize family resemblance:

$$\sigma(r, r^+) \leqslant \varepsilon + \rho, \qquad \sigma(r, r^-) \leqslant \varepsilon + \rho^{-1}$$

Combining these with the previous two inequalities gives bounds on the similarity of the response to the two possible correct responses:

$$\sigma(r, r^+) \in [\varepsilon + \rho, 1 - \rho^{-1}], \qquad \sigma(r, r^-) \in [\varepsilon + \rho^{-1}, 1 - \rho]$$

Hence, as expected, the response r reflects the similarity of the stimulus s to the positive and negative exemplars.

Note that ρ is defined in terms of the γ_i, which in turn depend on the similarities of the current stimulus s to the stored stimuli s_i and on the similarities of the current context c to the stored contexts c_i. Thus ρ is a function, $\rho(s, c)$, of both the current stimulus and the current context. In effect the positive and negative exemplars induce a *potential field* $\rho(s, c)$, which defines for each stimulus s in context c the extent to which it resembles the positive exemplars and is different from the negative exemplars (in their contexts). Thus ρ represents an abstraction from the examples, that is both indefinite boundaried and context-sensitive. Also note that the classification is not based on any specific features of the stimulus. On the contrary, since the state spaces are dense, we can "teach" an arbitrary abstraction by a suitable presentation of positive and negative cases.

We will say little else about learning at this time, except to note that present behavior can be habituated by adding each triple (s, c, r) to the memory at the end of the corresponding state transition.[51] Such a model is surely oversimplified; we investigate a slightly more sophisticated one below, in the context of continuous time.

C. *Continuous Time*

1. *Definitions*

We now turn to the case in which time is taken to be continuous and the contents of memory are taken to be a single continuous trace. That is, we take the input stimulus s_t to be a continuous function of the time t, as is the induced response r_t. Similarly, the memory trace (S_x, C_x, R_x) is taken to be a continuous function of a continuous index variable $0 \leqslant x \leqslant K$ (analogous to i, $1 \leqslant i \leqslant K$, in the discrete logic). We now must say what we mean by continuity.

Suppose that S and T are bounded metric spaces with similarity metrics σ_S and σ_T. Further suppose that f is a function from S to T. Then we say that f is continuous at a point $p \in S$ if and only if for all ζ with $0 \leqslant \zeta < 1$ there is an η with $0 \leqslant \eta < 1$ such that $\sigma_T[f(p),$

$f(q)] > \zeta$ whenever $\sigma_s(p, q) > \eta$ and $q \in f[S]$. That is, we can make $f(p)$ and $f(q)$ arbitrary similar by making p and q sufficiently similar.

For the sake of the continuous logic, I postulate that s_t and r_t are continuous at all $t \geq 0$, and S_x, C_x and R_x are continuous at all $x \geq 0$. Hence, I am assuming that the state of the neural systems cannot change instantaneously. This is a reasonable assumption for any physical system.

As before, define $\alpha_x = \sigma(s_t, S_x)$, $\beta_x = \sigma(r_t, C_x)$ and $\gamma_x = \Gamma(S_x, C_x)$. We define the total similarity of a response:

$$\sigma_{[a, b]}(r) = \int_a^b \sigma(r, R_x)\, dx$$

Similarly, the average is defined $\hat{\sigma}_{[a, b]}(r) = \sigma_{[a, b]}(r)/(b - a)$. Finally, the weighted similarity of a response to the memory trace is:

$$\varsigma(r) = \int_0^K \gamma_x \sigma(r, R_x)\, dx$$

These are just the continuous extensions of the previous definitions.[52]

The definition of the state transition function is the same: $s : c = \varepsilon r[\varsigma(r) = \max \{\varsigma(r) \mid r \in B\}]$. Note however that this defines the state *at the next instant of time*. That is, since $c_{t + dt} = r_t$, the new context is $c_{t + dt} = r_t = s_t : c_t$. This is the differential equation that defines the behavior of the system over time. Recall that we require r_t (and hence c_t) to be continuous.

2. Activation of a single trace interval

We can now derive results analogous to those for the discrete logic. For example, since the interval $[0, K]$ can be broken down into various size intervals $[0, x_1], [x_1, x_2], \ldots, [x_n, K]$ the discrete analysis can be applied to the subintervals. We consider a specific case.

Suppose there are a, b, c and d such that $0 < a < b < c < d < K$. Our intent is that the region $[b, c]$ of the trace is activated, the regions $[0, a]$ and $[d, K]$ are not activated, and the regions $[a, b]$ and $[c, d]$ are partially activated. Hence there are $\varepsilon \approx 0$ and $\zeta \approx 1$ such that $\zeta \leq \gamma_x$ for $b \leq x \leq c$, $\gamma_x < \varepsilon$ for $0 \leq x \leq a$ or $d \leq x \leq K$, and $\varepsilon \leq \gamma_x \leq \zeta$ for $a \leq x \leq b$ and $c \leq x \leq d$. We assume the concept is sufficiently sharp; that is, there is a $\delta \approx 0$ such that $b - a < \delta(c -$

b) and $d - c < \delta(c - b)$. We also assume that the activated responses are mutually similar; that is, $\mu[b, c] = \eta \approx 1$. Proceeding as in the discrete case,

$$\varsigma(r) \leqslant \varepsilon\sigma_{[0, a]}(r) + \zeta\sigma_{[a, b]}(r) + \sigma_{[b, c]}(r) + \zeta\sigma_{[c, d]}(r) + \varepsilon\sigma_{[d, K]}(r)$$
$$\leqslant \varepsilon a + \zeta(b - a) + \sigma_{[b, c]}(r) + \zeta(d - c) + \varepsilon(K - d)$$
$$< \varepsilon a + \zeta\delta(c - b) + \sigma_{[b, c]}(r) + \zeta\delta(c - b) + \varepsilon(K - d)$$

Conversely,

$$\varsigma(R_x) \geqslant 0 \cdot \sigma_{[0, a]}(R_x) + \varepsilon\sigma_{[a, b]}(R_x) + \zeta\sigma_{[b, c]}(R_x) + \varepsilon\sigma_{[c, d]}(R_x)$$
$$+ 0 \cdot \varepsilon\sigma_{[d, K]}(R_x)$$
$$= \varepsilon\sigma_{[a, b]}(R_x) + \zeta\sigma_{[b, c]}(R_x) + \varepsilon\sigma_{[c, d]}(R_x)$$
$$\geqslant \zeta(c - b)\,\mu[b, c] = \zeta(c - b)\,\eta$$

Noting that $\varsigma(r) \geqslant \varsigma(R_x)$ allows the inequalities to be combined as before. Simplifying and solving for $\sigma_{[b, c]}(r)$ yields:

$$\sigma_{[b, c]}(r) > \zeta\eta(c - b) - \varepsilon(K + a - d) - 2\zeta\delta(c - b)$$

Hence, the average similarity is:

$$\hat{\sigma}_{[b, c]}(r) > \zeta\eta + \varepsilon(K + a - b)/(c - b) - 2\zeta\delta$$

Hence,

$$\lim_{\substack{\delta \to 0 \\ \varepsilon \to 0 \\ \zeta \to 1}} \hat{\sigma}_{[b, c]}(r) > \eta$$

Hence, as the concept becomes sharp, the interference small and the similarity of the stimulus/context to their stored counterparts increases, the response approaches its stored counterparts.

3. *Learning*

So far we have described the memory trace as a function (S_x, C_x, R_x) defined for $0 \leqslant x \leqslant K$. How does the memory trace get extended? The simplest way is if we let all experiences be recorded with equal weight. That is, we set $K = t$, and let the memory trace be precisely the previous history of the system, (s_x, c_x, r_x), $0 \leqslant x \leqslant t$. The required modifications to the formulas are simple. The result is that the memory trace wanders through mental space under the influence of its own past history.

More realistically, we might assume that experiences are not equally likely to be recalled. Thus we can postulate a continuous function $\pi_x = \pi(R_x)$ that reflects the *inherent relevance* (such as pleasure or pain) of the mental experience. This is an indirect basis for other measures of relevance. The required modification is simple: $\varsigma(r) = \int_0^t \pi_x \gamma_x \sigma(r, r_x) \, dx$.

IV. CONCLUSIONS

I have had several goals in this paper. The first was to claim that AI is moving, and must of necessity move, into a new phase that comes to grips with nonverbal reasoning. My second goal has been to claim that traditional AI technology, based as it is on idealized verbal reasoning, is inadequate to this task, and therefore that the new AI requires a new logic, a logic that idealizes nonverbal reasoning. My final goal has been to show, by example, what such a logic might be like. This logic is at present in a very rudimentary form. My only consolation in this is that Boole's logic was a similarly rudimentary form of modern symbolic logic. I would certainly be very gratified if my logic were as near to the mark as his.

NOTES

[1] A critique from a (predominantly Heideggerian) phenomenological viewpoint of current AI technology can be found in H. Dreyfus (1979) and H. Dreyfus and S. Dreyfus (1986). In his (1982, pp. 3—27) Hubert Dreyfus claims that current AI technology is making the same fundamental error that Husserl made in his approach to phenomenology, and that it is facing the same "infinite task." Gardner (1985) provides a good overview of the strengths and limitations of cognitive science; much of this applies to contemporary AI. Haugeland (1985) likewise shows AI and cognitive science in their historical context. His (1981) collects important papers pro and con traditional AI methods.

[2] See Köhler (1947), p. 118. Dreyfus (1979) quotes Oettinger (1972): "Perhaps . . . in perception as well as in conscious scholarly analysis, the phoneme comes after the fact, namely . . . it is constructed, if at all, as a *consequence* of perception not as a step in the process of perception itself."

[3] For example, a speech recognition system may have to be trained to a specific voice, and may work poorly if the speaker's voice changes (e.g., when he's agitated).

[4] Peirce, a pioneer of symbolic logic, saw that logic machines would be "minutely analytical, breaking up inference into the maximum number of steps" (Goudge, 1969, p. 61). For a calculus of reasoning, efficiency is critical, so we should "seek to reduce the number of individual steps to a minimum" (Goudge, 1969, p. 61).

[5] As Wittgenstein (1958) says, "The question 'Is what you see composite?' makes good sense if it is already established what kind of complexity — that is, which particular use of the word — is in question" (§47, p. 22).

[6] Dreyfus (1982), p. 21.

[7] Heidegger (1962), p. 99. He continues: "That with which our everyday dealings proximally dwell is not the tools themselves. On the contrary, that with which we concern ourselves primarily is the work — that which is to be produced at the time; and this is accordingly ready-to-hand too. The work bears with it that referential totality within which the equipment is encountered." The latter observation is especially important with regard to the issues of context dependence and intentionality, discussed below.

[8] "This context and our everyday ways of coping in it are not something we *know* but, as part of our socialization, form the way we *are*" (Dreyfus, 1982, p. 21).

[9] M.I.T. Artificial Intelligence Laboratory Memo No. 299 (September 1973), p. 77; quoted in Dreyfus (1979), p. 34.

[10] Wittgenstein (1958), §81, p. 38.

[11] See H. Dreyfus and S. Dreyfus (1986).

[12] For this distinction, see Ryle (1949), chapter II (3).

[13] Wertheimer (1959), p. 239.

[14] Heidegger distinguishes ready-to-hand and present-at-hand as follows: "Original familiarity with beings lies in *dealing with them* appropriately. . . . The *whatness* of the beings confronting us every day is defined by their equipmental character. The *way* a being with this essential character, equipment, is, we call *being handy* or handiness, which we distinguish from being extant, [present] at hand" (Heidegger, 1982, p. 304). Nature itself can be considered either as ready-to-hand — the characteristic stance of technology — or as present-at-hand — the characteristic stance of science. "If its kind of Being as ready-to-hand is disregarded, this 'Nature' itself can be discovered and defined simply in its pure presence-at-hand" (Heidegger, 1962, p. 100).

[15] There is now ample evidence that nonverbal reasoning is an essential part of human (and animal) cognitive activity. See for example Arnheim (1971), Wertheimer (1959), Kosslyn (1980), Shepard (1971, 1975), and Gardner (1985). On the importance of nonverbal thinking in scientific creativity, see Miller (1986).

[16] In Peirce's scheme, the mathematics of logic is a subclass of mathematics, which is a science of discovery. Logic is an order within normative science, which is a subclass of philosophy, which is a science of discovery. History of science is a family within history, which is a suborder of descriptive psychics, which is an order in the psychical sciences, which is a subclass of idioscopy, which is a science of discovery. Psychology is also an order of the psychical sciences. "Mathematics studies what is and what is not logically possible, without making itself responsible for its actual existence." One of its branches is the mathematics of logic. "Nomological psychics [psychology] discovers the general elements and laws of mental phenomena." One of the divisions of normative science is logic. "Logic is the theory of self-controlled, or deliberate, thought; and as such, must appeal to ethics for its principles," since ethics "is the theory of self-controlled, or deliberate, conduct," (from *A syllabus of Certain Topics of Logic*, 1.180—192, quoted in Buchler (1955), p. 60—62). All of the above are sciences of discovery. Within the theoretical sciences of discovery there are three classes: mathematics, philosophy (including normative science) and idioscopy (a term Peirce borrowed from Bentham).

The latter is divided into physiognosy (the physical sciences) and psychognosy (the psychical sciences). The differentiae of the latter two are relevant: "Physiognosy sets forth the workings of efficient causation, psychognosy of final causation." (1.239—42, Buchler (1955), p. 67) We might say, physiognosy deals with mechanical laws, psychognosy with intentional laws.

[17] See papers by Johnson-Laird (1970), Wason (1966, 1972), Kahneman, Slovic, and Tversky (1982, 1984).

[18] See, for example, G. Polya's (1968).

[19] In Peirce's classification there are also Practical Sciences, which presumably include a science of practical logic. In his "Minute Logic" (1.239, quoted by Buchler (1955), p. 66), Peirce distinguishes two branches of science, "Theoretical, whose purpose is simply and solely knowledge of God's truth; and Practical, for the uses of life." (1.239) In particular, to the three normative sciences, logic, ethics and esthetics, there are three corresponding practical sciences, or arts: the art of reasoning, the conduct of life and fine art. The normative sciences, like the practical, "study what ought to be, i.e., ideals," but "they are the very most purely theoretical of purely theoretical sciences" (1.278—82, Buchler (1955), p. 69—70). For Peirce, esthetics is the primary normative science, for it is "the science of ideals, or of that which is objectively admirable without any ulterior reason" (1.191, Goudge (1969), p. 48). I deviate from Peirce's scheme in that I include the practical sciences under the normative. Thus I do not distinguish between logic as a normative science and logic as a practical science.

[20] See Johnson-Laird (1970), Wason (1966, 1972), and Kahneman, Slovik and Tversky (1982, 1984).

[21] See Miller (1986), pp. 125—183.

[22] See Wertheimer (1959), Chapter 10.

[23] See Gardner (1985), Chapter 13.

[24] It seems likely that the new AI will bring with it a rebirth of interest in analog computation. Current research on analog, molecular and hybrid optical computers is perhaps a harbinger.

[25] Wittgenstein (1958), §71, p. 34.

[26] Wittgenstein (1958), §69, p. 33

[27] See for example Zadeh (1965, 1975, 1983). Kichert (1978) has a good summary of the postulates of fuzzy set theory. See also Goguen (1969).

[28] See the 'pure water' example above. Note also that the context itself is indefinite boundaried. What if I'm serving guests at my campsite?

[29] Arnheim (1971), p. 37.

[30] Wittgenstein (1958), §65, p. 31.

[31] Wittgenstein (1958), §66, p. 31.

[32] Wittgenstein (1958), §79, p. 37.

[33] Arnheim (1971), p. 49.

[34] Rosch (1977), p. 30. See also Rosch (1978), and Armstrong, Gleitman and Gleitman (1983).

[35] Kuhn (1970), p. 200.

[36] Kuhn (1970), p. 192.

[37] We use this term in Brentano's and Husserl's sense, i.e., "the unique peculiarity of experiences 'to be the consciousness of something.'" (Husserl, 1962, §84, p. 223).

[38] Goudge (1969), p. 235; see also Tiercelin (1984).

[39] I am grateful to R. W. Hamming for alerting me to this limitation of analog computation.

[40] Paivio (1979), p. 8.

[41] Paivio (1979), p. 9. See also Miller (1986), especially chapter 4, for a discussion of the interplay of verbal and nonverbal reasoning.

[42] Arnheim (1971), p. 229.

[43] Arnheim (1971), pp. 231—232.

[44] The reader will observe that we use terminology inspired by neuropsychology. This should not be interpreted as implying that we are offering a theory of brain function. It is simply the case that we have found considerations of brain organization to be helpful in developing the theory.

[45] Although, as will become apparent later, our model permits the possibility that *all* processes have some effect on semipermanent information.

[46] Our definitions are inspired by, but, in the spirit of idealization, not identical with those common in neuropsychology. Kolb and Whishaw (1985), define afferent as "[c]onducting toward the central nervous system or toward its higher centers," efferent as "[c]onducting away from higher centers in the central nervous system and toward a muscle or gland," and interneuron as "[a]ny neuron lying between a sensory neuron and a motor neuron."

[46] Note that the minimum similarity is just 1 minus the radius of the set (i.e., maximum distance in the set).

[47] Actually, as will be explained shortly, it is a multiple-valued function and thus not, in the technical sense, a function.

[48] The number K is not fixed, but increases as new traces are made in the memory. However, for the analysis of state (i.e. transient information) changes, it can be taken as constant.

[49] There is no loss of generality in taking the index of the similar pair to be 1.

[50] Note that $|\bar{H}|/|H|$ is an (inverse) measure of the number of activated traces.

[51] Thus K increases with each state transition.

[52] Note that $\varsigma(r)$ is just the inner product $\gamma \cdot \sigma_r$ of γ and the $\sigma_r = \sigma(r, R_x)$.

REFERENCES

Armstrong, S. L., Gleitman, L. R., and Gleitman, H.: 1983, 'What Some Concepts Might Not Be', *Cognition* 13, 263—308.

Arnheim, Rudolf: 1971, *Visual Thinking*, University of California Press, Berkeley and Los Angeles.

Buchler, J. (ed.): 1955, *Philosophical Writings of Peirce*, Dover, New York.

Dreyfus, Hubert L.: 1979, *What Computers Can't Do: The Limits of Artificial Intelligence*, revised edition, Harper & Row, New York.

Dreyfus, H. (ed.): 1982, *Husserl, Intentionality and Cognitive Science*, MIT Press, Cambridge.

Dreyfus, H., and Dreyfus, S.: 1986, *Mind over Machine*, Macmillan, The Free Press, New York.

Gardner, Howard: 1985, *The Mind's New Science: A History of the Cognitive Revolution*. Basic Books, New York.

Goguen, J. A.: 1969, 'The Logic of Inexact Concepts', *Synthese* **19**, 325—373.

Goudge, Thomas A.: 1969, *The Thought of C. S. Peirce*, Dover, New York.

Haugeland, John (ed.): 1981, *Mind Design: Philosophy, Psychology, Artificial Intelligence*, MIT Press, Cambridge.

Haugeland, John: 1986, *Artificial Intelligence: The Very Idea*, MIT Press, Cambridge.

Heidegger, Martin: 1962, *Being and Time*, seventh edition, transl. J. Macquarrie and E. Robinson, Harper and Row, New York.

Heidegger, Martin: 1982, *The Basic Problems of Phenomenology*, transl. Albert Hofstadter, Indiana University Press, Bloomington.

Husserl, Edmund: 1962, *Ideas: A General Introduction to Pure Phenomenology*, transl. W. R. Boyce Gibson, Collier, London.

Johnson-Laird, P. N., and Wason, P. C.: 1970, 'A Theoretical Analysis of Insight into a Reasoning Task', *Cognitive Psychology* **1**, 134—148.

Kahneman, D. Slovic, P., and Tversky, A. (eds.): 1982, *Judgement under Uncertainty: Heuristics and Biases*, Cambridge University Press, New York.

Kahneman, D., and Tversky, A.: 1982 'The Psychology of Preferences', *Scientific American* **246**, 160—174.

Kahneman, D., and Tversky, A.: 1984, 'Choices, Values and Frames', *American Psychologist* **39**, 341—350.

Kanerva, Pentti: 1986, '*Parallel Structures in Human and Computer Memory*', RIACS TR 86.2, Research Institute for Advanced Computer Science, NASA Ames Research Center.

Kichert, Walter J. M.: 1978, *Fuzzy Theories on Decision-Making*, Martinus Nijhoff, Leiden.

Köhler, Wolfgang: 1947, *Gestalt Psychology*, New American Library, New York.

Kohonon, Teuvo: 1977, *Associative Memory: A System-Theoretical Approach*, Springer-Verlag, Berlin.

Kolb, Brian, and Whishaw, Ian Q.: 1985, *Fundamentals of Human Neuropsychology*, second edition, W. H. Freeman and Company, New York.

Kosslyn, Stephen Michael: 1980, *Image and Mind*, Harvard University Press, Cambridge.

Kuhn, Thomas: 1970, *The Structure of Scientific Revolutions*, second edition, University of Chicago Press, Chicago.

Miller, Arthur I.: 1986, *Imagery in Scientific Thought*, MIT Press, Cambridge.

Oettinger, Anthony: 1972, 'The Semantic Wall', in E. David and P. Denes (eds.), *Human Communication: A Unified View*, McGraw-Hill, New York, p. 5.

Paivio, Allan: 1979, *Imagery and Verbal Processes*, Hillsdale: Lawrence Erlbaum Assoc., 1979.

Polya, G.: 1986, *Patterns of Plausible Inference*, second edition, Princeton University Press, Princeton.

Ryle, Gilbert: 1949, *The Concept of Mind*, University of Chicago Press, Chicago.

Rosch, Eleanor: 1977, 'Human Categorization', in N. Warren (ed.), *Advances in Cross-cultural Psychology*, vol. I, Academic Press, London.

Rosch, Eleanor: 1978, 'Principles of Categorization', in E. Rosch and B. B. Lloyd (eds.), *Cognition and Categorization*, Lawrence Erlbaum Assoc., Hillsdale.

Shepard, Roger N.: 1975, 'Form, Formation, and Transformation of Internal Representations', in R. L. Solso (ed.), *Information Processing in Cognition: The Loyala Symposium*, Lawrence Erlbaum Assoc., Hillsdale.

Shepard, R. N., and Metzler, J.: 1971, 'Mental Rotation of Three-dimensional Objects', *Science* **171**, 701—703.

Tiercelin, C.: 1984, 'Peirce on machines, self control and intentionality', in S. B. Torrance (ed.), *The Mind and the Machine: Philosophical Aspects of Artificial Intelligence*, Chichester: Ellis Horwood Ltd. and New York: John Wiley, pp. 99—113.

Tversky, A., and Kahneman, D.: 1983 'Extensional vs. Intuitive Reasoning: The Conjunction Fallacy in Probability Judgement', *Psychological Review* **90**, 293—315.

Wason, P. C.: 1966, 'Reasoning', in B. Foss (ed.), *New Horizons in Psychology*, vol. 1, Penguin, Harmondsworth.

Wason, P. C., and Johnson-Laird, P. N.: 1972, *The Psychology of Reasoning: Structure in Context*, Harvard University Press, Cambridge.

Wertheimer, Max: 1959, *Productive Thinking*, Enlarged Edition, Harper & Brothers, New York.

Wittgenstein, L.: 1958, *Philosophical Investigations*, transl. G. E. M. Anscombe, third edition, Macmillan Company, New York.

Zadeh, L. A.: 1965, 'Fuzzy Sets', *Information and Control* **8**, 338—353.

Zadeh, L. A.: 1975, 'Fuzzy Logic and Approximate Reasoning', *Synthese* **30**, 407—428.

Zadeh, L. A.: 1983, 'Commonsense Knowledge Representation Based on Fuzzy Logic', *IEEE Computer* **16**, 10 (October), 61—65.

Computer Science Department
Naval Postgraduate School
Monterey, CA 93943, U.S.A.

PART II

EPISTEMOLOGICAL DIMENSIONS

CLARK GLYMOUR

ARTIFICIAL INTELLIGENCE IS PHILOSOPHY

1. INTRODUCTION

Artificial intelligence is philosophical explication turned into computer programs. Historically, what we think of as artificial intelligence arose by taking the explications provided by philosophers, and finding computable extensions and applications of them. Developments in artificial intelligence programming technology have tended to make the process of transforming certain sorts of philosophical explications into programs nearly automatic. Production rule systems, for example, exploit the ambiguity between conditional sentences and procedural rules, and permit one to turn theories consisting of a collection of conditionals into a simple program, with little explicit worry about control structure or algorithm design. Nowadays, most of the philosophical theories used in artificial intelligence are not taken from the philosophical literature directly, but that does not make them any the less philosophical. Computer science teaches us that there is more to philosophy than we might have thought, and it contains new and interesting philosophical contributions.

To see that artificial intelligence is philosophy of science and philosophical logic, or their application, one need only consider some of the work in AI[1]. What makes AI programs run? What are the *ideas* behind them? First consider some programs that make use of ideas from the philosophical literature.

- *The DENDRAL* and *META-DENDRAL Programs.* The task that the DENDRAL program addresses is the determination of molecular structure from data consisting of the molecular formula of a compound, and the mass spectrum of the compound. The program uses three key ideas: (1) An algorithm for computing the topological structures of organic molecules consistent with any given molecular formula; (2) Rules about which molecular bonds in an environment in a molecule are most likely to break; (3) Nicod's confirmation criterion and Hempel's theory of deductive nomological explanation.

195

James H. Fetzer (ed.), Aspects of Artificial Intelligence, 195—207.

There is, of course, a great deal more detail and, not least important, a control structure.

- The META-DENDRAL program is designed to improve the rules that the DENDRAL program uses. The program takes as data a description of a class of organic compounds (generally a restriction on the structure of the carbon skeleton and on the number and kinds of non-carbon and non-hydrogen atoms), the molecular structure, molecular formula, and mass spectrum of a number of instances of compounds of this class. The program outputs improved rules, for the class of compounds, regarding which molecular bonds will break in the mass spectrometer. META-DENDRAL uses the DENDRAL program plus the "prediction criterion of confirmation," discussed and rejected by Hempel, but championed by many other philosophers of science.

- Programs for medical diagnosis developed in the 1960s by Warner, and by Gorry and Barnett, use Bayesian procedures. The programs impose an *a priori* distribution on diagnoses and evidence, and update the distribution by conditionalization on the evidence. Gorry and Barnett's program uses simple utilities to implement Bayesian decisions. The ideas of course owe a great deal to professional statisticians, especially Savage, but also to Ramsey and Carnap.

- Ehud Shapiro's *Algorithmic Program Debugging* won the ACM Distinguished Dissertation award in 1982. Shapiro's book is dedicated to Sir Karl Popper, and the program can broadly be viewed as a computational implementation of Popper's methodology. A component program, The Model Inference System, uses conjecture and refutation to infer a finite axiomatization of the first order sentences true in any relational structure, provided that set has such an axiomatization and all theorems are computable from it in a bounded number of computational steps. The evidence provided to the program consists of basic sentences true in the target structure.

- Glymour, Scheines, Spirtes and Kelly have published a program, TETRAD, that helps to construct linear causal explanations of statistical data. The essential idea of the program is a philosophical theory of explanation which can be found in various forms in works by Causey, Glymour, Rosencrantz, Skyrms and elsewhere.

- If there is a theoretical side to machine learning, it is "formal learning theory." Formal learning theory investigates the existence of algorithmic procedures that will "identify" any object in a class of objects given an infinite sequence of evidence generated from the object. The objects may be recursive functions, languages, relational structures, or first-order theories. The subject was begun by a philosopher, Hilary Putnam, who used themes from Reichenbach to argue that, for every learning algorithm based on a Carnapian probability measure, there are hypotheses that the algorithm cannot learn, even given every positive instance of the hypothesis.

Many, many other examples could be given of ideas in artificial intelligence that had a previous hearing in the philosophical literature. Such examples perhaps only show that artificial intelligence is an activity that consumes philosophical results, not that it is itself a philosophical activity. But consider a few of the clearly *philosophical* contributions that artificial intelligence work has made.

- *Concept Learning Programs.* "Concept learners" are simply programs that infer a sentence of universal biconditional form, with an atomic formula on one side of the biconditional. The predicate in the atomic formula is the "concept" to be learned. The evidence from which the program is to learn the sentence consists of a list of conjunctions of singular sentences, with each conjunction containing a conjunct that is a basic sentence containing one of the predicates occurring in the sentence to be learned. The sentence to be learned might, for example, be of the form $(x)[Cx \leftrightarrow [(Ax \ \& \ Bx) \lor -Dx]]$ and the evidence might consist of a list such as $\langle (Ca \ \& \ Aa \ \& \ Ba \ \& \ Da), (-Ca \ \& \ -Aa \ \& \ Ba \ \& \ Da)...\rangle$. Items in the list are considered "positive" if they begin with "C", "negative" if they begin with "$-C$." Perhaps the earliest and best program of this kind is Earl Hunt's Concept Learning System. Hunt's program contains a simple algorithm that finds a biconditional for which all of the positive items, and none of the negative items, are instances of the formula on the right hand side. If the items all satisfy an appropriate biconditional, and they exhaust a structure in which the biconditional is true, Hunt's algorithm is guaranteed to converge to the correct hypothesis. Hunt's algorithm can be viewed as a simple theory of tentative hypothesis acceptance. If Hempel's work is philosophy, so is Hunt's.
- *Version Spaces.* Consider the concept learning problem again. Rather than trying to find a particular hypothesis to put forward as each new item is received as evidence, we might instead try to keep track of all hypotheses that are consistent with the evidence so far

received. That problem can be put in Hempelian terms: keep track of all of the biconditionals such that the evidence so far received has *some* consistent extension satisfying the biconditional. Can that be done? Tom Mitchell proposed a procedure, which he called computing "Version Spaces," to do it. (The procedure is so called because it keeps track of the space of all versions of the hypotheses consistent with the evidence.)

- *Nonmonotonic Reasoning.* Philosophers have known for a long time that inductive reasoning is "nonmonotonic": a conclusion derived inductively from a set of premises may not follow from consistent extensions of the premises. A great deal of reasoning outside of learning contexts has the same feature. Birds can fly, and Tweety is a bird, so I infer that Tweety can fly, but not if I also know that Tweety is an ostrich. What artificial intelligence work has added is a sense of how widespread such inference patterns are, and several attempts to articulate a formal structure for this sort of reasoning.

- *Accounts of Discovery.* The BACON programs, developed by Langley, Simon and Bradshaw, give a *reconstruction* of how various empirical laws might be (or might have been) discovered. The STAHL program, developed by Langley, Simon and Zytkow, reconstructs the reasoning of phlogiston chemists about the composition of matter. No one doubts that theories of argument forms (e.g., inference to the best explanation, or bootstrap testing), buttressed by historical cases, are philosophical; there seems no reason to doubt that theories of reasoning and discovery procedures are philosophical as well.

These are only a handful of examples, to which many others might be added. Even the work that directly applies previous philosophical theories often adds considerably to them. Shapiro's work, for example, is a better piece of philosophy than anything else I can think of in the Popperian tradition.

2. DESCRIPTION AND NORM

Philosophical theories usually equivocate between description and prescription. The equivocation is notorious in Bayesian epistemology, in logical empiricist philosophy of science, in ethics, and elsewhere. Rawls has made the practice into a principle: reflective equilibrium. If

artificial intelligence is philosophy, we should expect the same equivocation. In artificial intelligence the equivocation is between "cognitive
modelling" and not cognitive modelling.

Suppose we consider an algorithm, maybe implemented on a computer, that does something impressive, and will perform any of an
infinity of instances of some task that we think of ordinarily as requiring
intelligence. The program may or may not perform every instance of a
task *correctly*, but it does pretty well. Perhaps the algorithm when
implemented makes discoveries in chemistry or in social science, or
perhaps it identifies objects, or perhaps it reads and interprets English
text. Whatever the program does, there are several different claims that
might be made about it. Two of the most important are these:

- The program's performance on the task is within the range of human
  performances on the same task.
- People use the same algorithm to perform the same task.

If either of these claims is made on behalf of an algorithm, then the
combination, the claim and the algorithm, is a piece of *cognitive
psychology*. Instrumentalist psychologists, such as John Anderson, may
stop at the first claim; bolder psychologists, such as Herb Simon, may
make the second as well. In either case, an empirical claim is made, a
claim that can only be established by comparing programs and persons.

Instead of these claims, others might be made. For example:

- The algorithm performs a task correctly.
- The algorithm performs a task with a specified probability of
  success.
- The algorithm is a good way for a computationally bounded system
  to attempt the task.

Sometimes these two sorts of claims are clearly distinguished, and
one sort is made but not the other. In other cases the distinctions are
fudged. The fudging is as natural in artificial intelligence as it is in
philosophy. We equivocate over description versus prescription because
practice is a primary source of our intuitions about what is best or what
is obligatory. In attempting to get the computer to perform a task at
least as well as a human we must somehow reduce the task to a
computational problem, and we often have little guidance in that
reduction save by analyzing the procedures, judgments and criteria that

humans use. That humans seem to use certain procedures to accomplish difficult tasks sometimes suggests that they are good procedures to use.

3. LITTLE THEORIES AND LARGE INTELLIGENCES

Artificial intelligence comes with a philosophy of mind that provides the rhetoric of the enterprise and forms its public ambitions. In a phrase, cognition is symbolic processing, and symbol processors are (or can be) as cognizant as humans are. That philosophy of mind has been the focus of most of the critical attention that philosophers have given to AI. That focus is perhaps unfortunate, for the work that makes artificial intelligence part of philosophy is independent of the *details* of the philosophy of mind.

"Conventional" artificial intelligence programs, whether they are programs for computer vision, or general learning, or expert systems for a special domain, are little theories. They are theories about the confirmation and testing of propositions by other propositions, or theories about the geometry of edges of visual objects, or theories about what features make for a good combination of cutting oils in various machinery, and so forth. The more the theories look like theories of reasoning, the more the description of the program looks like a piece of philosophy.

Such theories always have a radical incompleteness. Computers executing programs based on the theories always have limitations beyond which their behavior is unintelligent. The limitation of the computer program parallels the limitation of the philosophical theory. I cannot think of an *effective*, i.e., computable, philosophical theory of anything for which there are not a multitude of counterexamples and limitations. Even excellent pieces of formal philosophy work only in appropriate contexts, and there is no formal theory of contextual variation. Hempel's account of explanation works very nicely for celestial mechanics, perhaps, but it is not so good on the heights of flagpoles. There are philosophical theories that do not admit ready counterexamples, but that is generally because they are vague, and therefore ineffective.

Now it may be that there is a grand theory that is not context dependent, and that can be executed on the computer, and that will show all of the resourcefulness and competence of human intelligence.

If there is and it can be found and programmed, then the vision of artificial intelligence championed by Simon and Newell and Pylyshyn and many others will be realized. That would be a *philosophical* accomplishment beyond all precedent, for it could only be done by constructing a philosophical theory of perception, learning, and reasoning that is adequate to all of human practice. Even if one does not expect any such thing. that need not lessen the interest in artificial intelligence work, any more than the conviction that any effective philosophical theory will have limitations and counterexamples lessens the interest of such theories. The interest derives from the fact that the theories are clear, if qualified, articulations of norms, and they do more to further our understanding than most unclear but unqualified articulations. Philosophical progress consists in pushing at the limitations without giving up the clarity. Progress in artificial intelligence consists in the same thing.

There are other lines of artificial intelligence work which are less clearly part of philosophy. These lines of work are different in that they do not depend on giving reconstructions of a domain of knowledge or form of reasoning. They are not philosophy. Instead, the reasoning patterns *emerge* from constraints on the computational system and from its interactions with the environment; they are not deliberately programmed by the system designer. Work of this sort includes "genetic" algorithms, in which structures compete with other another, tooth and claw, and competent structures evolve. It also includes computational systems that are based on brain architecture, and parallel distributed processing algorithms. Systems of the latter sort can be designed so that they learn, and the learning is very flexible. So far at least, it is also very slow. This sort of work is fascinating and promising, even though at present of much less technological use than more "classical" artificial intelligence efforts. It is also much less like philosophy of science and philosophical logic.

Whether it turns out that people execute algorithms that are founded on theories that look like explications, or instead turn out to be distributed processors without such representations, or something else altogether, the epistemological interest of AI remains. It remains, first, in the philosophical interest of the theories that are the basis of programs, and, second, in the thesis that no matter what sort of system humans turn out to be, they are *computationally bounded.*

4. HOW ARTIFICIAL INTELLIGENCE IS DIFFERENT

The main lines of artificial intelligence work consist in philosophical explications turned into computer programs. Turning explications into programs creates special problems that have not usually been addressed by philosophers. There is no clear intellectual reason why these further problems, which are largely computational, ought not to be considered a part of philosophy, and some philosophers are beginning to take part in the work.

A theory is a statement of relations and properties; a program is a specification of procedures. I can very well say, for example, that the best theories on any given body of data are those that are confirmed in Hempel's sense by that data, or those that have the highest posterior probability on the data. It is something else, however, to specify an algorithm that will compute the sentences in a language that are confirmed in Hempel's way by any body of data, or to specify a procedure that will compute the hypotheses with the highest posterior probability on any evidence. The philosophical task is expanded: one must provide a theory, and develop procedures to compute features in it. Several theoretical questions immediately arise.

- Given a theory and a property it describes, and an algorithm for computing the property, can it be shown that the algorithm does in fact compute the property, or that it computes some specifiable approximation to the property?
- Given an algorithm, what is its complexity? How does the time the algorithm requires (e.g., the number of steps it executes) increase as the size of instances to be computed increases?
- Given a theory and a property, what is the least complexity possible for any algorithm that computes the property? Complexity can be measured in various ways (e.g., worst case complexity, expected complexity), and generally, by whatever measure, better algorithms make lower demands on computational resources. But some properties may be intrinsically more difficult to compute than others.
- Given a property, what is the behavior of a system that computes that property when compared with some external standard? For example, under what conditions will a procedure that accepts increasing bodies of evidence and outputs one of the hypotheses with highest posterior probability eventually converge to the true hypothesis?

There are some recent striking contributions by professional philosophers to questions of this sort. Decision problems are put in complexity classes, according to the most efficient possible algorithms for deciding them. Problems are *NP complete* if they are equivalent in difficulty to any of a large class of problems for which no algorithm is known whose resource consumption is bounded by a polynomial function of the size of the problem. Kevin Kelly has proved that the problem of computing Version Spaces for simple hypothesis languages is NP complete. Discovery problems consist of a class of objects (which may be recursive functions, or languages, or relational structures, or theories). Features of each object in the class (e.g., finite segments of the graph of a recursive function, sequences of well-formed strings of the language, finite segments of the diagram of a relational structure, theorems of a first-order theory) constitute the *evidence* that is produced by the object. A discovery procedure is an algorithm that accepts (in any order) an infinite sequence of pieces of evidence, a sequence that eventually contains every piece of evidence. As each piece is received the discovery procedure outputs a conjecture as to the identity of the object, or withholds its conjecture until further evidence is received. A discovery procedure *solves* a discovery problem if, on every ordering of the evidence, after a finite sequence of evidence has been received the procedure ever after outputs the correct hypothesis. Scott Weinstein, in collaboration with Daniel Osherson, has proved that there are discovery problems that can be solved by some procedure or other, but cannot be solved by a Bayesian procedure that outputs at every stage one of the hypotheses with highest posterior probability. Both Kelly's work and Weinstein and Osherson's work are just some first steps in exploring the complexity and power of learning strategies, and many, many open questions remain.

5. WHY BOTHER?

Given a philosophical explication, why should one bother to develop procedures to compute properties the theory describes, and why should one investigate the theoretical questions described above concerning procedures for computing such properties? The short and complete answer is that procedures are at least as interesting as relations and properties. It is all very well to be told that one should infer the best explanation of a body of data; it is better still to be told what properties distinguish a best explanation; it is best to be told a procedure for finding the best explanation. The deeper answer is that the theoretical questions itemized above are really the issues of epistemological and normative interest. Ought implies can, or so I think, and any normative epistemological theory, like any ethical theory, bears the burden of showing how its imperatives can be fulfilled, or if not fulfilled, how they can be better or worse approximated. If, as I believe, people are computationally bounded, then ought implies can implies can compute. An artificial intelligence program is really a kind of normative theory. It says, "Such and such is a good thing to do, and here is a good way to do it." But to say that something is a good thing to do is otiose if there is in fact no way to do it, and there is no way to do computationally intractable problems. Further, making the case that such and such is a good thing to do consists at least in part in showing that doing it will lead to other desires (e.g., to discovering the truth). And finally, a way to do such and such may not be good if there are computationally better ways.

A lot of epistemology would seem very different if it were looked at in this way. So conceived, the urgent issues besetting Bayesian epistemology, for example, are transformed. Bayesians say that having degrees of belief that satisfy the requirements of probability measures and changing them by conditionalization is a good thing to do. Some also say that it is a good thing to be nondogmatic about hypotheses; i.e., not to assign contingent hypotheses degree of belief 1 or 0 (unless, of course, they are the results of observation). But to do as Bayesians recommend, one must be able to compute probability measures over languages, and it is known that there is a trade-off between dogmatism and the computational difficulty of probability measures: the less dogmatic you are, the more difficult it is to compute your probability measure. If you are nondogmatic about everything save logical truths, contradictions, and a decidable set of contingent propositions, then your probability measure (over a first-order language) is not computable at all. So an urgent question for computational Bayesianism is this: how can probability measures be approximated by computable measures? If nondogmatic but computationally intractable probability measures can be approximated by computable measures, will the computable measures converge towards the truth under conditionalization, or under some other transformation that approximates conditionalization?

None of this explains why one should bother implementing algorithms in programs and running the programs on a computer. It only shows why a theoretical investigation of procedures and their properties might be of philosophical interest, not why anyone serious should bother with a computer program. There are several reasons, not all of which may apply in any particular case.

- Sometimes, as with expert systems, there is a quite practical use for the computer program. There is no reason why philosophy ought not to be of practical use.
- Very often the relations between the computational procedure and some external standard are too complex or too vague to be untangled by mathematical analysis, and the only means to assess the procedure's performance is by implementing the procedure and running it.
- Sometimes the algorithm, when implemented, can serve as a research tool for discovering other properties of importance.

Perhaps only the third reason is opaque, and needs an illustration. In developing the TETRAD program, my collaborators and I were interested in statistical properties of linear statistical theories, or "causal models," that are determined entirely by a directed graph representing the causal hypotheses in such a theory. Kevin Kelly found and implemented an algorithm which we subsequently proved did indeed compute important statistical features of any such theory entirely from its associated directed graph. The statistical features computed are simple, and the algorithm is simple, but it is impossible to do the computations reliably by hand. Kelly's program, and a subsequent program written by Peter Spirtes, enabled us to classify directed graphs according to the statistical properties they entail. The classification in turn became the basis for a search procedure that helps find the theory whose statistical features best fit the statistical patterns found in any appropriate body of data. The applied work could not have been done without a computer implementation of the theoretical work, and implementing some of the theoretical work made possible other theoretical work.

6. GOOD AND BAD ARTIFICIAL INTELLIGENCE

The trouble with artificial intelligence work is that it can use the senses to overwhelm the intellect. If the cases considered are restricted, almost any inane theory can be programmed to produce output that looks

impressive. Oohs and Ahhs at the computer's behavior on carefully
chosen examples don't make the theory behind the program an accomplishment of any value if it is intrinsically trivial and not general, or if
the program breaks at the least variation in the examples, or the
computer grinds on forever if tasks of realistic size are imposed.
Programming technology has made it relatively easy to implement
theories of little or no value; one needs only a system of conditional
hypotheses, thought of as procedural rules, and the system can be
programmed with very little thought given to control structures.

My rules of thumb are these:

- Since AI is philosophy, the philosophical theory a program implements should be explicit. Any claim that a program solves some
  well-studied problem, whether finding the best explanation, or
  measuring simplicity, or whatever, but doesn't say how, should be
  disbelieved.
- If the procedures look simple enough for a theoretical analysis of
  their complexity and their concordance with some external standard,
  but none is given, the work should be suspect.
- Examples of program runs should be done not just to illustrate what
  the program can do, but also what it *can't* do.
- One needs to know whether a program is just an implementation of
  an old algorithm in a new context, or whether some new computational problem has been solved. One also needs some consideration
  of whether or not there are better ways to do the same thing.

By these standards there is a lot of pretty good work in artificial
intelligence, and some excellent work. The same could be said of
philosophy, and not surprisingly, for while artificial intelligence and
philosophy may be different academic disciplines, in the main they are
one subject.

NOTE

[1] One might also consider the people. Consider the backgrounds of two of the leading
contributers to artificial intelligence work, Bruce Buchanan and Herb Simon. Simon
studied with Carnap, and has contributed to the literature on philosophy of science
since the early fifties. Simon played a major role in the development of several
influential learning programs, notably the GENERAL PROBLEM SOLVER and the
BACON programs, and more than anyone else has articulated the view that there can
and should be computable, normative, philosophical theories of discovery *procedures*,

and that computers should implement them. Buchanan received a Ph.D in philosophy at Michigan State for a thesis on scientific discovery, supervised by Gerald Massey. He then went to work at the Stanford Heuristic Programming Project, where he helped to develop the DENDRAL and META-DENDRAL programs. Subsequently, he was a principal in the development of the MYCIN program. Between them, DENDRAL and MYCIN have served as models for much of the later work on expert systems.

REFERENCES

Skyrms, B.: 1980, *Causal Necessity*, Yale University Press, New Haven.
Causey, R.: 1981, *Unity of Science*, D. Reidel, Dordrecht.
Glymour, C.: 1980, *Theory and Evidence*, Princeton University Press, Princeton.
Glymour, C., R. Scheines, P. Spirtes and K. Kelly: 1987, *Discovering Causal Structure: Artificial Intelligence, Philosophy of Science and Statistical Modelling*, Academic Press, Orlando, forthcoming.
Hunt, E., J. Marin and P. Stone: 1966, *Experiments in Induction*, Academic Press, New York.
Kelly, K.: 1986, *The Automated Discovery of Universal Theories*. Ph.D Thesis, Department of History and Philosophy of Science, University of Pittsburgh, Pittsburgh.
Langley, P., H. Simon, G. Bradshaw, and J. Zytkow: forthcoming, *Scientific Discovery: An Account of the Creative Process*, MIT Press, Cambridge.
Lindsay, R., B. Buchanan, E. Feigenbaum and J. Lederberg: 1980, *Applications of Artificial Intelligence for Organic Chemistry*, McGraw-Hill, New York.
Mitchell, T.: 1983, 'Learning by Experimentation: Acquiring and Refining Problem-Solving Heuristics,' in R. Michalski, J. Carbonell and T. Mitchell (eds.), *Machine Learning*, Tioga, Palo Alto.
Osherson, D., M. Stob and S. Weinstein,: 1986, 'Mechanical Learners Pay a Price for Bayesianism,' submitted to the *Journal of Symbolic Logic*.
Rosenkrantz, R.: 1981, *Foundations and Applications of Inductive Probability*, Atascadero, CA: Ridgeview City.

Department of Philosophy
Carnegie-Mellon University
Pittsburgh, PA 15213 U.S.A.

BRUCE G. BUCHANAN

ARTIFICIAL INTELLIGENCE AS AN EXPERIMENTAL SCIENCE

1. INTRODUCTION

For many centuries the physical sciences have improved our understanding of the natural world through observation and experimentation. Although artificial intelligence (AI) does not focus on the same phenomena, there is no compelling reason to believe that the same methods will not work here as well. Although this study is largely a description, from one perspective, of how AI research is presently carried out, it is also a prescription (again from one perspective) of how AI can be more productive.

H. Simon's assertion that AI is an experimental science (Simon, 1969) is appealing to those in AI because of the comforting and legitimate sound of the term 'science'. Not all AI research has the form of experimental science, however, especially in its attention — or lack of attention — to data collected through observation and experimentation. This paper examines some of the consequences of saying that AI research is experimental and illustrates the process with examples of work done.

In calling AI an *empirical* science, we presuppose regularities in intelligent behavior of people and computers that can be discovered by observation and experimentation. We need also to presuppose that something like Mill's Canons of Induction will allow us to find regularities and assign causes and effects. And we presuppose a satisfactory answer to Hume's problem of induction. In calling AI an *experimental* science, we presuppose the ability to perform controlled experiments involving intelligent behavior. These are much more easily done with computer programs than with people. At the present time, AI has to be more concerned with qualitative statements of regularities than with statistical statements because the framework for being more precise does not yet exist, as we argue in Sections 2 and 3.

AI has both theoretical and experimental concerns. McCarthy (1983) distinguishes basic and applied research in AI and calls for more basic research on the premise that "reaching human level artificial

James H. Fetzer (ed.), Aspects of Artificial Intelligence, 209–250.
© 1988 *by Kluwer Academic Publishers.*

intelligence will require fundamental conceptual advances." Nilsson (1983) mentions two different research strategies in AI. The first he calls the "function-follows-form approach" — more popularly known as creating solutions in search of problems. He admits that "some of the enthusiasm for the use of logic in AI might also be explained by this function-follows-form approach." The second he calls the "form-follows-function approach" — which is the problem-driven approach followed in most AI research. Here, the researcher is motivated by a desire to find a way (any way) of making a computer or computer-controlled device solve some specific problem. What the researcher finds or invents to do then is less determined by enthusiasm for a methodology than by an enthusiasm for making things work.

Instead of a dichotomy of research paradigms, however, AI seems to contain a progression of steps from theorizing to engineering, from engineering to analysis, and from analysis back to theorizing. All seem important for progress. Upon examining individual pieces of research, however, we note that many researchers stop before completing the progression. Most work to date falls into either the theoretical or the engineering categories. Future progress requires additional work on analysis and generalization that is characteristically scientific.

There are six more-or-less separate steps in this cycle, as shown in Table I below. Some research includes one or two of these steps; the best experimental research involves all of them. One of the best and earliest examples of AI as an experimental science, illustrating all six of the steps in Table I, is Newell and Simon's research on GPS. it culminated in the detailed analysis with generalizations published in (Newell and Simon, 1972).

The first two steps are termed "theoretical" because they are often carried out with paper and pencil in the absence of an implemented program. The second two are termed "engineering" because they involve building an artifact (a program) and observing its performance, perhaps experimenting with it under different conditions. The last two are termed "analytical" to stand for both analysis of empirical data and formulation of general hypotheses that explain the data. The analysis steps are essential for progress in science. They also allow cycling back to theorizing or to engineering with new problems, sometimes with explicit hypotheses to test.

The six steps of the experimental approach can be illustrated by the

TABLE I.

Steps involved in experimental research in AI. Too many pieces of research to date stop after steps (2), (3), or (4).

Theoretical Steps:

    1. Identify the problem.
    2. Design a method for solving it.

Engineering Steps:

    3. Implement the method in a computer program.
    4. Demonstrate the power of the program (and thus of the method).

Analytical Steps:

    5. Analyze data collected in demonstrations.
    6. Generalize the results of the analysis.

research on the MYCIN program, a well-known AI program developed in the 1970s at Stanford. This account is necessarily oversimplified, since the research had many facets, which are more fully described in (Buchanan and Shortliffe, 1984). One AI *problem* that MYCIN addressed was how to use symbolic reasoning in a computer for medical diagnosis and therapy planning. Statistical approaches to diagnosis were only modestly successful and they failed to capture an explainable line of reasoning. Other medical AI work at that time depended on clever programming more than on a conceptually simple architecture for making the program easily extensible.

The *method* that was designed for MYCIN had two important components corresponding to the two major components of the architecture: the knowledge base and the inference methods. The knowledge base was designed to be simple and homogeneous, unlike the other complex programs in which medical knowledge was partly in parameters and mostly in procedures. The individual elements of the knowledge base were conditional sentences, called *rules*, in which premise conditions of a rule were asserted to be evidence for the conclusion of the rule. Also associated with each rule was a degree of evidential support (called a certainty factor, or CF). An example of one of MYCIN's rules is shown in Figure 1.

Rule 50

If: (1)   The morphology of the organism is rod, and
      (2)   The gram stain of the organism is gramneg, and
      (3)   The aerobicity of the organism is facultative, and
      (4)   The infection with the organism was acquired while the
            patient was hospitalized
Then: (1)   There is weakly suggestive evidence (.2) that the identity of
            the organism is pseudomonas, and
      (2)   There is suggestive evidence (.7) that the category of the
            organism is enterobacteriaceae

Fig. 1. A rule from the MYCIN program

The second component of the architecture was the inference procedure that interpreted the rules in the knowledge base. It evaluated them individually and in inference chains, to determine the degree of evidential support associated with plausible hypotheses for diagnosis and for treatment. Simply stated, the single inference rule was *modus ponens*. It was evaluated "backwards" by selecting a possible final conclusion, and inquiring about the degree to which premise conditions associated with the conclusion were satisfied. Often this led to chaining backward through several rules to primary data, because the degree of support of any one premise condition might itself be established by one or more rules in which that condition was named in the conclusion. Thus we speak of MYCIN's architecture as rule-based backward chaining, and also as evidence gathering.

The *implementation* of MYCIN was first done by E. H. Shortliffe over a two-year period, and was subsequently extended by a team of excellent programmers. Details of the implementation are not particularly relevant to this discussion. Suffice it to say that many parts of the program were added on top of the simple inference method to make the program easy to use, easy to understand, and easy to extend. Altogether, MYCIN was about 50 000 lines of LISP code, of which its knowledge base of about 500 rules was a small fraction.

The *demonstration* of MYCIN's ability to reason at the level of medical experts was a variation of Turing's test (Turing, 1950). MYCIN's knowledge base was refined (manually) so that MYCIN was correctly diagnosing 100 training cases of meningitis, and recommending appropriate treatment. Ten additional test cases of meningitis were then selected randomly from medical records, with guidelines to insure that the test set involved cases of several types of meningitis and that the cases were challenging. Summaries of these ten cases were shown to various persons at Stanford Medical Center, including several infectious disease faculty members, an infectious disease fellow (expert-in-training), a medical resident, and a medical student. Their therapy recommendations were collected, along with the actual therapy given at Stanford by the patients' own physicians. MYCIN's recommendations were determined and listed (unidentified) with the nine others for each of the ten test cases. Eight nationally recognized experts in meningitis were asked to read each of the ten patient summaries and to evaluate, for each patient, the therapy recommendations we listed. Their evaluation was binary: either a therapy recommendation was appropriate or it was inappropriate. MYCIN's recommendations were at the top of the list, in a statistically indistinguishable cluster of recommendations from Stanford's infectious disease experts. These results are shown in Table II.

TABLE II
Results of formal evaluation of MYCIN's performance (Buchanan and Shortliffe, 1984).

Prescribers	No. (%) of items in which therapy was rated acceptable* by an evaluator (n = 80)	No. (%) of items in which therapy was rated acceptable* by majority of evaluators (n = 10)	No. of cases in which therapy failed to cover a treatable pathogen (n = 10)
MYCIN	52 (65)	7 (70)	0
Faculty-1	50 (62.5)	5 (50)	1
Faculty-2	48 (60)	5 (50)	1
Infectious disease fellow	48 (60)	5 (50)	1
Faculty-3	46 (57.5)	4 (40)	0
Actual therapy	46 (57.5)	7 (70)	0
Faculty-4	44 (55)	5 (50)	0
Resident	36 (45)	3 (30)	1
Faculty-5	34 (42.5)	3 (30)	0
Student	24 (30)	1 (10)	3

* Therapy was classified as acceptable if an evaluator rated it as equivalent or as an acceptable alternative.

The *analysis* of why MYCIN worked as well as it did, and why it did not work better, is contained in many research papers and summarized in a 700-page book (Buchanan and Shortliffe, 1984). This included some quantitative studies but was mostly qualitative. For example, two facts enabling us to write an interactive editing program for MYCIN's rules were that the syntax of the rules was simple and all the rules were in the same syntax. We noted a tradeoff, however, between simple syntax and ability to express all of the interrelationships we would like. Temporal relationships, for instance, are not easily represented in MYCIN's syntax.

The *generalization* of our results was expressed as several hypotheses about rule-based systems outside the narrow domain of medicine. Many of these were tested by generalizing the MYCIN program itself into a framework system, called EMYCIN (for "essential" MYCIN). EMYCIN has the same architecture but contains no domain-specific knowledge. We hypothesized that this architecture would be appropriate for many diagnostic and classification problems. This has been confirmed by many persons at Stanford and elsewhere using EMYCIN successfully.[1]

An alternative view to the six-step experimental approach shown in Table I — that AI is a theoretical discipline — emphasizes formalism in describing methods, skips the two engineering steps entirely, and formalizes the analysis to proofs about the strengths and limits of the method. The results of the analysis are presumably as general as they can be, so the final step of generalization is not needed. The modified steps, (1), (2′), and (5′), thus constitute the essence of the theoretical approach. It would be naive to argue that theoretical studies do not advance our understanding of physical phenomena: Einstein's mathematics provide a compelling counterexample to anyone so inclined. While this paper attempts to show that we lack an adequate vocabulary for describing our methods and argues in favor of the six-step experimental approach in AI, it is not an argument against theoretical methods. When they are successful they can revolutionize a discipline. In the absence of a theoretician to do this, however, the rest of the scientific community can be making advances experimentally (Kuhn, 1962). Moreover, even theoreticians must link their terminology with concepts that are generally considered important.

AI is not entirely different from physics and chemistry in these six steps, but is closer to economics and psychology in its theoretical development and its dependence on computers to demonstrate consequences of ideas. Although the implementation of ideas in a computer program is not as integral a part of other disciplines at it is in AI, each one has some idiosyncratic experimental apparatus. Moreover, computer simulation in science is playing an ever-increasing role in confirming hypotheses. AI is somewhat different, however, in that the computer program is not only a simulation, it is the sole experimental apparatus.

1.1. *What is AI?*

The term 'artificial intelligence' was introduced in 1956 (McCorduck, 1979) and is now in general use, although some universities prefer to call the discipline 'machine intelligence' or 'cognitive science' and the Soviets call it 'cybernetics'. Its goal is to understand the nature of intelligent behavior. It is often loosely defined as encompassing all those activities of programming computers to perform tasks that we generally would say require intelligence when people perform them. Gnats don't play chess, for example, and persons who are good players are undeniably intelligent.

Within computer science, AI has a somewhat more technical characterization. AI is considered to be the branch of computer science that studies the representation and use of *symbolic information* as (opposed to using only numbers) and studies *heuristic processes* (as opposed to algorithmic processes in which there is a guarantee of success in a finite time). Chess, for example, requires reasoning about *plausible* moves and requires representing *symbolic* features of a game, such as threats and control.

Expert systems are one well-known class of AI programs that reason at the level of specialists, that can explain their reasoning, and whose knowledge can be easily extended. MYCIN, mentioned above, is one of the first expert systems. Work on expert systems is often characterized as solely engineering because of its emphasis on performance of a running program. But in a larger context, expert systems benefit as much as any program from a sound design; and they can be used to further the scientific progress of AI if they are used for experimentation (Buchanan, 1986).

1.2. *Theoretical Steps in AI*

Theoretical work in AI is largely pursued in the framework of symbolic logic. Mathematics is the language of choice for theoretical work in the so-called "hard" sciences of physics and chemistry. However, almost by

definition, mathematics is inappropriate for describing the phenomena
that AI is concerned with because those include methods of symbolic
reasoning, or reasoning without expressing everything numerically. The
only candidate for a formal framework suitable for formalizing symbolic
reasoning is symbolic logic. McCarthy wrote (McCarthy, 1983) "AI
badly needs mathematical and logical theory, but the theory required
involves conceptual innovations — not just mathematics."

One appeal of a formal framework, of course, is the *guarantee* that
accompanies the notion of proof. Once the axioms describing a system
are established, deductive consequences establish the theoretical limits
of that system. Another appeal is the formalism itself. By expressing our
axioms concisely and in standard form, we gain in *understanding* our
beliefs and their relations to one another. Both of these are desirable.

While its formal notions of proof and its simple syntax make first-
order predicate logic an obvious candidate, it lacks the expressive
power to represent reasoning about necessity, beliefs, or other mo-
dalities. It also fails to account for, or deal with, changes in truth values
of propositions, e.g., as new information is introduced. And second-
order constructs are needed to reason about strategies in which first-
order propositions (or classes of them) are described. There may be no
simpler or more understandable formal descriptions of programs than
the code itself.

In a theoretical approach to AI problems, the first two steps are (1)
identify a problem; and (2) design a method.

Unfortunately, there is considerable work in AI in which the method
is designed first and the problems for which the method offers a
solution are identified second. This is usually justified as "pure research"
and undoubtedly has an important place in every science. But it is
undirected by data and guided by the same notions of simplicity and
elegance that brought us windowless monads and other rationalistic
constructs. They are not necessarily wrong, but they are speculations
about how intelligent programs ought to behave.

One of the more successful pieces of theoretical work in AI was
Doyle's work on truth maintenance (Doyle, 1979). In the absence of an
implemented program, Doyle designed, and argued for, a method for
maintaining a true description of the world in the face of new axioms
that contradict previous information. (This is called the problem of
non-monotonic reasoning (McDermott and Doyle 1980.)) The formal
solution is logically satisfying — linking premises and conclusions

(much as in natural deduction) so that removing a premise leads easily to removing the deductions depending on it. It is conceptually powerful and simple enough to encourage others to attempt implementations.

Theorizing in the absence of data can be successful in AI because AI programs are artifacts designed to run under the control of man-made languages and operating systems on man-made computers. Insofar as the layers of artifacts — and their interactions — are well understood, it is possible to design programs that behave in specified ways.

In complex problem areas the number of interactions leaves the result of theorizing uncoupled from interesting instantiations of the problem. As Simon notes (Simon, 1969), the number of variables in problems of interest is too great to allow us to understand the complexity of the interactions. Carnap's logic of confirmation (Carnap, 1950) is an example of theorizing independently of the engineering considerations involved in implementing the theory in a computer program. Computational issues were not Carnap's concern, of course. But there is a need for a computationally efficient calculus of degree-of-confirmation, for use in programs such as MYCIN, that weigh uncertain evidence for and against alternative hypotheses, and propagate a degree of evidential support through inferences that are themselves uncertain (Shortliffe and Buchanan, 1975). Carnap's measure of strength of confirmation, as a logical relation between hypothesis and evidence, depends on having a formal language capable of expressing every attribute of the individuals in the universe of discourse. It also depends on having the means to examine the 2^n state descriptions formed by conjoining elementary propositions or their negations. Except for trivially small languages, power set calculations are impractical in programs.[2]

Another difficulty with theorizing in the absence of data is oversimplification of the initial problem. Bayesians, for instance, believed that MYCIN's one-number calculus of confirmation could be replaced by Bayes' Rule. It had, after all, been used in medical diagnosis programs for years. But MYCIN's calculus was developed partly in response to a need to work with a single number associated with premise-conclusion pairs (a measure of *increased* support of a conclusion (Horvitz *et al.*, 1986), not with prior and posterior probabilities, because that was the way that our medical collaborators defined the problem. The ideal conditions required by most theories hold only in simplified problems, which are often so unrelated to their complex instantiations that the principles for solving them bear little resemblance to the original theory.

1.3. *Engineering Steps in AI*

Work in building expert systems and other applications programs is very clearly seen as engineering work. As such, it engenders considerable controversy over its part in scientific research because it is seen as stopping with the engineered product. In addition to expert systems, applications of AI, include robotics and vision, automatic programming, and natural-language understanding. In all of these areas, AI ideas and methods have matured enough to be applied outside the research labs. There is little controversy about the benefits of transferring and developing research ideas, when they are done well, but there is no universal agreement about the necessity of steps (3) and (4), implementation and demonstration, in AI research.

Omitting the implementation certainly saves time and trouble. It often takes graduate students 12—24 months to construct and debug a working prototype, and it requires substantial equipment resources beyond paper and pencil. The running program itself is only an existence proof that a computer program *can* be written to solve a few problems. It doesn't tell us how well the program works or how much of its performance is due to the design (and how much is a result of the programmer's cleverness or a happy choice of test problems). So why bother with the implementation? Much of the answer lies in the nature of the evidence that compels one to accept the ideas proposed in the design.

As stated above, the complexity of the problems and of the solution methods precludes making a convincing theoretical argument that a designed method is sufficient for solving problems of a type — unless the problems are reduced to trivial proportions. The implementation thus allows demonstrations of the method working (or not) on problems of varying complexity. "Only when it [a bridge] has been overloaded do we learn the physical properties of the materials from which it is built," Simon wrote (Simon, 1969).

The most elementary question to be answered by the demonstration is whether the method works at all. Another question is the scope of problems for which it works. Both the problem description and the design are often open-ended, in the sense that they allow endless argument over these two questions. An implementation provides specifics to argue about: we can examine what the program does and how it does it.

Demonstrations can take many forms and can provide different levels of support for claims about methods. Several fine pieces of AI research culminated in implementations that ran only on single problems. MOLGEN (Stefik, 1981a; 1981b) and TEIRESIAS (Davis, 1982) are two of those. Such demonstrations show that the methods will work on at least one large problem, more convincingly than if there had been no implementation at all because of their complexity and the complexity of both the programs and the problems. MYCIN was run on many medical cases and its performance was demonstrated with three formal studies (Shortliffe, 1974; Yu et al., 1979b; Yu et al., 1979a). DENDRAL (Lindsay et al., 1980) analyzed molecular structures of many sets of challenging chemical molecules convincingly enough to warrant publication in the chemistry literature. Meta-DENDRAL (Buchanan and Mitchell, 1978) discovered rules of mass spectroscopy for a previously unreported family of chemical structures, and those rules were published (Buchanan et al., 1972) as a contribution to the science of mass spectrometry.

Section 3.1 below discusses the question of *what* we demonstrate. As McCarthy wrote (McCarthy, 1983) "We have to think hard about how to make experiments that are really informative. At present, the failures are more important than the successes, because they often tell us that the intellectual mechanisms we imagined would intelligently solve certain problems are inadequate."

1.4. *Analytical Steps in AI*

Many experimental researchers stop after the demonstration step in the hope that the power of their ideas is now obvious. This is characteristic of an engineering approach to questions in which the product speaks for itself. However, the needs of science are not met — and progress in AI is impeded — until there is considerable work on steps 5 and 6 of the process: analysis; and generalization.

Progress in science is only achieved through careful analysis of methods and their power. Designing without analysis is idle speculation. Implementation without analysis is tinkering. Alone they have no research value. All too often we read of major pieces of AI work that stop with the engineering steps. But we need to know how the implemented program embodies the designs and that the program works well *because of* the design.

In Lenat's work on the program AM (Lenat, 1976), for example, there was considerable puzzlement among researchers about what and how much Lenat had shown with this program. It had successfully rediscovered some important concepts in mathematics, such as prime numbers, but other researchers were unable to understand well enough why the program worked. A subsequent paper (Lenat and Brown, 1984) analyzed AM's methods carefully and provided a better understanding of them than is provided by Lenat's first description of the program with examples.

Writing on scientific method emphasizes the importance of analyzing data, with the whole weight of mathematical statistics thrown into the task. AI has not taken advantage of these analytical methods for two related reasons: it is tedious to collect data, and it is not obvious what is worth measuring. These are discussed in Section 2 (measurement).

Generalization is added as an explicit step, because science progresses beyond the strict boundaries of what has actually been observed with the formulation of *general* hypotheses. With natural phenomena there is no logical justification for generalizing beyond the data at all, as Hume noted. Nevertheless, a clearly formulated statement of a claim provides a target for attempts at refutation, which in Popper's view (Popper, 1959), at least, is a necessary element of scientific progress. As in the rest of the literature, we do not say how the hypotheses are discovered, although some work in AI has focused on hypothesis formation and discovery (Buchanan, 1983; Buchanan, 1985; Lenat and Brown, 1984).

Since programs are artifacts, one might be tempted to argue that there can be logical justifications for generalizing beyond examples, from the fact that the layers from program to operating system to computer are describable in every detail, thus allowing prediction of future behavior. The operation of the computer follows physical laws, however, so the logical argument rests on assumptions open to Hume's objection. More importantly, as mentioned above, the complexity of interactions within any layer — and across layers — precludes our making predictions with certainty. AI needs, as much as the physical sciences, the insightful but unproven generalizations that are based on data, but that cover more than the data.

McDermott (1983) has noted that the increased number of expert systems "has not been accompanied by even a modest growth in our understanding of the relationship between knowledge and search. The problem is lack of data. Though the number of expert systems that has been developed is now sufficient to allow us to begin to deepen our understanding, the information available about those systems is inadequate." He advances four dimensions of knowledge in expert systems that are necessary (but not necessarily sufficient) for understanding the role of knowledge in problem solving. These are:

1. Compartmentalizability of the task knowledge — the decomposition of knowledge into smaller pieces, as into subtasks.
2. Uncertainty of the task information — the extent to which data are unavailable or unreliable (and the extent to which we know they are).
3. Applicability of the task knowledge — the extent to which problem solving knowledge is relevant to finding a correct solution (and the extent to which we know it is).
4. Thickness of the task knowledge — the variety of responses allowed by the knowledge base in a problem solving situation.

Davis (1982) states four simple architectural principles for building expert systems which he generalized from the collective literature and experience of early work:

1. separate the inference engine and knowledge base;
2. use as uniform a representation as possible;
3. keep the inference engine simple;
4. exploit redundancy.

These remain good, easy-to-remember guiding principles even after some years. But, as Davis pointed out, they are a bit oversimplified. Separating the inference engine and the knowledge base is not always as easy or as desirable as it sounds in cases where (a) the grain size of relevant knowledge is large or (b) the knowledge encoded in the knowledge base is strongly sequential. Also, there are many instances where one abandons a uniform representation in favor of specialized representations. If specialized representations are employed, however, it is often necessary to abandon the principle of keeping the inference procedure simple.

Davis correctly notes that performance is not the only criterion for measuring the design and implementation of an expert system; in addition he cites explanation and knowledge acquisition.

Criteria such as these are discussed in more detail in Section 3, for these are the kinds of terms needed in generalizations.

2. MEASUREMENT

In every science we see measurements of the systems under study, with m :ch of routine science devoted to making increasingly more precise measurements. For instance, precise values of theoretical constants such as the speed of light are of intense interest — or frequencies of expressed genetic traits in a population, or correlations among proposed chemical causes and biological effects. Sometimes, better measurements provide better understanding. AI, too, has a need for measurement in order to increase our understanding of the methods we desgin and the programs we build. We talk about the size and complexity of a program, or about the simplicity and expressive power of a knowledge base representation, but we can't measure them. And we talk about the efficiency and inferential power of reasoning methods without being at all precise. Even if we don't ever agree on units for absolute measures, we still need a way to place programs on relevant comparative scales.

A theory of AI would allow, among other things, predicting the consequences of modifying a program. It is of extreme interest to know, for example, how a program's execution speed will change as its knowledge base grows.

In a simple production-rule architecture, for example, a program will examine *every* rule in order to determine which rules are applicable and then decide which one to apply (Davis and Buchanan, 1977). That changes the description of the situation, then, and on every reasoning cycle the same things must happen. As more rules are added the program has more to do on every cycle. Thus adding knowledge to improve the quality of the answer has a predictable effect on the increased time a program will take, if it is implemented in this simple architecture.

In many, more complex programs, we observe the paradox of increased knowledge — a disappointing exponential slow-down as the program's knowledge increases. In other programs, a richer organization of the knowledge base keeps the slow-down linear. In some, however, we see a positive effect of adding more knowledge. The scientific disappointment is that we rarely can predict whether the effect of more knowledge will be positive or negative, let alone predicting the magnitude of the effect.

In MYCIN, for example, Davis (Davis and Buchanan, 1977) added strategy rules (called meta-rules) to guide MYCIN's reasoning. Our belief at the time was that an additional layer of reasoning about *which* medical rules to invoke, and in which *order*, would make the domain-level reasoning about the medical problem more efficient. When we tested that hypothesis by implementing a few strategy rules, however, we were surprised. Meta-rules carry an overhead, as does all reasoning about strategy. And we were able to find implementations of our simple strategy considerations in the domain-level rules themselves. Therefore the test showed less efficiency when meta-rules were used, in the few situations we tested, than when domain rules were modified to reflect the same strategies. Although our initial belief still seems reasonable, MYCIN's rule set was not complex enough to justify the extra layer of reasoning about strategy. We are currently examining this belief in a more complex problem area, the determination of molecular structures of proteins, described below (Section 3.1).

Toulmin (1953) contrasts descriptive and prescriptive phases of science. Biology before Mendel was descriptive, where much effort was expended on collecting specimens and describing them. Until there was an adequate vocabulary for a theory, however, the science remained descriptive. With Mendel's theory of inheritance, and even more with biochemical models of biological processes, biology has entered a prescriptive phase. AI is still in a descriptive phase. Observation of natural objects and phenomena, however, has been replaced in AI by observation of computer programs and their behavior. We have many specimens and we publish lengthy descriptions of them. Unfortunately, we don't use our descriptive terms consistently, nor do we even know all of the features of a program that are worth noticing.

Data collection in science can be accomplished either by passive observation or by active experimentation, or by variations on both. In AI there has been little effort expended on collecting sets of data points, although considerable expense goes into the construction of each program. Each program can be seen as a single data point, or as the experimental apparatus with which we can generate many data points. Either way, there is little activity that we call active data collection in AI.

Pairs of implementations of the same programs can provide interesting data for comparisons. Too seldom are there records of initial implementations that failed to achieve desired performance, and almost never are there records comparing discarded and saved methods or written notes on why methods were changed. The primary reasons for these lapses are probably (a) the complexity of the methods (and their implementation) and (b) our present inability to sort out essential from nonessential attributes of problems and solution methods. In addition, the process of writing and debugging a large program encourages making hundreds of incremental changes — some small and some large — and using the resulting program to stand as the best — sometimes only — documentation of the details of the method.

A few of the pairs of observations that have been noticed are listed in Table III below. It is beyond the scope of the present paper to analyze their differences, even if the relevant attributes could be identified. It is worth noting, however, that research methods in the social sciences may tell us how to be systematic about observing complex systems even in the absence of controlled experiments in the sense of the laboratory sciences.

Active experimentation in AI is rare. The existence of pairs of observations listed above is mostly accidental. Sometimes researchers construct a second implementation as a way of seeing if a second method will work, or work better, as in the case of DART, MOLGEN and INTERNIST. Sometimes they are responding to fundamental dissatisfactions with the quality or speed of performance from one or another of the versions.

Programs, when compared at all, are usually described without a detailed analysis of the prior version and without systematic experimentation and measurement. But what do we measure? What are the relevant terms for describing programs and intelligent behavior? We don't know the answers to these questions, but we can look at some of the attributes researchers already are measuring and some of the generalizations they are making.

3. VOCABULARY

Formulating general hypotheses about AI methods and programs either

TABLE III.

Some pairs of implementations of programs using different AI methods. In each case at least two major versions were implemented for the same task and have been described. Few detailed analyses have been published from such pairs.

PROGRAM	MAJOR VERSIONS
MYCIN	(a) Semantic network representation of medical knowledge (b) Rule-based representation of knowledge (Buchanan and Shortliffe, 1984)
DENDRAL	(a) Knowledge of mass spectrometry in Lisp code (b) Knowledge in explicit rule form (Lindsay *et al.*, 1980)
DART	(a) Rule-based representation of knowledge for troubleshooting electronic circuits (Bennett and Hollander, 1981) (b) Logic-based representation (Genesereth, 1984)
MOLGEN	(a) Planning experiments in molecular genetics from first principles (Stefik, 1981a, 1981b) (b) Planning from pre-stored schemas (Friedland, 1979)
PROTEAN	(a) Determining 3-dimensional structure of proteins by optimization methods (Frayman, 1985) (b) Determining structures by heuristic refinement (Hayes-Roth *et al.*, 1986)
INTERNIST	(a) "Flat" representation of associations between diseases and their manifestations (Miller *et al.*, 1984) (b) Structured representation of causal associations (Pople, 1982)

theoretically or experimentally requires a vocabulary containing names of relevant features of those procedures, as well as names of effects we want to predict. That is, a general form of the sort of hypothesis in AI that would be testable is:

Under background conditions (C), methods or programs with attributes $A_1 \ldots, A_n$ will result in effects (E).

Schema H_1: $C \,\&\, A \,\&\, \ldots \,\&\, A_n \to E$

A variation on this form that would be easier to formulate and test is:

Under background conditions, including specification of many relevant attributes, a change in one or more relevant attributes will result in a change in effect E.

Schema H_2: $C \,\&\, \Delta A \to \Delta E$

In this section, we examine the kinds of terms that have been used in AI to describe effects and those used to describe programs and methods.

Niwa *et al.* (1984) undertook an experimental comparison of four representation schemes for the same task, risk management of large construction projects. The four were a simple production system, a structured production system, a frame-based system, and a logic-based system. Their careful comparison led to many interesting observations, which they summarized in three general hypotheses:

1. "In a poorly understood domain whose knowledge structure cannot be well described, modular knowledge representations, e.g., simple production and logic systems, should be used. However, this causes low run time efficiency.
2. The use of structured knowledge representations, e.g., structured production and frame systems, increases run time efficiency as well as reducing the effect of the knowledge volume on run time. However, system implementation is more difficult.
3. Mathematical completeness makes logic systems more difficult to implement and less efficient in run time. Our problem was too simple to adequately demonstrate the advantages of logic representation."

These hypotheses represent the kinds of associations that are being made at present, when they are explicitly stated at all. We examine below some of the studies from which generalizations have been formed and examine some of the terms that are used in the formulation of hypotheses.

3.1. *Outcome Variables*

The convincingness of a demonstration depends on stating clearly, in advance, what the demonstration will "prove." With AI programs, designers often advance many interdependent claims at once. And, too often, they do not state explicitly what those claims are. This is reprehensible scientific behavior.

The claims usually center on efficiency of methods because measures of efficiency are the coin of the realm of computing. This assumes that the quality of solutions is good enough that we even want to measure efficiency. For example, an extremely fast program with dubious performance for identifying shore birds is the one line program:

PRINT   'It is a sandpiper.'

Several categories of claims are discussed in this section. Whatever the claim, it is important that it focus on attributes of programs that are well defined and measurable in principle. The capabilities of human problem solvers that lead to intelligent performance are generally capabilities that we want AI programs to have, especially if we cannot reproduce that quality of performance otherwise. Intuitive problem solving is one of those capabilities. But it doesn't advance the science of AI to argue whether a program has the capability for reasoning intuitively unless we can define what that means. For that reason, measuring quality of *performance* is likely to tell us more about progress than trying to determine if some *modes* of human thinking are reflected in a program.

Some suggested criteria for evaluation expert systems were discussed in (Buchanan, 1983b). These included: accuracy and reliability, correctness of the lines of reasoning, appropriateness of input-output content, characteristics of the hardware environment, efficiency of the program, and cost effectiveness of the whole system. This — like other similar sets of suggestions — is a disparate list that can be structured under the four headings suggested below.

3.1.1. *Productivity*

AI research aims at many subgoals as means to the larger end of understanding intelligent behavior. Expert systems, for instance, are constructed commercially to provide assistance within an organization but can also serve as data points in our collection of examples that behave more or less intelligently. Their ability to provide assistance can be taken as one measure of how successfully the system designer has captured the important aspects of problem solving in a narrow area. Insofar as an expert system saves times or money within the organization it is commercially successful.

This measure of *productivity* is too gross, however, for the science of AI. It is akin to using volume of traffic over the Golden Gate Bridge as a measure of the builder's design principles. If it carries no traffic we still don't know if the design was sound or faulty, and we don't know if volume is high *because of* a good design or *in spite of* flaws. Nevertheless, the financial success of numerous expert systems constructed on the same design lends some evidence that their design is sound and worth emulating (See Buchanan, 1986).

3.1.2. *Human factors*

Measures of *human factors* are more important than productivity for judging some aspects of the merit of designs. References to ease-of-use, familiarity, and understandability are sometimes published. Davis charted the approximate time in person-years that was needed to construct several expert systems (See Figure 2). Decreased times were taken to be a measure of increased understanding of design principles, but also reflects better judgment about selecting smaller problems to work on.

In the context of chess, Wilkins wrote the PARADISE program (Wilkins, 1979) to examine the strength of some methods of reasoning about plans. One of his claims was that the methods were extensible, that is, that new chess knowledge could easily be given to the system when PARADISE stumbled over new problems. He demonstrated this by showing that he could add new chess knowledge within a few minutes to solve failed problems. (The time — about ten minutes — was long enough to allow analysis of what went wrong and to conceptualize the new rule that was needed, but short enough to disallow major changes in representation or inference procedures.) In this way, a claim about ease of modification was quantified.

3.1.3. *Computational requirements*

Computational requirements are also used as a measure of design

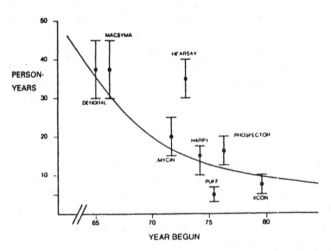

Fig. 2. Approximate times to construct eight early expert systems (from Davis, 1982).

principles. We prefer methods that run in time linearly proportional to the number of elements being reasoned about to methods that run in exponential time. We also prefer methods that use less memory in the computer. The unconstrained DENDRAL algorithm generated huge numbers of molecular structure graphs with chemical compositions of only 10—20 atoms (Smith, 1975). With heuristics as constraints on the generator, problems 5—10 times that large could be solved in "reasonable" amounts of time.

In addition to time and space requirements, two other computational characteristics of AI programs that are of interest are portability and extensibility. Portability refers to the technical characteristics that allow a program to run on a variety of different computers. Extensibility refers to characteristics that allow the program to grow without severely straining the time and space limits. (Another aspect of extensibility — the ease of making extensions — was discussed under human factors above.)

3.1.4. *Performance*

The final set of outcome measures used to substantiate claims about AI methods is the set of *performance* measures. Performance can be defined in many ways, often including the measures of computational resources mentioned above. It seems informative to separate quality of performance from efficiency; thus we mean here measures of quality. Performance is a fundamental measure: unless two programs are performing at about the same levels of quality, it doesn't make sense to compare their efficiency. This can be difficult to measure when problems have no decision procedure for determining with certainty whether or not a purported solution is a solution. Also, many problems have more than one correct solution. In MYCIN's problem area, diagnosis and treatment of bacterial infections, the problem was defined to be finding appropriate therapy before all the relevant information was available to identify the bacteria causing the infection. In the absence of complete and certain information, defining a single best solution was not possible.

When there is no easily determined "correct" solution, performance can be measured against a "gold standard" (Gaschnig *et al.*, 1983). In the case of MYCIN, the gold standard was the therapy recommendation of Stanford specialists in infectious diseases. With expert systems this kind of comparison is now routinely mentioned (even if not frequently enough used) to validate the strength of the methods. In other contexts outside of expert systems, performance is rarely measured. AM (Lenat, 1976), for example, is cited for its methods of discovering new concepts in mathematics, even in the absence of demonstration beyond a few examples. It accomplished something that had not been done before — that was sufficient. Both accuracy and precision are important components of performance. DENDRAL was demonstrated to be accurate by showing for hundreds of known problems that the correct solution was contained in DENDRAL's set of best five solutions. Its precision was measurable in terms of the mean and variance of the rank of the correct solution. The accuracy and precision of PROTEAN's reasoning have been measured (Lichtarge *et al.*, 1987) in the context of using NMR data to determine the three-dimensional structures of proteins. In both cases, analysis shows that precision can be improved with more, or better, chemical data about the problems. That is, imprecision came less from the methods than from the data provided to the programs.

3.2. *Some Additional Terms*

In addition to the outcome variables discussed in Sec. 3.1., various criticisms of "artificially intelligent" systems focus on dimensions that are not as easily measured. People seem to reason using common sense, intuition and causal models. And people often behave intelligently. Thus there are grounds for believing that machines should also reason in these ways if they are to be called "truly" intelligent. It doesn't follow that machines *must* use these modes if they are to behave *as if* they were intelligent, nor is it clear that we could maintain a distinction between being intelligent and acting all the time as if one were intelligent.

The following three subsections discuss three terms that have been proposed as important, or necessary, for describing intelligent systems. These terms, and others like them, fail to provide operational tests of intelligence and thus become more problematic than helpful.

3.2.1. *Common sense*

AI programs — expert systems in particular — have been criticized because they lack common sense. Insofar as this is a criticism of performance it is dealt with in the same way — by adding more knowledge. But this is not satisfactory to the critics because of the potentially vast number of relevant pieces of common sense knowledge. They would like to see more general representations and inference methods and a disposition on the part of the program to motivate its behavior by its common sense knowledge.

Common sense seems to guide all of us in many situations — often successfully, but often poorly. We speak of common sense sometimes as facts and relations that "everyone knows." Some examples are, perhaps, "Water runs downhill," and "Ice is slippery." We also speak of common-sense reasoning as a method for using general facts appropriately. For instance, we would expect "any fool" to slow down when driving on icy roads, unless special circumstances render them not slippery.

But is there anything special about the propositions just cited? They are general and they have many exceptions. That is equally true of many propositions in technical, uncommon domains as well. Is it common sense in chess that one should try to control the center, or in chemistry that atomic nuclei do not touch, or in programming that naming frequently used procedures makes debugging easier? The point of noticing that people use such propositions is not to *define* something that people do better than computers. Rather, the point is to ask *how* computers can do something similar.

The only consistent definition of common sense is that it is whatever we commonly appeal to when we want to criticize someone else's actions. Everyone knows, for example, that he who hesitates is lost and that a stitch in time saves nine. In different circumstances, however, we can point out that haste makes waste, or that fools rush in where angels fear to tread, or that one should look before leaping.

Common-sense reasoning may be captured as a very large set of facts together with their exception conditions.[3] Water runs downhill *unless* it is under pressure, or it is in a non-liquid phase, or . . . . In the development of the INTERNIST program at the University of Pittsburgh, the project's expert, Dr. Jack Myers, has added about 100 000 "common sense" facts of medicine to improve the program's performance and appearance (Miller *et al.*, 1984). For example, males don't get pregnant, aspirin obscures the results of thyroid tests, and so on. Each time the expert notices that INTERNIST asks a "stupid" question, he adds the commonsense facts that allow it to avoid that question in similar cases.

Another way of capturing the sense of common sense reasoning may be to represent only *some* of the most general rules, but to augment them with new modes of reasoning. So-called default reasoning is one such mode in which a program assumes that the general fact holds in a situation unless it is contradicted by a more specialized piece of knowledge. Taxonomic hierarchies play an important role in default reasoning because general rules can be written once to apply to all members of a class, with special exception clauses being attached to subclasses or individuals. For example, we would not need to represent the fact that water flows downhill if we had already represented more general facts about gravity and liquids flowing and the fact that water is a liquid.

If we cannot build programs that behave intelligently with *assistance* from intelligent users, we will not be able to build completely *autonomous* intelligent systems. Thus the most common method for using common sense in expert systems is to expect the users of the system to supply it. This is a pragmatic approach that completely sidesteps this research issue in favor of pursuing others. For the most part it is safe to make some assumptions about the knowledge that a user can bring to an interactive program. Users of MYCIN, for example, were expected to be physicians or nurses. In the design of MYCIN's knowledge base, choices about what to ask users and what to expect MYCIN to infer were made with respect to how much the users would know. Even medical students know that males don't get pregnant.

3.2.2. *Intuition*

The most outspoken and persistent critic of artificial intelligence is Hubert Dreyfus (Dreyfus and Dreyfus, 1986). For twenty years, Dreyfus has been making impossibility claims based, partly, on the premise that human thought and problem solving cannot be reduced to rule-governed behavior. AI programs have obviously been successful at capturing some elements of good problem solving performance — on tasks that require intelligence when people do them. So Dreyfus's claim must be modified to

(1) Not all human problem solving mechanisms can be explained as rule-governed behavior, i.e., implemented in computer programs.

The first response from the AI community is "So what?" Even if the *mechanisms* underlying human performance are inexplicable in the vocabulary of computer programs, the *behavior* of human problem solvers may be adequately reproduced by programs. For example, the DENDRAL program systematically enumerates descriptions of chemical structures to find structures that explain some experimental data. Human chemists are unable to generate correct lists of alternative explanations systematically when the lists are at all long. Also, even if they did search for explanations this way, they would not use DENDRAL's double-coset algorithm (Brown *et al.*, 1974) without computer assistance because of its complexity. This is an instance where we don't need to know *how* a person interprets experimental data in chemistry in order to find an alternative method that is as good or better.

The second response to claim (1) is to challenge proponents to describe at least one reproducible cognitive activity that necessarily cannot be reproduced in computer programs. Chess is often named by Dreyfus as an activity that is not rule-governed at the level of grand masters. While chess-playing programs have become very good, they have not, in fact, achieved performance levels of the world champions. But what evidence do we have that they cannot — in principle — achieve those levels? Induction from past performance doesn't establish a necessity claim. We do not know what computer programs will be able to do in the future, nor — with a few mathematical exceptions — do we have any arguments that let us conclude with certainty what they can or cannot do.

Dreyfus's argument for proposition (1) involves two unsupported premises:

(2) Some human problem solving activities, at their highest levels, require intuition.

(3) Intuition is necessarily not describable as rule-governed.

Granted these premises, proposition (1) does follow. But (2) requires introducing a mechanism or mode of intuitive thought which, by assumption (3), has no precise description. Intuition is everything that is left over after all the mechanisms that may be described as rule-governed have been factored out. Assumption (2) asserts that there is a non-null residue.

It is always tempting to claim that people have some unique elements not shared by other animals, plants, or machines. One must distinguish, however, examples of the many tasks where people currently outperform machines from examples of performance that *only* humans can ever achieve. The present performance levels of machines are empirical questions. Their inability to reason intuitively, however, has been made definitional. Thus a scientific approach to AI research demands a more precise formulation of such an objection.

3.2.3. *Causal models*

Another suggested necessary criterion of success of AI programs, especially expert systems, is that their reasoning be motivated by causal models whenever appropriate. As with common sense and intuition, this criterion is dispositional: the program should be disposed to act out of regard for certain kinds of knowledge.

Current representation and reasoning methods in AI are weak in dealing with theoretical models describing the behavior of a complex system. Most of the successful expert systems for diagnosis, for example, use a set of diagnostic rules that describe relations between manifestations of a problem and its cause. In MYCIN, excellent performance was achieved without the program's having anything more than a superficial causal model of the infectious disease process. In one rule, shown in Figure 1, the program related some factors associated with an infection with pseudomonas bacteria as a possible cause. Part of that rule is reproduced as:

> *Rule 50-a* — Pseudomonas should be considered as a possible cause of infection for every patient who has a hospital-acquired bacterial infection, if there is evidence that the infecting organism is a gram negative facultative rod.

We were aware of many arguments justifying this association, some of them appealing to quasi-theoretical causal laws in medicine. Several of these justifying steps are shown in Table IV below.

It seems clear that many of these steps could themselves be expanded ten- or twenty-fold. Yet we chose to stop with the association expressed in Rule-50 because that was considered to be the level of most *clinical* relevance, as opposed to the levels of more theoretical interest to a researcher. That is, questions of performance drove the formulation of the knowledge base, with potentially relevant items omitted if they did not contribute to performance.

In domains where practical knowledge can be easily derived from theoretical knowledge or domains where the systems of interest change faster than the practical knowledge accumulates, it may make better

TABLE IV.

Some of the assumptions and causal reasoning that justify an association
used in MYCIN

---

1. A patient is a person.
2. Almost every patient in a hospital is sick.
3. Pseudomonas is a bacterium.
4. Sick persons cannot mobilize defenses sufficiently to combat transient infections.
5. Pseudomonas has been isolated from fruits and vegetables, water around cut flowers, water faucets, ventilation tubing, Foley bags, and hands.
6. Hospital patients frequently eat fruits and vegetables
     OR have cut flowers in their rooms
     OR are around water faucets, ventilation tubing, Foley bags, or people with hands.
7. If a bacterium can be isolated from x and a patient has close contact with x
     then that patient may ingest that bacterium.
8. Therefore, hospital patients frequently ingest pseudomonas.
9. Bacterial flora in hospitalized patients are altered, allowing ingested bacteria to colonize the mouth and lower intestine.
10. Colonized bacteria in the mouth or lower intestines may enter the bloodstream through daily activities such as brushing teeth.
11. People normally brush their teeth, etc.
12. Therefore, colonized bacteria may enter the bloodstream of hospitalized patients.
13. Pseudomonas may colonize the mouth or lower intestines.
14. Therefore, pseudomonas may enter the bloodstream of hospitalized patients and cause significant infection.
15. If an organism is causing a significant infection in patients who are otherwise sick, then it is risky to ignore that organism.
16. Therefore, it is risky to ignore a pseudomonas-caused infection (unless it has been ruled out).
17. Pseudomonas is a member of the family of gram-negative, facultative rods.
18. Evidence for a family of organisms constitutes some evidence for each member of the family.
19. If there is some evidence for a risky organism then it is important to consider it as a possible cause of a significant infection.
20. Therefore, Rule 50.

---

sense to encode the theory rather than the specialized associations that
are instantiations of the theory. Troubleshooting (diagnosing faults in)
electronic circuits is one such domain. Electronic equipment is changing

faster than field engineers can build a set of troubleshooting rules from experience. Moreover, all electronic equipment is constructed from just a few elementary components, such as transistors and wires, that are well understood theoretically. Thus it makes sense to use a wiring diagram and fundamental principles to reason about faults in a circuit. Unfortunately, even single faults cannot be diagnosed for circuits of interesting size yet. But the method of reasoning "from first principles" when they exist is certainly one type of reasoning we want future computer programs to be able to use.

Claims made for AI methods have involved all of the classes of outcome variables mentioned above. Only the first four — productivity, human factors, computational requirements, and performance — are precisely enough defined at this time to be measured, and these only with difficulty. Other capabilities suggested for AI programs such as reasoning with common sense, intuition, or causal models are better dealt with as performance issues at a behavioral level than at a dispositional level.

There is no reason, or desire, to use only a single measure of success; evidence favoring a method can take many forms. Yet if AI is to be seen as an experimental science, the claims must first of all be explicit and clear. Second, the claims must be testable, with reproducible experiments. And third, it is desirable, although not necessary, that these claims be measurable.

3.3. *Measurable Attributes*

There are many analyses of algorithms that relate attributes of an algorithm to computational resources. Algorithms for sorting have been well analyzed (Knuth, 1973), for example. The attributes are simple quantities, such as the number of items in a list to be sorted. The analyses relate these quantities to the time a particular sorting algorithm will take or the amount of computer memory that will be required to store intermediate and final results. Whereas small algorithms can be analyzed, complex programs cannot and must be studied empirically. With AI programs, especially, there are so many potentially relevant variables that attempts to name the one or two *most relevant* ones in simple hypotheses have not been satisfactory. It is also extremely troublesome to argue that two sets of background conditions are the same "for all practical purposes." Thus it is necessary to do controlled experiments in which some attributes are held constant and others are varied. The two parts of this section discuss large and small variations. In the former, gross attributes of programs are varied, with qualitative (or sometimes quantitative) observations made on effects. In the latter, considerably more attributes are held constant, with small variations in only a few, carefully studied attributes.

3.3.1. *Observations on large variations*

Two classes of attributes of AI programs that are often mentioned are system architecture and task type. Both are abstract concepts that are ill defined and not measurable. In the pairs of implementations of programs listed in Table III, architectural attributes were varied while the task (and thus the task type) was held constant. These include primarily the methods for representing knowledge in the system and the methods for making inferences. (See (Chandrasekaran, 1986 and Buchanan *et al.*, 1983a) for discussion of several major architectural choices.)

Descriptions of task type have been unsatisfactory in that they do not settle the question of when two problems (tasks) are of the *same* type. We distinguish tasks better by the methods that can be used to solve them than by intrinsic features of the tasks themselves. To some extent, it is helpful to view all problem solving as search (Simon, 1969; Newell and Simon, 1976; Buchanan, 1983a), so to that extent all tasks fit into a single category. Broadly speaking, we can distinguish construction (or formation) methods from decomposition methods (Amarel, 1971), i.e., (a) methods that successively combine primitives to construct a solution from (b) those that iteratively decompose or refine a whole into primitive parts. DENDRAL is a prime example of the former, in which descriptions of molecular structures were constructed by systematically adding one chemical atom at a time into partial structure descriptions until all atoms in the specified chemical composition were accounted for. MYCIN is a good example of the latter, in which the final goal of recommending appropriate therapy for a patient with a bacterial infection was decomposed into subgoals. One of the subgoals was to identify, as well as possible on the available evidence, the organism causing the infection. That subgoal, in turn, was decomposed further. At the "bottom" MYCIN reached sub-subgoals whose truth or falsity, if known at all, could be established by asking questions that were factual and observational.

Two programs may be compared at a gross level if either task type or system architecture is held constant. Additionally, some experiments have varied the task, within the same task type. If all vary, no comparisons can be made. If all are constant, we can compare differences in implementation or other variations at a finer level of detail. The remaining four possibilities are shown in Table V.

TABLE V.

Some types of large variations from which comparisons of AI methods have been made. See text for other combinations.

	Task Type	Task	Architecture	Examples
1.	constant	constant	varying	Pairs of programs in Table III
2.	constant	varying	constant	EMYCIN, other shell systems
3.	constant	varying	varying	MYCIN-PROSPECTOR comparison
4.	varying	varying	constant	Prolog, other languages providing an architecture

In each of four system architectures, Niwa *et al.* (1984) implemented two small prototype expert systems that solved the *same* task. The major architectural classes were simple production system, structured production system, frame system and logic system. For each they implemented both forward chaining and backward chaining inference procedures. They measured a number of static attributes of each program including the size of the knowledge bases and of the inference procedures (e.g., in number of characters used to express them). They also measured some dynamic attributes of the programs, including average inference time (in seconds) required to reach (presumably correct) solutions to three problems with three different sizes of knowledge bases. One set of results is shown in Figure 3 below.

Some experiments in AI have held a program's architecture and task *type* constant and varied the tasks. EMYCIN's architecture is rule-based backward chaining, with inexact inferences. It was used — with different knowledge bases — to solve many problems of a type we call classification or evidence-gathering problems. BBI (Hayes-Roth *et al.*, 1986) is another framework system for solving many similar problems with the same architecture, which is opportunistic reasoning in what is

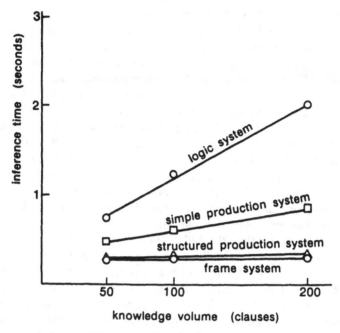

Inference Time and Knowledge Base Volume
for Backward Reasoning.

Fig. 3. From Niwa *et al.* (1984)

called the Blackboard Architecture. It has been used to solve several problems of a type we call assembly problems. Other frameworks have been used in similar ways. (See, for example (Harmon and King, 1985; Kulikowski, *et al.*, 1982). Other framework systems make less of an architectural commitment to a class of problems, e.g., (Brownston *et al.*, 1985; Genesereth, 1982; Novak, 1981; Smith, 1983; Stefik *et al.*, 1983), and serve more as general system-building environments.) By keeping the architecture and task *type* fixed, these systems allow experimentation with different aspects of tasks of the same type. These are closer to controlled trials than the grosser ones in which either task type or architectural attributes vary.

The task of the PROSPECTOR system (Duda, 1978) — advising on the likelihood of significant mineral deposits in a geographical area — shares many features with MYCIN's task of medical diagnosis. The

designers of PROSPECTOR used an architecture with several differences from MYCIN's: for example, they added more structure to the
rule set (linking rules in a network instead of using a hierarchy of rule
sets as in MYCIN) and they added capabilities to allow users of
PROSPECTOR to guide the line of reasoning. One of the most
significant differences is in their calculus for uncertain reasoning, which
they analyzed and compared to MYCIN's (Duda *et al.*, 1976). Insofar
as the tasks are of the same type and the architectures are roughly the
same (both are rule-based), their analysis shows a better theoretical
grounding for PROSPECTOR's (essentially Bayesian) calculus of confirmation with equal success in building their knowledge base of rules.

In these kinds of gross experiments with variations in either system
architecture or task type, it is extremely difficult to say what has varied
and what has remained constant. In most experiments — including
those by Niwa, or the PROSPECTOR-MYCIN comparison — new
procedures or new knowledge bases are written to make different
systems. All who are familiar with computer programming, however,
will recognize the difficulty in comparing effects of code written by
different persons or in different languages. Or, if the effects can be
compared, it is impossible to sort out the contributions of the programmer from those of the method in these large-variation experiments.

3.3.2. *Controlled experiments on smaller variations*

The gross changes of attributes mentioned in the previous section allow
qualitative statements about characteristics of programs. Too many
things may vary when one is only concerned about holding the architecture or the task type constant. But there is still a hope that quantitative
attributes can be found.

There have been many attempts in AI (e.g., (Gaschnig, 1979)) to
measure smaller, relevant attributes of AI programs, such as the size of
their knowledge bases, and make meaningful statements relating these
attributes to performance (or some other outcome variable). Mostly this
search for relationships must take place, as in other disciplines, in very
controlled situations in which it is possible to vary one input parameter
at a time and measure effects.

Except in very simple situations, controlled experiments in AI are
extremely tedious and difficult to interpret. Small problem areas, sometimes called "micro-worlds", serve as the rat mazes or the Drosophila of
AI (McCarthy, 1983). The two most widely used micro-worlds have

been board games (specifically chess) and the world of children's blocks. In the latter, programs have been constructed for answering questions in English about the arrangements of colored blocks on a table (Winograd, 1972), for planning the steps involved in arranging blocks into a specified goal state (Sussman, 1975), and learning classification rules for arches (Winston, 1975). In principle, micro-worlds offer the opportunity for controlled experiments; regrettably, the opportunity has seldom been exploited. A primary reason is that we lack a good set of descriptive attributes of methods. That is, we don't know very well how to simplify our descriptions of methods to identify just a few relevant attributes to vary systematically while holding everything else constant.

Size of Knowledge Base
Simple counts of the number of lines of code in a program are misleading because of the variability of how much goes in a line, what concepts are represented, how efficiently a programmer writes code, and so forth. Similarly, simple counts of the number of statements in an AI program's knowledge base are unenlightening. A decade ago, rule-based expert systems were compared on the number of rules in their knowledge bases as if the quality of their performance depended more on the number of rules than on their content. Nevertheless, computational requirements are nearly always sensitive to sizes, as the graphs shown in Figure 3 show. The three sizes of rule sets used in Niwa's experiments are 50, 100, and 200 rules (or axioms in the logic-based systems), and these are related to problem solving time.

Conditional statements are not the only measurable quantity in a knowledge base, however. Each conditional rule mentions objects and their attributes. For example, INTERNIST's knowledge base contains associations among diseases and manifestations. It represents 571 diseases with an average of about 80 relevant manifestations per disease. Another measure of "how much" a program knows might be the size of its vocabulary. One such measure is a count of objects, attributes and values known to the system — perhaps expressed as the sum shown in Table VI. This measure is problematic when it comes to continuous ranges of values and object-types that can be instantiated in arbitrarily many ways. For example, MYCIN used some continuous values of numerical attributes such as fever, weight, and concentrations. And it used names of object-types, such as Culture and Organism in rules which were then instantiated for each culture of bacteria growing from specimens taken from a patient and each organism growing in the culture.

TABLE VI

Vocabulary sizes of some well-known expert systems. This descriptive measure of an AI program is too simple to be useful.

	Size	$= \# obj + \# attrib + \# vals$
MYCIN	715+	$=(17 + 257 + 441+)$
INTERNIST	4674	$=(571 + 4100 + 3)$
XCON	934+	$=(94 + 840 + ??)$
XSEL	408+	$=(79 + 329 + ??)$
XFL	326+	$=(74 + 252 + ??)$

Note: Many attributes take continuous numerical values.

XCON (McDermott, 1981), XSEL (McDermott, 1982), and XFL are members of the same family of systems designed to assist in con-figuring computer systems for customers. (These data were provided by McDermott (personal communication).)

In an object-oriented architecture such as a frame-based system, the number of objects may be more indicative of the scope of a program's knowledge. Depending on whether a rule-based or object-centered architecture is used, a designer generally increases the number of rules or objects respectively with equivalent quality of performance. (See Table VII.) So there is no special significance to these counts.

Size of Solution Space

The power of a method is somehow related to its ability to find correct (or acceptable) solutions to large problems, where the size of the problem is the number of possible solutions in the search space. MYCIN gathered evidence for and against 120 organism identities in its diagnostic procedure and then collected the likely identities. Since there were rarely more than six likely causes of infection in a patient, MYCIN's search space of diagnoses could be said to be all subsets of 120 identities with six or fewer items — a space of millions. The size of DENDRAL's search space was exponentially related to the number of atoms in the chemical compound of interest and grew to tens of millions with 15—20 atoms. Some other examples are shown in Table VIII.

TABLE VII

The numbers of rules and objects in some rule-based and object-centered expert systems. The ratio of the two separates the two paradigms. The Dipmeter Advisor (Smith, 1984) assists in interpretation of data from oil wells; the Car Advisor (Plotkin, personal communication) assists in diagnosing automobile problems. This ratio is interesting but not useful for describing the size or complexity of a knowledge base.

		($\#$ Rules / $\#$ Object names)
Rule-based systems:		
MYCIN	62.3	≈(1059/17)
XCON (McDermott,		
Personal comm.)	61.0	≈(5739/94)
XSEL (*Ibid.*)	27.1	≈(2148/79)
XFL (*Ibid.*)	21.8	≈(1618/74)
Object-centered systems:		
INTERNIST	5.2	≈(2600/500)
SCHLUMBERGER DIPMETER		
ADVISOR	1.4	≈(90/65)
TEKNOWLEDGE CAR		
ADVISOR	0.4	≈(1242/3317)

TABLE VIII

The size of the solution spaces for several expert systems. This number is important in describing the complexity of a problem. But it must be interpreted carefully since there are so many arbitrary ways to choose the numbers used.

MYCIN: combinations of 1—6 organisms from list of 120 organisms	≈ 10^9
Plausible ones	≈ 6×10^6
INTERNIST: combinations of 1—3 diseases from list of 571	≈ 31×10^6
DIPMETER ADVISOR: any one of 65 geological categories for an arbitrary number of depth levels say, 500 ten-foot zones	≈ 65^{500}
XCON: a configuration of 50—150 computer system components selected from 20000 types	≈ billions

Complexity of Solution Space: Number of Inferences

Another measure of a problem solving method is the way it organizes its search of the solution space. Exhaustive search requires less intelligence than guided search. Thus the number of nodes in the search space that are *actually* examined is as interesting as the total number. An approximate measure of this is a function b^d, where d is average depth of search and b is the average branching factor. For instance, MYCIN's knowledge base contained chains of at most six rules, with the average length of an inference chain (depth of search) about 4. At each intermediate conclusion in the inference chains, about 5.5 rules contributed evidence to the conclusion. Thus it actually examined about a thousand (5.5^4) nodes in the search space for each single case, out of a very much larger implicit space of inferences.

Contents of Knowledge Base
Some experiments vary the contents of the program's knowledge base to measure its effects on quality of performance. In these, the semantic force of the elements in the knowledge base is the issue, rather than an objective measurement like the cardinality of a set. All other performance being equal, we may prefer smaller sets to larger ones, but even that is not assured since the understandability of the program's knowledge base is also an issue.

DENDRAL was run many hundreds of times with small changes to a single list, called BADLIST. The intent of these informal experiments was to find the rules of chemical structure that accurately predicted instability of chemical compounds under normal conditions. Peroxides, for example, can be identified in a structural description by the presence of two oxygens covalently bonded. Manual examination of DENDRAL's lists of structures by a chemist revealed structures that were unstable and which should have been excluded by BADLIST (while excluding no stable structures). The rules on BADLIST were formulated gradually through this sequence of controlled trials. When we attempted to vary many things at once (e.g., changing the procedures by which structures were generated or filtered), we were easily confused.

Greiner (1985) introduced the term "ablation experiments" into AI to describe systematic excision of parts of a program with subsequent analysis of failure (a variation on Mill's Method of Differences). This requires careful design of the program so that pieces can be excised with resulting gradual degradation of performance. Usually when we attempt this with programs, we cause catastrophic failures because the relevant pieces are so interdependent. At the time of design, then, the experiments must be envisioned so that the program can be made to be robust enough and the knowledge base made to be modular enough to avoid these catastrophes.

In many expert systems, knowledge bases are separate from inference procedures and elements of the knowledge base are modular, i.e., nearly independent. In these cases, ablation experiments can be easily performed on the knowledge bases to determine how much knowledge is needed for good performance. It is not easy to know which are the meaningful experiments, however, nor how to interpret the results.

The size of the vocabulary used for constructing a knowledge base is an important consideration in comparing AI programs. But it is not unambiguously described, as discussed above, nor is it the only consideration. The complexity of the inference network, including the size of the total space of possible solutions, is equally important. Finally, no comparison of numbers can be as illuminating in comparing AI programs as comparison of the semantic *content* of knowledge bases. The power of a program to reason effectively in a large combinatorial space lies in the knowledge it can use to guide the search and to prune implausible solutions.

Both large-scale and small-scale variations in programs can provide meaningful data when extreme care is taken in controlling unwanted variations and in interpreting the results. Some patterns and principles are emerging from the experiments already performed. But they are sparse. Considerably more experimentation is needed to advance the science of AI.

4. CONCLUSION

AI is a new discipline, so it is unreasonable to expect all the same methodological characteristics found in the more mature disciplines. In looking at research in AI we see theoretical (often mathematical), engineering, and analytical concerns, as in other experimental disciplines. Each has its advantages, but the greatest advantage to AI comes from a coordinated effort in all three styles.

AI's struggle to develop a coordinated, coherent methodology may provide an opportunity for researchers in other empirical disciplines to take stock of their own methods. One important lesson to be learned from the varied and many successes of the sciences is that theorizing *after* collecting data has a far better track record than theorizing in the *absence* of data. But we see in AI, once again, the fundamental difficulty of deciding on the vocabulary of attributes that we *can* measure and that are *worth* measuring. As we learn what aspects of AI programs to describe and measure, we will realize great benefits from the experimental paradigm.

NOTES

[1] See (Buchanan and Shortliffe, 1984). Two commercial systems which are close copies of EMYCIN, Personal Consultant and KS300, have also been used to construct many expert systems.

[2] See also (Putnam, 1963), p. 773.

[3] McCarthy (1984) refers to this as Feigenbaum's hypothesis, because Feigenbaum has postulated that a very large knowledge base of individual facts and rules will suffice to capture everything needed for good performance.

REFERENCES

Amarel, S.: 1971 'Representations and Modeling in Problems of Program Formation', in D. Michie (ed), *Machine Intelligence 6*, pp. 411—466, American Elsevier, New York.

Bennett, J. S. and C. R. Hollander: 1981, 'DART: An Expert System For Computer Fault Diagnosis', in *Proc. IJCAI-81*, pp. 843—845. IJCAI,Vancouver, B.C.

Brown, Harold, L. Hjelmeland, and L. Masinter: 1974, 'Constructive Graph Labeling Using Double Cosets', *Discrete Mathematics* 7, pp. 1—30.

Brownston, L., Farrell, R., Kant, E. and Martin, N.: 1985, *Programming Expert Systems in OPS5: An Introduction to Rule-Based Programming*, Addison-Wesley, Reading, MA.

Buchanan, B. G.: 1983, 'Mechanizing the Search for Explanatory Hypotheses', in *PSA-II*, Philosophy of Science Association, East Lansing, MI.

Buchanan, B. G.: 1985, 'Steps toward Mechanizing Discovery', in K. F. Schaffner (ed.), *Logic of Discovery and Diagnosis in Medicine*, chapter 4, University of California Press, Berkeley, CA, pp. 94—114.

Buchanan, B. G.: 1986, 'Expert Systems: Working Systems and the Research Literature', *Expert Systems* 3, 32—51.

Buchanan, B. G., E. A. Feigenbaum, and N. S. Sridharan: 1972, 'Heuristic Theory Formation: Data Interpretation and Rule Formation', in Bernard Meltzer and Donald Michie (eds.), *Machine Intelligence 7*, pp. 267—290. John Wiley & Sons, New York.

Buchanan, B. G. and T. M. Mitchell: 1978, 'Model-Directed Learning of Production Rules', in Waterman, D. A. and Hayes-Roth, F. (eds.), *Pattern-Directed Inference Systems*, Academic Press, New York, pp. 297—312.

Buchanan, Bruce G. and Richard O. Duda: 1983a, 'Principles of Rule-Based Expert Systems', in M. Yovits (ed.), *Advances in Computers*, Academic Press, New York.

Buchanan B. G. *et al.*: 1983b, 'Constructing an Expert System', in F. Hayes-Roth, D. Waterman, and D. Lenat (eds.), *Building Expert Systems*, Addison-Wesley, New York.

Buchanan, B. G., and E. H. Shortliffe: 1984, *Rule-Based Expert Systems: The MYCIN Experiments of the Stanford Heuristic Programming Project*, Addison-Wesley, Reading, MA.

Carnap, R.: 1950, *The Logical Foundations of Probability*, University of Chicago Press, Chicago.

Chandrasekaran, B.: 1986, 'Generic Tasks in Knowledge-Based Reasoning: High-Level Building Blocks for Expert Systems Design'. *IEEE Expert* 1, 23—30.

Davis, R.: 1982, 'Teiresias: Applications of Meta-Level Knowledge', in R. Davis and D. B. Lenat (eds.), *Knowledge-Based Systems in Artificial Intelligence*, McGraw-Hill, New York.

Davis, R. and Buchanan, B. G.: 1977, 'Meta-level knowledge: overview and applications', in *Proc. IJCAI-77*, pp. 920—927. IJCAI, Cambridge, MA, August, 1977.

Doyle, J.: 1979, 'A truth maintenance system'. *Artificial Intelligence* 12, 231—272.

Dreyfus, H. and Dreyfus, S.: 1986, 'Why Computers May never Think Like People', *Technology Review*: 42—61.

Duda, R. O., Hart, P. E., and Nilsson, N. J.: 1976, 'Subjective Bayesian methods for rule-based inference systems', in *Proc. 1976 National Computer Conference (AFIPS Conf. Proc.)*, pp. 1075—1082. AFIPS Press, New York.

Duda, R. O., Hart, P. E., Nilsson, N. J., and Sutherland, G. L.: 1978, 'Semantic Network Representations in Rule-Based Inference Systems'. In Donald A. Waterman and Frederick Hayes-Roth (eds.), *Pattern-Directed Inference Systems*, pp. 203—221. Academic Press, New York.

Frayman, F.: 1985, *PROTO: An Approach for Determining Protein Structures from Nuclear Magnetic Reasonance Data: An Exercise in Large Scale Interdependent Constraint-Satisfaction*, Ph.D. thesis, Northwestern University.

Friedland, P.: 1979, *Knowledge-based Hierarchical Planning in Molecular Genetics.* Ph.D. thesis, Computer Science Department, Stanford University, Report CS-79-760.

Gaschnig, J.: 1979, 'Preliminary Performance Analysis of the Prospector Consultant System for Mineral Exploration'. In *Proc. IJCAI-79*, pages 308—310. IJCAI, Tokyo, Japan.

Gaschnig, J., Klahr, P., Pople, H. and Shortliffe, E.: 1983, *Evaluation of Expert Systems: Issues and Case Studies*, in F. Hayes-Roth, D. Waterman, and D. Lenat (eds.), *Building Expert Systems*, Addison-Wesley, Reading, MA, pp. 241—280.

Genesereth, Michael R.: 1982, *An Introduction to MRS for AI Experts*. Technical Report HPP-82-27, Stanford University.

Genesereth, Michael R.: 1984, *The Use of Design Descriptions in Automated Diagnosis.* Technical Report HPP 81—20, Stanford University.

Greiner, Russell: 1985, *Learning by Understanding Analogies*, Ph.D. thesis, Stanford University. Report CS-85-1071.

Harmon, P. and King D.: 1985, *Expert Systems: Artificial Intelligence in Business.* John Wiley & Sons, New York.

Hayes-Roth, B., Buchanan, B. G., Lichtarge, O., Hewett, M., Altman, R., Brinkley, J., Cornelius, C., Duncan, B. and O. Jardetzky: 1986, PROTEAN: Deriving Protein Structure from Constraints., in *Proc. AAAI-86*, pp. 904—909, AAAI, Philadelphia, PA.

Horvitz, E. J., Heckerman, D. E. and Langlotz, C. P.: 1986, 'A Framework for Computing Alternative Formalisms for Plausible Reasoning', in *Proc. AAAI-86*, pp. 210--214, AAAI, Philadelphia, PA.

Knuth, D. E.: 1973, *The Art of Computer Programming*, Volume 3: *Sorting and Searching*. Addison-Wesley, Reading, MA.

Kuhn, T. S.: 1962, *The Structure of Scientific Revolutions*. Chicago Univ. Press, Chicago, IL.

Kulikowski, C. and Weiss, Sholom M.: 1982, 'Representation of Expert Knowledge for Consulation: The CASNET and EXPERT Projects', in P. Szolovits (ed.), *Artificial Intelligence in Medicine*, Westview Press, Boulder, CO, pp. 21—55.

Lenat, D. B.: 1976, *AM: An Artificial Intelligence Approach to Discovery in Mathematics as Heuristic Search*, Ph.D. thesis, Computer Science Department, Stanford University. Reprinted with revisions in Davis, R. and Lenat, D. B., *Knowledge-Based Systems in Artificial Intelligence*. New York: McGraw-Hill, 1982.

Lenat, D. B., and Brown, J. S.: 1984, 'Why AM and EURISKO Appear to Work', *Artifical Intelligence* 23, 269—294.

Lichtarge, O., O. Jardetzky, C. Cornelius, and B. G. Buchanan: 1987, 'Validation of the First Step of the Heuristic Refinement Method for the Derivation of Solution Structures of Proteins from NMR Data'. Forthcoming in *PROTEINS: Structure, Function and Genetics*.

Lindsay, R. K., Buchanan, B. G., Feigenbaum, E. A., and Lederberg, J.: 1980, *Applications of Artificial Intelligence for Organic Chemistry: The DENDRAL Project*. McGraw-Hill, New York.

McCarthy, J.: 1983, 'President's Quarterly Message: AI Needs More Emphasis on Basic Research', *AI Magazine* 4(4), 5.

McCarthy, J.: 1984, 'What is Commonsense?' (Presidential Address). In *Proc. AAAI-84*, AAAI, University of Texas at Austin, TX.

McCorduck, P.: 1979, *Machines Who Think*, W. H. Freeman and Co., San Francisco.

McDermott, J.: 1981, 'R1: The Formative Years', *The AI Magazine* 2(2), 21—29.

McDermott, J.: 1982, 'XSEL: A Computer Sales Person's Assistant', in J. E. Hayes, D. Michie and Y-H Pao (ed.), *Machine Intelligence 10*, Ellis Horwood Ltd., Chichester, England, pp. 325—337.

McDermott, J.: 1983, 'Extracting Knowledge from Expert Systems', in *IJCAI-83*, IJCAI, Karlsruhe, West Germany, pp. 100—107.

McDermott, J.: Personal communication, Carnegie-Mellon University.

McDermott, D. V. and Doyle, Jon: 1980, 'Non-Monotonic Logic I', *Artificial Intelligence* 13, 41—72.

Miller, R. A., H. E. Pople, Jr., and J. D. Myers: 1984, 'INTERNIST-1: An Experimental Computer-Based Diagnostic Consultant for General Internal Medicine', in W. J. Clancey and E. H. Shortliffe (ed.), *Readings in Medical Artificial Intelligence: The First Decade*, Addison-Wesley, Reading, MA.

Newell, A. and Simon, Herbert: 1972, *Human Problem Solving.*, Prentice-Hall, Englewood Cliffs, N.J.

Newell, A. and H. A. Simon: 1976, 'Computer Science as Empirical Inquiry: Symbols and Search, The 1976 ACM Turing Lecture'. *Communications of ACM* **19**, 113–126.

Nilsson, N. J.: 1983, 'Artificial Intelligence Prepares for 2001', *AI Magazine* **4**, p. 7.

Niwa, K., Sasaki, K. and H. Ihara: 1984, 'An Experimental Comparison of Knowledge Representation Schemes'. *AI Magazine* **5**, pp. 29–36.

Novak, G. S.: 1981, 'GLISP: An Efficient, English-Like Programming Language', in *Proc. 3rd Annual Conf. of Cognitive Science Society.* University of California at Berkeley.

Plotkin, Barry. Personal communication, Teknowledge Corp.

Pople, H. E., Jr.: 1982, 'Heuristic Methods for Imposing Structure on Ill-structured Problems: The Structuring of Medical Diagnostics', in P. Szolovits (ed.), *Artificial Intelligence In Medicine*, Westview Press, Boulder, Colo., pp. 119–190.

Popper, K. R.: 1959, *Logic of Scientific Discovery*, Hutchinson, London.

Putnam, H.: 1963, 'Degree of Confirmation', in P.A. Schlipp (ed.), *The Philosophy of Rudolf Carnap*, Open Court, La Salle, IL.

Shortliffe, E. H.: 1974, *MYCIN: A Rule-based Computer Program for Advising Physicians Regarding Antimicrobial Therapy Selection*, Ph.D. thesis, Stanford University. Reprinted with revisions as Shortliffe, E. H., *Computer-Based Medical Consultations: MYCIN.* New York: American Elsevier, 1976.

Shortliffe, E. H., and Buchanan, B. G.: 1975, 'A Model of Inexact Reasoning in Medicine', *Mathematical Biosciences* **23**, 351–379.

Simon, H. A.: 1969, *Sciences of the Artificial*, Massachusetts Institute Technology Press, Cambridge, MA.

Smith, D. H.: 1975, 'The Scope of Structural Isomerism'. *J. Chem. Inf. & Comp. Sci.* **15**, 203–207.

Smith, R. G.: 1983, 'STROBE: Support for Structured Object Knowledge Representation', in *Proc. IJCAI-83*, IJCAI, Karlsruhe, West Germany. pp. 855–858.

Smith, R. G.: 1984, 'On the Development of Commercial Expert Systems', *AI Magazine* **5**, 61–73.

Stefik, M. J.: 1981a, 'Planning with Constraints (MOLGEN: Part 1)', *Artificial Intelligence* **16**, 111–140.

Stefik, M. J.: 1981b, 'Planning and Meta-Planning (MOLGEN: Part 2)', *Artificial Intelligence* **16**, 141–170.

Stefik, M., Bobrow, D., Mittal, S. and Conway, L.: 1983, 'Knowledge Programming in Loops', *AI Magazine* **4**(3), 3–13.

Sussman, G. J.: 1975, *A Computer Model of Skill Acquisition'*, American Elsevier, New York. Based on Ph.D. thesis, MIT, Cambridge, MA.

Toulmin, S.: 1953, *The Philosophy of Science*, Hutchinson, London.

Turing, A. M.: 1950, 'Computing Machinery and Intelligence', *Mind* **59**. Reprinted in *Computers and Thought*, Feigenbaum and Feldman (eds.), McGraw-Hill, New York, 1963.

Wilkins, D. E.: 1979, *Using Patterns and Plans to Solve Problems and Control Search.* Ph.D. thesis, Stanford University. Report CS-79-747.

Winograd, T.: 1972, *Understanding Natural Language.* Academic Press, New York.

Winston, P. H.: 1975, 'Learning Structural Descriptions from Examples', in Winston, P. H. (ed.), *The Psychology of Computer Vision*, McGraw Hill, New York, chapter 5, pp. 157–209.

Yu, V. L., Buchanan, B. G., Shortliffe, E. H., Wraith, S. M., Davis, R., Scott, A. C., and

Cohen, S. N.: 1979a, 'An Evaluation of the Performance of a Computer-based Consultant', *Computer Programs in Biomedicine* **9**, 95—102.

Yu. V. L., Fagan, L. M., Wraith, S. M., Clancey, W. J., Scott, A. C., Hannigan, J., Blum, R. L., Buchanan, B. G., Cohen, S. N., R. Davis, J. Aikins, W. van Melle, E. Shortliffe, and S. Axline', 1979b, 'Antimicrobial Selection By A Computer: A Blinded Evaluation By Infectious Disease Experts', *JAMA* **242**, 1279—1282. Also in Buchanan, B. G. and Shortliffe, E., *Rule-Based Expert Systems*, Addison-Wesley, 1984.

Knowledge Systems Laboratory
Department of Computer Science
Stanford University
Stanford, CA 94305

DONALD NUTE

DEFEASIBLE REASONING:
A PHILOSOPHICAL ANALYSIS IN PROLOG

We reason defeasibly when we reach conclusions that we might be forced to retract when faced with additional information. I contrast this with both invalid deductive reasoning and inductive reasoning. This reasoning is defeasible, but its defeasibility is not because of incorrectness. Nor is it ampliative as is inductive reasoning. It is the kind of "other things being equal" reasoning that proceeds from the assumption that we are dealing with the usual or normal case. Conclusions based on this kind of reasoning may be defeated if we find that the situation is not usual or normal.

I will develop a philosophical analysis of defeasible reasoning. This analysis will be primarily descriptive, although it may justify certain pieces of reasoning that would not be performed in ordinary circumstances. My goal is to capture accurately those bits of defeasible reasoning we observe in the thought and speech of ordinary people. This will produce a unified account that exceeds the boundaries of actual practice in some respects.

The method of my analysis will be formal, but it will not be any of the usual methods of the philosopher logician. Instead, I will take a different approach and formulate my analysis of defeasible reasoning as a computer program. The program will be written in Prolog, a recently-developed programming language used in artificial intelligence research. In this way, I will apply some of the tools of artificial intelligence to the task of philosophical analysis. I think that this will not be the last time that such an exercise is attempted.[1]

PHILOSOPHICAL ANALYSIS AND LOGIC

Contemporary philosopher logicians often use methods of formal logic in philosophical analysis. The two most familiar approaches are to use either a formal system or a formal semantics to capture the structure of some philosophically puzzling concept or set of concepts. Modal logics were developed to clarify the concepts of possibility and necessity and to explain how we reason with them. Deontic logics are similarly

251

James H. Fetzer (ed.), Aspects of Artificial Intelligence, 251–288.
© 1988 *by Kluwer Academic Publishers.*

motivated by a desire to clarify normative concepts, and tense logics aim at an explanation of the roles tense constructions, temporal adverbs, etc., play in our thought and speech. Possible-worlds semantics has been used to elucidate all of these features of our conceptual scheme. The formal method I will demonstrate here is an alternative to building formal systems or formal semantics in the usual sense, but it also complements these methods as they complement each other.

By a formal system, I mean an artificial language and a set of axioms and inference rules defining a logical calculus or a proof theory for that language. In the formal language, we focus on a particular part of our conceptual scheme and ignore the rest. We temporarily hide the features of our conceptual scheme we don't want to consider within the primitive "nonlogical" symbols of the artificial language we construct. These details are hidden, for example, within the p's and q's of a sentential logic or the F's and G's of a predicate logic. We assume a ready supply of these and concentrate our attention on the special "logical" symbols in the formal language. The axioms and rules of inference of the formal system focus on patterns for manipulating expressions in the formal language according to the role their natural counterparts are supposed to play in our ordinary thought and speech. These manipulations constitute a notion of derivability within the formal system.

The formal system is tenuously connected to our everyday reasoning, our ordinary speech, or our actual conceptual scheme through certain conventions for translating natural thought and speech into the formal language, and vice versa. These conventions are ordinarily far less precise than what goes on within the system. We can have an excellent formal system, one that can hardly be faulted in the abstract for its rigor and elegance. But the "translation lore" that connects it with ordinary thought and speech may be vague and poorly understood.

If our motive for building the formal system is to explain and clarify some piece of our conceptual scheme, or our ordinary way of thinking and speaking, how do we determine its success? We do this in two ways.

First, we look at examples of ordinary reasoning where we draw certain conclusions from certain starting points. We translate both the conclusions and the starting points into the formal language and see if we can use the axioms and rules of the formal system to derive the translations of the conclusions from the translations of the starting points. Every success is a point for and every failure is a point against the formal system.

Second, we look at proofs we get in the formal system and try translating the starting points and conclusions of these proofs back into natural language. If all goes well, we would ordinarily accept the translations of the conclusions if we accept the translations of the starting points. Every success is a point for the theory. Failures can occur in two ways: either we ordinarily would reject the translation of the conclusion when we accepted the translations of the starting points, or we simply would neither accept nor reject the translation of the conclusion when we accepted the translations of the starting points. The first discrepancy is usually considered more serious than the second.

What do we do when our formal system together with its translation lore doesn't exactly fit our ordinary thought and speech? We can revise our translation lore, our formal system, or *our ordinary usage*. The first two options are obvious, but it is the third that really distinguishes philosophical analysis from formal linguistics or cognitive psychology.

The most interesting conceptual problems, the ones most likely to draw the attention of a philosopher, are those involving concepts that are vague or unclear. Often people are unsure or even inconsistent in their thinking and speaking when they use these concepts. While the first goal of the philosopher logician is descriptive, his or her ultimate goal is to offer an analysis that resolves some of the difficulties that arise in ordinary usage. Where there are perceived problems with ordinary usage, the philosopher anticipates that at some point his or her analysis will take a prescriptive or normative turn. Either it will suggest ways to fill gaps where ordinary usage is uncertain, or it will reject certain usages that lead to contradiction, or both.

Before we begin the formal analysis, we can not decide how the analysis should reform ordinary usage. There are no predetermined criteria for when misfits should prompt a revision of the formal system, a revision of the translation lore, or a revision of ordinary usage. There are many desiderata affecting this decision that would take us beyond our present concerns. In any case, these desiderata are neither absolute nor clearly arranged in some priority scheme.

Similar remarks apply to the role of formal semantics in philosophical analysis. Just as with formal systems, the development of a formal semantics usually begins with the definition of a formal language. There is no reason in principle why we cannot develop a formal semantics directly for a natural language and this is sometimes done, but we usually resort to the formal language so that we can once again pack that part of the language or conceptual scheme we don't want to consider inside the "nonlogical" primitives of the formal language. As before, we require a translation lore that gives us a way to move between ordinary thought and speech on the one hand and the formal language on the other.

A formal semantics provides a way of interpreting the expressions of its formal language in terms of some set of abstract objects. Usually these objects are set-theoretical models of some sort. Typically, an interpretation function maps the expressions of the language to some components or substructures of the model. The most common kind of formal semantics, although not the only kind, is truth conditional. It provides a definition of truth for an expression in the formal language relative to a model in terms of the properties of the objects the expression or its subexpressions are mapped to in the model. Some property or properties other than truth could play a similar role in a formal semantics. These are the *designated properties* of the formal semantics.

Whatever the designated properties of a formal semantics are, they correspond to some natural properties we would attribute to expressions in our ordinary thought and speech. These could be truth, necessary truth, obligatoriness, or something else, and they are among the concepts the formal semantics is supposed to clarify. We evaluate the semantics by considering complex natural expressions where we have opinions about the natural designated properties possessed by the expression and its subexpressions. We then translate the complex natural expression into the formal language and see if the semantics will allow the same pattern of formal designated properties for the formal expression and its subexpressions.

As with formal systems, a formal semantics may fit more or less closely the ordinary thought and speech it models. With imperfect fit, we have the same three options as before: we can revise our translation lore, our formal semantics, or our ordinary usage. Where the ordinary or natural concepts involved are vague or unclear, the semantics can suggest ways to fill gaps where there is uncertainty, or ways to resolve inconsistencies. Like formal systems, our first goal in developing a formal semantics should be descriptive even when we are confident that our final results will be prescriptive or normative.

The two techniques of formal systems and formal semantics are often used together. Then we have a single formal language for which we provide both a proof theory and a model theory. We will be concerned to establish certain relationships between the proof theory and the model theory. Ideally, we will be able to show that there is a one-to-one correspondence between proofs or theorems (things that can be proved from the axioms alone) and some model-theoretic property. The attempt to establish this relation often prompts revisions in the formal system, the formal semantics, or both. There are, however, some interesting examples from philosophical logic where such a tight relation has not been or cannot be established.

I will not use either formal systems as they are ordinarily presented or formal semantics in my analysis. Instead, I will use a computer program. This approach will be closer to the formal systems approach than to the formal semantics approach. I have spent so much time discussing formal systems and formal semantics to clarify their purpose in philosophical analysis and the ways we evaluate how well they accomplish this purpose because I propose to use a computer program for the same purpose and to evaluate the success of the program in the same ways.

PHILOSOPHICAL ANALYSIS AND LOGIC PROGRAMMING

We can think of the data structures for a particular computer programming language as a formal language. The programming language itself is then, among other things, a metalanguage we can use to talk about and manipulate the expressions in this formal language. In the same way, set theory is the usual metalanguage used to talk about the formal languages usually developed as the starting points for formal systems and formal semantics.

For the purposes of philosophical analysis, we can confine our attention to some subset of the data structures for some programming language. In fact, the structures we consider could be the expressions of some more familiar formal language such as classical propositional logic, or some set of structures that stand in a natural one-to-one correspondence to these expressions. Just as before, we will need a translation lore for moving between the structures of ordinary thought and speech and our data structures/formal language. This translation lore will play exactly the same role it would play if we were constructing a formal system or a model theory of the usual sort.

A program or algorithm can play a role similar to that of a set of axioms and inference rules in a formal system. With given data structures as inputs, it will provide certain data structures as outputs. With some additional requirements, this is a kind of proof theory. Provided the program will output a certain structure when provided with certain other structures as inputs (while meeting other computational requirements we might want to impose), we could say that the computation that gives this result is a proof of the first structure (the output) from the set of starting structures (the inputs).

Work in automated theorem proving has not proceeded in this way. Here, we typically begin with some formal system and try to develop a program that will construct objects corresponding to proofs in that preexisting system. The approach I propose is quite the reverse. Rather than writing a program that complies with some predefined notion of proof, I suggest that we take the program itself as defining what is and what is not derivable from some set of starting points. Turning the usual way of looking at things entirely around, we might then try to develop a formal system of the more familiar type that can provide a proof in exactly those cases where the computation succeeds.

It could be argued that there is no real difference between what I am proposing and automated theorem proving. Automated theorem provers of the sort we find familiar attempt to implement formal systems while the kind of program I am suggesting tries to implement the natural reasoning of ordinary people. But in either case, we are just trying to develop a program that behaves according to certain preestablished criteria for what does and does not constitute a good piece of reasoning. But this argument points to what is really the most striking difference between automated theorem proving and philosophical analysis via logic programming. We begin our analysis by trying to model ordinary usage, but we expect at some point that our developing model (in this case a program) will provide advice for modifying ordinary usage. An automated theorem prover in the usual sense that didn't fit the theory it was patterned after would never be taken as providing a reason for changing the preexisting theory.

An advantage we might claim for the formal system, and a corresponding disadvantage for the computer program, is that formal systems are usually developed in simple and precise language one step at a time, while computer programs are often highly complicated and difficult to trace. Thus, as a descriptive and analytic tool, formal systems clarify and explain while computer programs obscure and are often more difficult for most people to grasp than the vague and unclear ways of thinking and speaking that they might be intended to clarify. It is not helpful to build what most people will see as a "black box" that cranks out conclusions from starting points. We want to see what is in the box, and we want what is in the box to help us understand why we arrive at the conclusions we do.

Here declarative computer languages offer relief. In a declarative computer language, a program does not consist of many minute steps telling the computer in "infinite" detail how to perform some task. Instead, a program consists of some set of statements describing a problem and defining the conditions that an acceptable solution to the problem must satisfy. Computation then becomes an automated process for finding one or more solutions that satisfy this set of conditions.

Some declarative programming languages use a variant of predicate logic for their syntax. Computation for these languages amounts to proving in some fragment of predicate logic that some statement is a theorem derivable from some other statements taken as axioms and using inference rules that are valid for predicate logic. Logic programming is the development and use of such programming languages and of other languages based on extensions of predicate logic or alternatives to classical logic. If we use one of these languages for our programming, the charge of obscurity is alleviated. Programs in these languages can look very much like the definitions we give in set theory when we develop a formal system. With the advent of logic programming, the familiar formal systems approach to philosophical analysis and the computer program approach I am proposing begin to converge. There will always be important differences, though, since derivability for a computer program will be defined as successful computation.

If languages for logic programming provide a reasonable metalanguage for discussing and manipulating formal languages, the next question to address is the evaluation of programs offered as philosophical analyses. When we translate ordinary thought and speech into a formal language, run a program, and translate the resulting conclusions back into ordinary thought and speech structures, we may get a better or worse fit. Where the fit is imperfect, we can modify the translation lore, modify the program, or modify ordinary usage. The choices are exactly the same, and the admittedly ill-defined methods we use in deciding individual cases will also be the same.

PROLOG

Probably the best-known logic programming language is Prolog, developed by Alain Colmerauer and his colleagues in the early 1970s. This language is growing in popularity among artificial intelligence researchers and expert systems developers. Although there are problems with understanding exactly what a statement in a Prolog program means, I will ignore these. Just as we must assume that we understand set theory or English supplemented with set theory when we set out to build a formal system or a formal semantics, here I will assume that we have a primitive understanding of the Prolog statements used in our analysis. Since many readers really will not be familiar with Prolog, I will summarize the most important features of Prolog here and other features as they are needed. For a detailed discussion the reader is referred to Clocksin and Mellish (1981) or Covington, Nute and Vellino (1987).

A Prolog program is made up of a set of *clauses*. A clause is either a *fact* or a *rule*.

Facts have the form of an atomic formula in predicate logic. Examples of Prolog facts are **f(a)**, **likes(Anyone, cake)** and **between(usa, canada, mexico)**. Notice that we can use any expression we choose as a predicate, a constant, or a variable, except that we begin constant expressions with lower-case letters and we begin variables with upper case letters.

A rule has a head and a body. The head of a rule is an atomic formula and the body of a rule is a combination of atomic formulae. The rules we will consider will all have conjunctions of atomic formulae as their bodies. The head of the rule and the body of the rule are connected with the symbol :—, and the atomic formulae conjoined in the body of the rule are connected with commas. Examples of Prolog rules are **father_of(Father, Son) :— parent_of(Father, Son), male(Father)** and **father(Person) :— father_of(Person, Anyone)**.

We can think of Prolog facts as axioms in a formal system, and Prolog rules as inference rules that say that the formula on the left of the symbol :— (the head of the rule) is to be inferred from the formulae on the right of the symbol :— (the body of the rule). The variables in a

Prolog clause can be thought of as falling within the scope of universal quantifiers at the beginning of the clause. One thing we should be careful not to do is to think of the symbol :— as representing material implication. It would be much better to think of it as the logician's "turnstile" or derivability symbol.

Prolog computation takes place when we present Prolog with a clause or "query" for it to try to derive from the available program or database. (There is no sharp distinction between program and data in Prolog). Queries begin with a question mark and a hyphen. Examples include ?— **father_of(jimmy, amy)** and ?— **between(Place, canada, mexico)**. If a query does not contain any variables, Prolog will answer **yes** it can derive the query or **no** it cannot. If a query contains a variable, Prolog interprets the variables as bound by existential quantifiers. Thus, the query ?— **between(Place, canada, mexico)** is interpreted as 'Can you prove that there is a place between Canada and Mexico?' If Prolog cannot prove this, it answers 'no.' If it can prove it, it gives values for the variables for which it can prove it. For example, it would respond to our sample query with **Place = usa.** We are given the option of rejecting this solution. If we reject it, Prolog will try to derive the query again, possibly producing a new solution.

Nearly all Prolog implementations include a built-in predicate **not**. Just as :— is not material implication, **not** is not the negation most of us are familiar with. In Prolog, **not** really means 'not provable.' Thus, Prolog considers a formula **not f(a)** to be proven if it is unable to prove **f(a)**. In our example, Prolog will answer **yes** to the query ?— **not between(georgia, canada, mexico)** because it cannot prove that Georgia is between Canada and Mexico using only the facts and rules we have provided.

We do not have any truth-functional connectives in Prolog. I said earlier that we used the comma to conjoin formulae in the body of a Prolog rule, but even this conjunction is not ordinary truth-functional conjunction. For one thing, we can have rules like **bachelor(Person) :— male(Person), not married(Person)**. Since **not** has a distinctly metalinguistic character, we would confuse the object language/metalanguage distinction if we thought of the comma as ordinary conjunction.

Variables in Prolog can occur in the position of a formula. This allows us to express the classic dictum that knowledge is justified true belief as **knows(Person, P) :— justified_in_believing(Person, P), P, believes(Person, P).** This means that Prolog is a self-referential fragment of higher-order predicate logic. A Prolog program can reason about itself or selected parts of itself. If we build one part of a Prolog program that reasons about another part, we can arbitrarily reestablish the object language/metalanguage distinction. We will need other features of Prolog that allow a Prolog program to examine itself, but we will discuss these as they are required.

This is the tool I will use in my analysis of defeasible reasoning. I will develop a core Prolog program that clarifies and analyzes defeasible reasoning. This core program will also define an extension of Prolog syntax suitable for expressing the kinds of facts and rules appropriate for defeasible reasoning. When additional facts and rules of the appropriate sort are added to this core program, we produce a new program that can determine what is defeasibly derivable from the added facts and rules. The core Prolog program will represent the promised analysis of defeasible reasoning. The sets of additional facts and rules and the conclusions the core program derives from them will provide the tests we need to evaluate this analysis.

DEFEASIBLE REASONING

Now that our method of analysis is at least provisionally specified, we can turn to the particular task of philosophical analysis to be attempted. First we will consider some examples of the kind of reasoning I am calling "defeasible reasoning."

I recently invited my friend Clyde to meet me for lunch at a Mexican restaurant called *El Azteca* and he accepted. At noon, I showed up at *El Azteca* fully expecting Clyde to meet me. He never did. Of course, I gave up my belief that Clyde would join me for lunch at *El Azteca* when I clearly saw that he did not. Later Clyde told me that he had been thinking of another restaurant, *Gus Garcia's*, when he agreed to meet me at *El Azteca*. The point, of course, is that we often conclude that something has or will happen because somebody tells us that it has or will happen. These conclusions are risky, and we often must abandon them in light of definite evidence to the contrary.

I recently saw a fine example of defeasible reasoning illustrated in Fulton (1984). An amateur astronomer is examining magazine advertisements to decide whether to buy a certain kind of telescope from Precision-Eye Optics or Betelgeuse. First she notices that Betelgeuse offers a free poster with their telescope. Based on this, she decides to order from Betelgeuse. Then she sees that the Betelgeuse telescope costs $19.65 more than the Precision-Eye telescope. Since the free poster isn't worth $19.65, she changes her mind and decides the Precision-Eye telescope is the one to buy. Looking even more closely, she sees that Precision-Eye charges $17 more for their tripod, an essential accessory. The difference in total price, then, is only $2.65. The "free" poster is worth this much, so the two companies are on even terms. However, the Precision-Eye ad explains that their "bi-pentafical wertner guards the sympathetic stop from excessive genuflexure" and points out that their competitor's telescope doesn't even have a wertner! Too bad for Betelgeuse! But wait; Betelgeuse claims that "the inclusion of any type of wertner tends to encourage photon splatter."

This back-and-forth reasoning is common when we must make a decision and pertinent information becomes available a bit at a time. Indeed, our most difficult and important deliberations are typified by just such a series of contradictory decisions each replacing the last as a result of new evidence.

Defeasible reasoning is so common that there is a whole genre of humor based on it. This is the familiar "good news-bad news" story. "I submitted a paper for a meeting of the Hawaiian Philosophical Association. Unfortunately, they didn't accept it. But they invited me to comment on another paper. My department didn't have money for my travel expenses. Luckily the dean came up with the money. By that time, all the flights to the meeting were booked. But I don't really mind. They were meeting in Cleveland." With apologies to Cleveland, I offer this as a good example of defeasible reasoning in humor.

These extended examples are provided only to show how common defeasible reasoning is. In what follows, we will look at simple pieces of defeasible reasoning. We will base our analysis on a careful examination of the patterns to be found in the small steps from which these extended chains of defeasible reasoning are built.

DEFEASIBLE WARRANTS AND RULES

Often when we make a claim, we can provide some particular reasons for making that claim and some warrant or rule connecting the reasons with the claim. This pattern of explanation is found in Toulmin, Rieke and Janik (1984) where it is explicitly applied to defeasible reasoning (although Toulmin, Rieke and Janik do not use this term.) In such cases, we can often locate the defeasibility of the reasoning in the warrant or the rule used.

Consider a simple example. I claim that Tweety can fly. You ask me why, and I tell you that Tweety is a bird. When you ask, "So what?" I respond that (typically, normally) birds fly. You can list atypical or abnormal cases including ostriches, penguins, and emus, but this will only prompt me to add a qualifier to my claim and say, for example, that *presumably* Tweety can fly. You would actually have to show me that Tweety is an atypical or abnormal bird before I am likely to give up my opinion that Tweety can fly.

That birds fly is the kind of common-sense rule that we use constantly, often without even being consciously aware of it. It is a rule with known exceptions; hardly any adult doubts that there are birds that do not fly. Yet we use the rule with some confidence unless we know that a particular bird is atypical or abnormal. By using such "rules of thumb," we simplify our reasoning and we allow ourselves to draw conclusions on the assumption that we are dealing with a typical or normal case even when we do not have certain information that the case is typical and normal. This is of particular advantage when we must act on incomplete information or when it is too bothersome to acquire complete information. The warrant or rule we use is defeasible since there are instances where its antecedent condition is true but we do not detach the consequent of the rule because there is reason to think the case at hand may be exceptional.

Other examples of defeasible rules can be extracted from the extended examples of defeasible reasoning considered earlier. In the aborted luncheon example, I assume that my friends say what they mean and that they do what they say they will do. Several rules were involved in the telescope example. One is that if two items are of equal quality, then the lower-priced item is the better value. Another is that if two items requiring accessories are of equal quality, then the one with the lower total cost for the item and the necessary accessories is the better value. The Hawaiian conference example also involves several rules such as the rule that regional societies normally hold their conferences within their regions. These rules are defeasible as the examples show, but it would have been appropriate to offer these rules as warrants to justify our various judgments.

Reasoning based on a defeasible rule like '(Typically) birds fly' is not deductive, since the premises can be true even though the conclusion is false. Defeasible reasoning is risky. But neither is defeasible reasoning ampliative. The conclusion that 'Tweety presumably flies' contains no information not already contained in our premise that Tweety is a bird and our rule that birds typically fly. Of course the reasoning process that led us to accept the defeasible rule that birds fly is very likely both ampliative and inductive, but that is another issue.

As was mentioned before, Toulmin, Rieke and Janik recognize that many of the warrants and rules we use in ordinary thought and speech are general rules with exceptions. In their terms, specific information might *undercut* a rule or *undermine* its authority. They give some attention to the problem of deciding what is and what is not the normal case (particularly in Section 10), but they do not consider in detail the different ways these kinds of rules and warrants can be undercut, the ways we can use indefeasible rules and warrants in defeasible reasoning, and the ways that defeasible rules and warrants can conflict. These are some of the issues we want to get clearer about through our analysis.

ABSOLUTE RULES, DEFEASIBLE RULES AND PRESUMPTIONS

Many of the rules we use in our everyday thought and speech are defeasible rules, but by no means all the rules we use are defeasible. The contrasting kind of rule I will call an *absolute* rule. These rules play a crucial role in defeasible reasoning at several levels.

Absolute rules fall into several categories. Among these are rules of categorization, rules based on definitions, lawlike rules from science, and rules stating geographic regularities.

Our categories are not absolute in any metaphysical sense, and there are nontrivial controversies in taxonomy. Nevertheless, many of our categorizations are entirely unexceptional given the categories we use. Absolute rules of categorization include 'Penguins are birds' and 'Whales are mammals.' A penguin that is not a bird or a whale that is not a mammal is an impossibility. This is not to say that our categories are not chosen for their convenience, or that we could not devise alternative categories given good enough reason. But given the way we do categorize, certain subordinations of one category to another are absolute.

The meanings of our terms are conventional, but again certain rules are absolute given the meanings that we assign to certain terms. Because of the meanings we give to our words, the rule 'No bachelor is married' and the rule 'Any two siblings have a parent in common' are absolute. These rules are similar in some respects to rules for categorizing things, yet they are different enough that they are worth mentioning separately.

A moderately sophisticated view of scientific law will treat the lawlike statements of science as more or less highly confirmed hypotheses, not as absolute truths. But outside the context of theory-testing, we often use such lawlike statements as though they were absolute. Examples are the laws of thermodynamics or Newton's principle of gravitational attraction. The ultimate scientific status of these principles is that they are useful hypotheses still open to test, but in engineering or even more mundane contexts they may function as absolute rules.

A final example of an absolute principle is a geographical regularity. Anything in Atlanta is in Georgia and is south of Canada. Of course, these relations may change over time as the example of European history testifies. (Is anything in Alsace-Lorraine in France today, or in Germany?) But at a given time many such rules are in principle exceptionless.

These examples may suggest that nothing is really absolute, but they also point out that there are many kinds of rules that are absolute in certain contexts or domains. By this I mean that nothing could undercut or defeat them when they are used in the right context or domain.

Prolog treats all rules as absolute. Only atomic formulae that begin with the Prolog predicate **not** are defeasible at all, and then only in the sense that something may become provable when we add new data that was not provable before. Since **not** cannot occur in the head of a Prolog rule, Prolog rules as complete statements are not even defeasible in this weak sense. So long as all the clauses in the body of the rule are provable, the head of the rule is taken as proven no matter what else is in the database.

Even though Prolog rules are essentially absolute, I will adopt another way to represent absolute rules in Prolog to maintain a distance between the object language I will develop and the Prolog rules I use to talk about this language. I will represent absolute rules using the connective $\rightarrow$ and defeasible rules using the connective $\Rightarrow$. Either kind of rule will have a list of literals (atomic formulae and negations of atomic formulae) as antecedent and a single literal as consequent.

An example of an absolute rule as formulated in this language is
[penguin(X)] → **bird(X)** and an example of a defeasible rule is **[bird(X)]**
⇒ **flies(X)**.

I will use the predicate **neg** for ordinary negation. Then we can
represent the rule that wet matches don't normally light when struck as
[match(X), wet(X,T), struck(X,T)] ⇒ **neg lights(X,T)**, where **T** is a
variable ranging over times. I will not mark the literals in our object
language in any special way to distinguish them from ordinary Prolog
facts. Very few facts, as distinct from rules, will occur in our analysis of
defeasible reasoning. We should have no trouble distinguishing the few
facts occuring in our analysis from the literals we take to be part of our
object language.

Before ending this section, I want to draw attention to what amounts
to a special kind of defeasible rule. Sometimes we *presume* that
something is true even when we recognize that certain information
would defeat this presumption. In American jurisprudence, the accused
is *presumed* innocent until proven guilty. Ideally, this would mean that
in the mind of a juror there exists a presumption of innocence that
prevails unless evidence is presented that forces the juror to surrender
this presumption. We can represent such presumptions as defeasible
rules of the form **[ ]** ⇒ **Presumption** where **[ ]** is the empty antecedent. We symbolize our juror's presumption as **[ ]** ⇒ **innocent(accused)**. This means that the juror will conclude that the accused is
innocent unless this presumption is somehow defeated. Exactly how
this and other defeasible rules may contribute to our conclusions or
may be defeated will be explored in the remainder of this paper.

ABSOLUTE DERIVABILITY

Since every Prolog rule is an absolute rule, any query that Prolog can
prove from a given database is absolutely derivable from that database.
Of course if the Prolog predicate **not** is involved in any of the rules
used in the derivation, the same query may not be derivable from a
strictly larger database. Because of this feature, we say that Prolog is
non-monotonic.

We won't allow **not** to appear in expressions of our object language,
but we will use **not** in the Prolog rules we develop to define certain
kinds of derivability for our object language. Having excluded **not** or
anything like it, we will be able to characterize a relation of absolute

derivability for our object language that is *monotonic*. This means that any literal in our object language that is absolutely derivable from any set of literals and rules in our object language will also be absolutely derivable from any larger set.

A literal is absolutely derivable from a set of literals and rules if we can derive it using only the literals and the absolute rules in the set. We use an absolute rule by first showing that all the literals in its antecedent are absolutely derivable and then detaching the consequent of the rule. To simplify things, we define absolute derivability so that both literals and lists of literals (the antecedents of rules) are absolutely derivable. We need only four Prolog rules to define the absolute derivability relation we want.

> **absolute_derivability([]).**
>
> **absolute_derivability([First|Rest]) :-**
>   **absolute_derivability(First),**
>   **absolute_derivability(Rest).**
>
> **absolute_derivability(Literal) :-**
>   **not list(Literal),**
>   **Literal.**
>
> **absolute_derivability(Literal) :-**
>   **not list(Literal),**
>   **((Antecedent -> Literal)),**
>   **absolute_derivability(Antecedent).**

The first two clauses in our definition tell us when a list of literals is absolutely derivable. The Prolog expression **[First|Rest]** represents a list where **First** is the first member of the list and **Rest** is a list of remaining members. The clause says that a list is absolutely derivable if its first member is absolutely derivable and the list of remaining members is absolutely derivable. By testing each member of the list, we eventually run out of members and are left with an empty list of remaining members. Our first clause says that the empty list [ ] is absolutely derivable. This clause provides the terminating condition for our recursive procedure.

The third and fourth clauses in our definition of absolute derivability tell us when an individual literal is derivable. First, a literal is derivable if it succeeds as an ordinary Prolog query. This can only happen if the literal is a fact in the Prolog database since we are not putting any ordinary Prolog rules in our object language. Second, a literal is derivable if it is the consequent of some absolute rule whose antecedent is derivable. Both of these clauses apply only if the value of the variable **Literal** is an individual literal and not a list. For Prolog to understand this condition, we need to define a list as the empty list or anything that has a first member and a list of remaining members.

> **list([]).**

> **list([X|Y]).**

What we have done with these definitions of the predicates **absolute_derivability** and **list** is to build a Prolog interpreter within Prolog. This interpreter is different from the usual Prolog interpreter, though, in that it replaces Prolog's usual derivation procedure with our absolute derivability for a language having a *positive* negation rather than the *negation as failure* implemented in Prolog. Our **neg** means a positive denial rather than the *can not prove* found in Prolog.

DEFEASIBLE DERIVABILITY

Our goal is not to write a Prolog interpreter within Prolog or even to add a true negation to Prolog. These are just steps toward constructing an analysis of defeasible reasoning within Prolog. Now we will offer a first approximation of what it means for a literal to be defeasibly derivable from a set of facts, absolute rules and defeasible rules.

We will need to define defeasible derivability for both individual literals and lists of literals just as we did for absolute derivability. So we begin our definition of defeasible derivability with two clauses for lists.

> **defeasible_derivability([]).**

> **defeasible_derivability([First|Rest]) :-**
> **defeasible_derivability(First),**
> **defeasible_derivability(Rest).**

Any single literal will certainly be defeasibly derivable if it is absolutely derivable.

```
defeasible_derivability(Literal) :-
    not list(Literal),
    absolute_derivability(Literal).
```

Next we need to consider the role absolute rules should play in our analysis of defeasible reasoning. If we can conclude tentatively that something is a penguin, then we should also tentatively conclude that it is a bird. The only thing that should interfere with this inference might be that we knew with certainty that it was not a bird. In Prolog it is easy to describe this role of absolute rules in defeasible reasoning.

```
defeasible_derivability(Literal) :-
    not list(Literal),
    ((Antecedent -> Literal)),
    defeasible_derivability(Antecedent),
    not contradicted(Literal).
```

This looks much like the final clause in our definition of absolute derivability, except for the condition that the formula that is defeasibly derivable must not be contradicted.

We will want to say that a formula is contradicted if some contrary of the formula is absolutely derivable. A contrary of a formula would be its usual negation, but a formula might also have other contraries. For example, 'Tommy is blind' and 'Tommy can see' are what we might call *semantic* contraries. This is certainly part of our background understanding of English and would play a role in our defeasible reasoning. We will represent information about contrary properties in our object language using a new predicate **incompatible**. In our example we will say

```
incompatible(sees(X),blind(X)).
```

We will also define a two-place predicate **negation**.

```
negation(neg Atomic_formula,Atomic_formula).
```

```
negation(Atomic_formula,neg Atomic_formula) :-
    not functor(Atomic_formula,neg,1).
```

If a formula is already a negative literal, its negation is the result of stripping off the **neg**. If a formula is atomic, if it does not already begin with a **neg**, we form its negation by adding a **neg** to it. Thus, we simplify our object language by prohibiting double negations. Prolog has a built-in predicate **functor** that we can use to test to see if a formula begins with **neg**.

Now we can easily define both the predicate **contradicted** and its subordinate predicate **contrary**.

```
contradicted(Literal) :-
    contrary(Literal,Opposite),
    absolute_derivability(Opposite).

contrary(Literal,Opposite) :-
    negation(Literal,Opposite).

contrary(Literal,Opposite) :-
    incompatible(Literal,Opposite).

contrary(Literal,Opposite) :-
    incompatible(Opposite,Literal).
```

The most interesting case to consider, of course, is the case where our defeasible reasoning depends on the use of a defeasible rule. A general statement of the way such rules are used is not difficult.

```
defeasible_derivability(Literal) :-
    not list(Literal),
    ((Antecedent => Literal)),
    defeasible_derivability(Antecedent),
    not contradicted(Literal),
    not defeated(Literal,Antecedent).
```

The defeasible detachment of the consequent of an absolute rule whose antecedent is defeasibly derivable is prevented only if the consequent is contradicted, but the defeasible detachment of the consequent of a defeasible rule whose antecedent is defeasibly derivable is prevented if the consequent is contradicted or if the antecedent of the rule is undercut or defeated for this particular consequent.

Notice what happens when our defeasible rule is a presumption. Then its condition is the empty list [ ] which is defeasibly derivable. One way that such a presumption can be defeated is if it is contradicted, i.e., if we can absolutely derive some contrary of its consequent. In the next section we will consider other ways that presumptions and other defeasible rules may be defeated.

DEFEATING RULES

We are prevented from drawing a conclusion using either an absolute or a defeasible rule if the consequent of the rule is absolutely contradicted by our other knowledge. This is the only way we can circumvent a defeasible derivation using an absolute rule, but there are other ways to defeat a proposed inference based on a defeasible rule.

Let's consider a simple case where two defeasible rules have contradictory consequents. We know that birds fly and that penguins don't. We could represent these two bits of common knowledge in our developing object language as follows.

[bird(X)] => flies(X).

[penguin(X)] => neg flies(X).

Now suppose we know that Chilly is both a bird and a penguin.

bird(Chilly).

penguin(Chilly).

If we can ignore for a moment our knowledge of the relationship between birds and penguins, we should find ourselves in a dilemma. Does Chilly fly or doesn't he? Intuitively, each of our two defeasible rules should undermine the other.

Let's add the following absolute rule.

[penguin(X)] -> bird(X).

This is a piece of background knowledge we were ignoring before. With only this much information, we can conclude that the first of our defeasible rules does not undercut or defeat the second after all. With this information, we draw the tentative conclusion that Chilly does not fly.

What happens when we add the piece of information that absolutely all penguins are birds? Then we can see that the defeasible rule about penguins is *more specific* than the rule about birds. In some sense, the rule about penguins looks at our information more closely than does the rule about birds. Knowing that something is a penguin tells us that it is also a bird, but not *vice versa*.

The particular example we were considering depends on our scheme of categorization for certain kinds of animals, but similar cases do not. Other kinds of relations between antecedent conditions that can be expressed using absolute rules can also help us decide which of two competing defeasible rules we should use. As an example, let's look at a geographical case. Suppose that there are two geographical areas, one called 'Washington' and the other called 'Jefferson.' I will tell you that one is a town and the other is a county, and that the town is in the county. I will also provide the following two rules.

> [lives_in(Person,washington)] =>
>   immediate_neighborhood_urban(Person).

> [lives_in(Person,jefferson)] =>
>   immediate_neighborhood_rural(Person).

Suppose we know that

> lives_in(george,washington)

and

> lives_in(george,jefferson).

Is George's immediate neighborhood urban or rural? We can't tell. Maybe Washington is a county that falls almost entirely within the limits of some large city, but Jefferson is a small farming community on the edge of Washington County. Or maybe Jefferson County is a rural county and Washington is its urban county seat. (Compare Atlanta, Georgia and Elmer, New Jersey.) Of course, we see the conflict only because we know that someone's *immediate* neighborhood couldn't be both urban and rural.

> incompatible(immediate_neighborhood_urban(X)),
>   immediate_neighborhood_rural(X)).

Our problem is resolved if we are provided with either of the following two absolute rules.

> [lives_in(Person,washington)] -> lives_in(Person,jefferson).

> [lives_in(Person,jefferson)] -> lives_in(Person,washington).

These examples show two things. First, when we have two defeasible rules with contrary consequents, one may defeat the other. Of course, this will only happen if the antecedent of the defeating rule is itself defeasibly derivable. Second, this does *not* occur when we can derive the antecedent of one of the competing rules from the antecedent of the other using only absolute rules in our derivation. So we have some indication both of when a defeasible rule should be defeated and when it should not.

When we can resolve conflicts between competing defeasible rules using absolute rules to compare the antecedent conditions of the defeasible rules, we might say that the winning rule is *better informed* than the losing rule. This means that the antecedent of the winning rule can only be satisfied if we have enough information to conclude that the antecedent of the losing rule can also be satisfied, but it would be possible to know that the antecedent of the losing rule was satisfied without knowing that the antecedent of the winning rule was satisfied. Another way to put this would be to say that one of the antecedents, the antecedent of the losing rule, is a list of *relative consequences* (relative to the absolute rules available) of the other. We need to define this notion of relative consequence.

```
relative_consequence([],Premises).

relative_consequence([First|Rest],Premises) :-
    relative_consequence(First,Premises),
    relative_consequence(Rest,Premises).

relative_consequence(Literal,Premises) :-
    not list(Literal),
    member(Literal,Premises).

relative_consequence(Literal,Premises) :-
    not list(Literal),
    ((Antecedent -> Literal)),
    relative_consequence(Antecedent,Premises).
```

Our first two clauses provide the now-familiar conditions for a list of literals to be relative consequences of a list of premises. Our other two clauses tell us when an individual literal is a relative consequence of a list of conditionals. The recursive definition of our relative consequence relation looks exactly like our definition of absolute derivability except that it is relative to some specified set of literals instead of being based on all the literals available to us.

Our definition of the relative consequence relation uses the notion of a literal being a member of a list. We define this notion recursively as follows.

member(X,[X|Y]).

member(X,[Y|Z]) :-
    member(X,Z).

This looks right except for one difficulty. Suppose our literal is **lives_in(george, jefferson)** and our list of premises is **[lives_in(X, jefferson)]**. Strictly speaking our formula is not a member of our list of premises. The question is whether

member(lives_in(george,jefferson),
    [lives_in(Person,jefferson)])

will succeed in Prolog. Prolog reinterprets this question as: Is there a value for **Person** for which I can show the above? And of course there is. Our goal will match the first rule for the definition of **member** if we let **Person** be **george**, let **X** be **lives_in(george, jefferson)**, and let **Y** be the empty list. Prolog does all necessary universal instantiations for us when it tries the rules for **member**.

Now let's make our first try at specifying when a defeasible rule is defeated. First, there will have to be a competing rule, a rule with a consequent that is a contrary of the rule we are defeating. Second, all the antecedent conditions of this competing rule must themselves be defeasibly derivable. (We wouldn't care that penguins don't normally fly if we didn't have reason to think that Chilly is a penguin.) Finally, we have to determine that the rule we are defeating is not better informed than its competitor.

defeated(Literal,Antecedent_1) :-
    contrary(Literal,Opposite),
    ((Antecedent_2 => Opposite)),
    defeasible_derivability(Antecedent_2),
    not better_informed(Antecedent_1,Antecedent_2).

better_informed(Antecedent_1,Antecedent_2) :-
    relative_consequence(Antecedent_2,Antecedent_1),
    not relative_consequence(Antecedent_1,Antecedent_2).

DONALD NUTE

I think this analysis is correct, but there is one objection we should consider before we proceed. Why, we might ask, should we restrict ourselves to the use of absolute rules in showing that a literal is a relative consequence of some list of premises? Suppose, for example, we know that Squeeky is a mammal. Suppose we also know that absolutely all mammals with wings are bats. Given that we know Squeeky is a mammal and we know mammals with wings are bats, doesn't the information that Squeeky has wings bring along with it the information that Squeeky is a bat? Shouldn't 'Squeeky is a bat' be a consequence of 'Squeeky has wings' relative to our prior knowledge? I think not, at least in the sense of 'relative consequence' important to the comparison of the antecedents of competing defeasible rules.

Let's look at the example of Squeeky more carefully. We have the following rules and facts.

mammal(squeeky).

has_wings(squeeky).

[mammal(X), has_wings(X)] -> bat(X).

[bat(X)] -> mammal(X).

[mammal(X)] => neg flies(X).

[bat(X)] => flies(X).

Does Squeeky fly? Presumably, he does. And this is the answer we get using our current analysis of defeasible derivability. But what happens if we allow the use of the information that Squeeky has wings in deciding what is a relative consequence of a list of premises? Then we find that relative to the information given in our example, **bat(squeeky)** is a relative consequence of **[mammal(squeeky)]** and **mammal(squeeky)** is a relative consequence of **[bat(squeeky)]**. This means that the antecedent of neither of our defeasible rules is more informative than the other and, given our definition of **defeated**, both defeasible rules for Squeeky are defeated. Contrary to our intuitions, we could draw no conclusion.

I think this example tells us something about the structure of our scheme of defeasible principles. We often have one principle that takes precedence over another in some circumstances. We can only understand this order of precedence by comparing the rules and their antecedents independently of any other information available to us. The information given in absolute rules, however, has a special status in this regard. It is a part of our conceptual scheme in a way that simple facts are not. It represents unalterable connections between our concepts, connections that must already be taken into account when we form our defeasible rules and that implicitly give our rules the order of precedence they have. This is most obvious for absolute rules based on our systems of categorization and on definitions, less obvious for rules with other bases such as scientific laws and geographical regularities. Nevertheless, this account seems to fit the evidence of our everyday thought and speech.

MIGHT CONDITIONALS AND DEFEATERS

In English we sometimes object to a general principle or even a specific assertion by voicing an opposing *might* conditional. Suppose someone told you that rental property was a good investment. We might reformulate this principle as 'If you invest in rental property, you will enjoy a significant return.' You might object to this piece of advice by saying "If you invest in rental property, you might have trouble keeping the units rented, or you might get tenants who damage the property." Both of the possibilities proposed present obstacles to seeing a good return on your investment.

These might conditionals play a special role in our reasoning. Normally a conditional provides us with a justification for believing its consequent when its antecedent conditions are satisfied. Even defeasible rules play this role although they can be defeated by additional information. Might conditionals don't work in the same way. Look again at our example, 'If you invest in rental property, you might have trouble keeping the units rented.' Suppose you accept this conditional but you buy rental property anyway. Do you now conclude that you will have trouble keeping your rental units occupied? No, you don't. You continue to hope that you will not have this problem.

In Pollock (1976), an analysis of might conditionals as the denial of the corresponding ordinary conditionals is proposed. On this view, to assert the conditional 'If I strike this match, it might not light' is equivalent to *denying* the conditional 'If I strike this match, it will light.'

This shows that the role of the might conditional in English is to *prevent* an inference rather than to warrant one. Here the might conditional is intended to prevent you from inferring that the match will light if you strike it while the ordinary conditional expresses a license to make the same inference.

We do not need to concern ourselves with the details of the analysis of conditionals offered in Pollock (1976) and elsewhere or with the connection between recently developed conditional logics and defeasible reasoning to appreciate the importance of the might conditional for both. Defeasible rules may compete, but we may also want to express conditions where a defeasible principle is unreliable without committing ourselves to a contrary inference. We want the potential investor in rental property to realize that he might not profit from his investment, not to conclude that he will not.

I will call a principle that plays this purely negative role a **defeater**. I will represent defeaters in our object language with the operator ? → as in **[bird(X), sick(X)] ? → neg flies(X)** which expresses the warning that a sick bird might not fly. The only role such rules will play is that they may be substituted for a competing defeasible rule in showing that some other defeasible rule is defeated. This means that we add a second rule to our definition of the predicate **defeated**.

```
defeated(Literal,Antecedent_1) :-
    contrary(Literal,Opposite),
    ((Antecedent_2 ?-> Opposite)),
    defeasible_derivability(Antecedent_2),
    not better_informed(Antecedent_1,Antecedent_2).
```

While there are similarities between English might conditionals and defeaters, there is also an important difference. Simply because a defeasible rule is defeasible, there is always some justification for saying that the consequent of such a rule might not be true even when the antecedent is true. 'Rental property is a good investment' is a defeasible principle because you might lose your shirt investing in rental property. If there is always some justification for accepting the corresponding defeater whenever we also accept a defeasible rule, how will we ever arrive at any defeasible conclusions?

The answer to these concerns, of course, is that we must take care when we adopt a defeater. The defeasible rule is intended to represent the typical or normal case. The place of the defeater is to *identify* cases that are not typical or normal, not merely to emphasize that there are exceptions *of some kind or other*. This essential defeasibility is already expressed by the defeasible rule itself.

Putting these observations together, let's take another look at our real estate example. I say "If I invest in rental property, I will make a good profit." It is not enough for you to object by saying "If you invest in rental property, you might not make a profit." If you really want to deter me, you will need to say something like "If you invest in rental property and your renters damage your property, you might not make a profit." You might then solidify your objection with further information like "This is a college town, and college students are notoriously destructive tenants." Now you have given me specific cases where my defeasible principle might be defeated, and you have given me reason to think my case will be of this sort.

Recognizing the existence of defeaters and representing them in our program strengthens our analysis of defeasible reasoning.

CHAINING

English conditionals are not universally transitive. By this I mean that from English conditionals of the forms 'If *A*, then *B*' and 'If *B*, then *C*' we can not reliably infer the conditional 'If *A*, then *C*.' Several logics for nontransitive conditionals have been developed since the seminal Stalnaker (1968). But these logics do not solve a closely-related problem I will call *the chaining problem*.

We all agree that Herbert's health would improve if he would stop smoking. And we also agree that Herbert would be forced to stop smoking if he contracted emphysema. Having agreed to all this, suppose we learn that Herbert has contracted emphysema. Then we must conclude first that he stops smoking and second that his health improves. But this is absurd. Conditional logics prevent us from inferring the conditional 'If Herbert contracts emphysema, his health will improve,' but they do not prevent us from detaching the consequents of our two original conditionals one after the other. Our only alternative is to decide that one of our conditionals was, after all, false.

The situation is even more interesting when we consider general defeasible principles. In the normal case, the health of any heavy smoker who stops smoking will improve. Furthermore, anyone who develops emphysema will normally stop smoking. Now what happens if we learn that Herbert, a heavy smoker, has developed emphysema? We tentatively conclude that Herbert stops smoking, but we don't want to conclude that his health improves. Nor do we want to reject either of our defeasible rules, the equivalent I suppose of deciding a conditional is false. What we want to do instead is to defeat the second defeasible rule in Herbert's case and for anyone else who develops emphysema. More generally, we could add the defeater that the health of a person who contracts any serious disease might not improve. Add that emphysema is a serious disease and we are done.

Conditional logics may offer another way to avoid the chaining problem. While these are logics for nontransitive conditionals, they do usually include a restricted transitivity principle. From conditionals of the forms 'If A, then B' and 'If A and B, then C' we are allowed to infer 'If A, then C.' We could emulate this principle in our analysis of defeasible derivability.[2] To do this, we would need to remember which defeasibly derivable rules each defeasibly derivable formula depends on and the conditions of all those rules. In this example, we would need to remember that 'Herbert stops smoking' comes from the rule 'Normally anyone who contracts emphysema stops smoking' and depends on the fact that 'Herbert contracts emphysema.' Then we can't use 'Herbert stops smoking' to satisfy the condition of any other rule unless that rule also has 'Herbert develops emphysema' as a condition. This will prevent us from getting the wrong result in our example.

One problem with this approach is that some chains involve absolute rules as their first links. Normally Mother is happy when Baby isn't crying, but it is an absolute rule based on the physiology of babies that Baby doesn't cry when it is choking. Suppose Baby is choking. Then Baby isn't crying. This conclusion isn't tentative, didn't follow from any defeasible rule, and doesn't depend defeasibly on the fact that Baby is choking. The technique outlined in the last paragraph won't work in this case. We might try to extend our technique to absolute derivability, but I see no easy way to do this.

But there is another reason to reject this partial solution for the chaining problem. The simple fact is that chaining is usually reliable. Just as we use defeasible rules routinely unless we believe that we are faced with an exception to the rule, we also chain rules routinely unless we have reason to believe that a particular chain of reasoning is leading us to an absurd result. We know that a person's health doesn't improve when he or she contracts emphysema. And we know that mothers aren't normally happy when their babies choke. This information is what tells us that the chains of reasoning in our examples are unreliable, not the form of the arguments themselves.

The better way to handle chaining is to allow it to happen unless there is a fact, defeasible rule or defeater available to prevent it. Reliable chaining is the norm, and defeasible reasoning is committed to the norm. Just as we have exceptions to particular defeasible rules, we have exceptions to particular chains of rules. These should be handled in similar ways. This is the way they are handled in the analysis we have developed here.

SUMMARY OF THE ANALYSIS

We have put our analysis together a piece at a time. This is one advantage of developing an analysis as a Prolog program. We can write a few rules and run the program, knowing that there are special considerations that our analysis does not yet handle properly. We refrain from giving our program examples of this kind until we have added the extra rules these examples require. Meanwhile, we can run the program on other examples that we think our analysis can handle to see how successful our analysis is so far. But now we are at the end of this process and it is time to collect our results.

Let's begin by looking at the object language we have created for our analysis. We have only two kinds of formulae in this language: atomic formulae and their negations. Besides our formulae, we have three kinds of rules in our language and three different operators we can use to represent them. Absolute rules are represented using the operator $\rightarrow$, defeasible rules are represented using the operator $\Rightarrow$, and defeaters are represented using the operator $? \rightarrow$. The consequent of a rule is always a formula, and the antecedent of a rule is always a finite list of formulae taken conjunctively as conditions of the rule.

It is important that we realize the rules of our language are not formulae in the language itself. Instead, they are similar to inference rules in formal systems. They tell us that we are entitled to infer certain formulae from others under appropriate conditions. Defeaters tell us that we are not entitled to make such inferences if conditions are wrong in certain ways. Thus our object language is two-tiered, with literals as the bottom tier and rules and defeaters as the top tier.

Our definitions of the predicates **absolute_derivability** and **defeasible_derivability**, together with their subordinate predicates, provide the core of a proof theory for our formal language. We will say a formula in our object language is derivable in the appropriate sense from a set of formulae and a set of rules provided Prolog can show that it is, but we must add a few constraints to this test.

First, a Prolog database is always finite even though we could have infinite sets of formulae and infinite sets of rules. Unlike other kinds of reasoning, we can not hope for a compactness result for defeasible reasoning. Suppose we had an infinite set of defeaters all contradicting one defeasible rule we wanted to use. We would have to examine every one of these defeaters to determine whether it actually defeated our rule. That some finite subset of the defeaters fail to actually defeat our rule means nothing. Fortunately, real defeasible reasoning as performed by humans does not depend on infinitely large sets of rules. Since I am interested in an analysis of something people actually do, I am not embarrassed to have produced a proof theory with similar limitations. Our analysis of defeasible derivability can only tell us whether a formula is derivable from a finite set of formulae and rules since it would be impossible for Prolog to complete a derivation in the case I have described.

Second, the order that facts and rules are written into a Prolog database can affect what is or is not provable. We don't want order to make a difference, so we will say that a formula is derivable in the appropriate sense from some finite set of rules and facts if there is some sequence of this set from which Prolog can prove the derivability of the conclusion. The problem here is not that we get different results if we apply the rules in a different order. The problem is that we can write rules that will cause the Prolog inference engine to enter an infinite loop. This can happen before Prolog tries the rule that would allow it to reach a conclusion. This is the kind of failure we are trying to avoid by allowing different permutations of the list of rules and formulae.

Third, even finite databases can be very large, and some derivations may take longer than any machine will last. I do not count these as serious problems. We make our notion of derivability abstract in a further way by positing a *very* big machine that will last a *very* long time.

So we have an analysis of defeasible reasoning expressed in terms of a Prolog program. A formula is defeasibly derivable from some finite set of formulae and rules if our Prolog program can perform such a derivation from some perturbation of the set on a machine of suitable size. My reasons for thinking that this analysis is adequate are to be found in the discussions of examples of defeasible reasoning that motivate the different pieces of the program.

Finally, we list the complete program for defeasible reasoning. At the head of this listing we need a few lines that tell Prolog the order of precedence for the logical symbols of our object language.[3, 4]

```
init :-
    op(900,fx,neg),
    op(1100,xfy,->),
    op(1100,xfy,=>),
    op(1100,xfy,?->).

:- init.

absolute_derivability([]).

absolute_derivability([First|Rest]) :-
    absolute_derivability(First),
    absolute_derivability(Rest).

absolute_derivability(Literal) :-
    not list(Literal),
    Literal.

absolute_derivability(Literal) :-
    not list(Literal),
    ((Antecedent -> Literal)),
    absolute_derivability(Antecedent).

defeasible_derivability([]).

defeasible_derivability([First|Rest]) :-
    defeasible_derivability(First),
    defeasible_derivability(Rest).

defeasible_derivability(Literal) :-
    not list(Literal),
    absolute_derivability(Literal).
```

```
defeasible_derivability(Literal) :-
   not list(Literal),
   ((Antecedent -> Literal)),
   defeasible_derivability(Antecedent),
   not contradicted(Literal).

defeasible_derivability(Literal) :-
   not list(Literal),
   ((Antecedent => Literal)),
   defeasible_derivability(Antecedent),
   not contradicted(Literal),
   not defeated(Literal,Antecedent).

contradicted(Literal) :-
   contrary(Literal,Opposite),
   absolute_derivability(Opposite).

contrary(Literal,Opposite) :-
   negation(Literal,Opposite).

contrary(Literal,Opposite) :-
   incompatible(Literal,Opposite).

contrary(Literal,Opposite) :-
   incompatible(Opposite,Literal).

negation(neg Atomic_formula,Atomic_formula).

negation(Atomic_formula,neg Atomic_formula) :-
   not functor(Atomic_formula,neg,1).

defeated(Literal,Antecedent_1) :-
   contrary(Literal,Opposite),
   ((Antecedent_2 => Opposite)),
   defeasible_derivability(Antecedent_2),
   not better_informed(Antecedent_1,Antecedent_2).

defeated(Literal,Antecedent_1) :-
   contrary(Literal,Opposite),
   ((Antecedent_2 ?-> Opposite)),
   defeasible_derivability(Antecedent_2),
   not better_informed(Antecedent_1,Antecedent_2).
```

```
better_informed(Antecedent_1,Antecedent_2) :-
   relative_consequence(Antecedent_2,Antecedent_1),
   not relative_consequence(Antecedent_1,Antecedent_2).

relative_consequence([],Premises).

relative_consequence([First|Rest],Premises) :-
   relative_consequence(First,Premises),
   relative_consequence(Rest,Premises).

relative_consequence(Literal,Premises) :-
   not list(Literal),
   member(Literal,Premises).

relative_consequence(Literal,Premises) :-
   not list(Literal),
   ((Antecedent -> Literal)),
   relative_consequence(Antecedent,Premises).

list([]).

list([X|Y]).

member(A,[A|B]).

member(A,[B|C]) :-
   member(A,C).
```

EXPRESSIVE POWER AND MEANING OF THE DEFEASIBLE LANGUAGE

I am sure that there will be many objections to both the general method of philosophical analysis and the particular details of the analysis of defeasible reasoning offered here. I will not try to anticipate all objections, but there is one that I will discuss. One apparent weakness of my analysis of defeasible reasoning is the expressive poverty of the object language provided. I will show how this problem can be at least partially alleviated. My solution will depend on a particular view about what kind of semantics is appropriate for rules and their close kin.

The only formulae that appear in our object language are literals. Besides **neg**, we have no sentence connectives in the object language in the strict sense. Of course we have the three operators $\rightarrow$, $\Rightarrow$ and $? \rightarrow$ used to form rules, but these are best viewed as occuring at a higher level than the operator **neg**. In particular, we can not negate a rule. (A defeater comes close to representing the negation of a defeasible rule, but we don't have even this possibility for absolute rules.) How are we to represent the kinds of complex statements that can occur in English?

We do have a kind of conjunction in our rule formation. A rule can have several literals as antecedent conditions and these are treated conjunctively. And we can use two rules to do the work of a single rule with a disjunctive antecedent. '*A* if *B* or *C*' can be represented as '*A* if *B*' and '*A* if *C*.' We can not negate a conjunctive antecedent directly, but here again we can use two rules to do the same job. '*A* if not both *B* and *C*' says the same thing as '*A* if not *B*' and '*A* if not *C*' taken together. The real difficulty comes in trying to represent a disjunction outside the context of an antecedent.

Before we can make any suggestions about how to represent disjunction at the highest level, we need to think about what '*A* or *B*' means. Classically 'or' is taken to represent a truth functional connective. A sentence '*A* or *B*' is true unless both *A* and *B* are false. We have already abandoned the truth functional analysis of the English 'if-then' as it is used to express our conditional warrants or rules. Indeed, it is not clear that our account will fit a truth *conditional* semantics for English 'if-then.'[5] I propose that we also think of at least some occurrences of English 'or' as providing justification for inferences rather than as statements having truth values.

What does it mean for someone to have a belief that he or she would express by saying '*A* or *B*'? The classical view, as we have just noted, is that such a person believes that *A* or *B* is true without believing that A is true or that B is true. Instead of this view, I suggest that such a person is prepared to accept *A* on learning that *B* is false and is prepared to accept *B* on learning that *A* is false. He or she is committed to certain strategies for revising beliefs about atomic propositions without already holding distributive beliefs about these propositions. Accepting '*A* or *B*' is equivalent to accepting the two rules '*A* if not *B*' and '*B* if not *A*'.

This account of disjunctive belief has an interesting immediate consequent. The two rules that are supposed to replace the disjunction might be either absolute or defeasible. This means that there might be both an absolute and a defeasible 'or' in English if the proposal is correct. The analysis of 'or' is discussed at greater length in Nute and Covington (1986) and the *inferential* semantics for defeasible reasoning is developed in Nute (1986). The point I want to make here is that there are alternatives to the truth conditional semantics with which we are familiar and that far from the expressive poverty we seem to detect at first, the formulae and rules of our analysis may provide an expressive richness not available in classical systems.

ADVANTAGES OF THE METHOD AND THE ANALYSIS

I claim no theoretical advantages for the method of analysis I have used in this research, although I do think there are important practical benefits. As for the account of defeasible reasoning offered here, I believe it to have obvious advantages from the perspectives of the cognitive scientist, the artificial intelligence researcher, and the philosopher.

One advantage of using logic programming for philosophical analysis is that it extends the community of inquirers in two important ways. First, it provides a common ground for exchange between cognitive scientists, artificial intelligence researchers, and philosophers. Formal systems and formal semantics may seem too abstract to many cognitive scientists and AI researchers. They often prefer models of a more concrete kind, specifically models that actually *work* in some way comparable to the natural process being modeled. At the same time, the advent of logic programming languages like Prolog should make the use of computational models more attractive to philosopher logicians. Second, using logic programming allows anyone, even someone who does not understand the details of the analysis, to propose test cases for the analysis. I have found that this leads to much faster evolution of an analysis. Just as Prolog allows quick prototyping and testing of "expert systems," it also allows quick testing and revision of a piece of philosophical analysis.

The particular account of defeasible reasoning I have proposed is natural and fits our ordinary experience well. Compared to other accounts that depend on subjective confidence factors or degrees of belief, this account finds far greater support in our introspective experience. Simply put, we usually are not aware that we associate a confidence factor with our different beliefs and warrants. I do not deny that something like probabilities plays some role in our reasoning, but I don't think our experience supports the view that all of our reasoning should be explained in this way. A very common pattern of reasoning is captured by the idea of a defeasible rule that we rely on until we have specific reasons for believing it is defeated in a particular case. Because the account is very natural and fits our common experience well, it should be very attractive to philosophers and cognitive scientists.

Artificial intelligence researchers are not only interested in cognitive modeling. Many are also interested in applying techniques developed in the artificial intelligence labs to the solution of practical problems. The area of expert system development is an example of this effort. Here again our account of defeasible reasoning has promise. Because we are not consciously aware of any confidence values we place on our beliefs or warrants, it is difficult to justify any particular values that might be assigned if we try to use a probabilistic inference engine for an expert system. Systems built around probabilistic approaches are also difficult to maintain because we may have to adjust probabilities for existing rules in a system whenever we add new rules. Defeasible reasoning as I have characterized it offers an alternative approach.[6]

NOTES

[1] This research was conducted with the support of National Science Foundation Grant No. IST-8505586. Support from the Advanced Computational Methods Center, University of Georgia; the Department of Computing, Imperial College of Science and Technology, London University; and the Forschungstelle für natürlich-sprachliche Systeme, Universität Tübingen is also gratefully acknowledged.

   I am deeply indebted to Marvin Belzer and Dov Gabbay for many helpful discussions and penetrating criticisms. I also thank Michael Covington, Franz Guenthner, Michael Lewis, Clyde Melton, and Andre Vellino for their helpful comments.

[2] This approach is implemented in an earlier version of the analysis, both in a program called Prowis (for PROgramming With Subjunctives) and in a formal system called LDR1. For details, see Nute (1984) and Nute (1985).

[3] The program listed here is written to run using the Arity Prolog Interpreter. With a few changes, it will run in any Prolog that supports a syntax close to that found in Clocksin and Mellish (1981). In particular, the numbers used in most Prolog implementations for the precedence of operators cannot exceed 256.

   A program similar to the one listed, but supporting many more features, is described in Nute and Lewis (1986). This program, called d-Prolog (for defeasible Prolog) is available to run under both the Arity Prolog Interpreter and the Prolog 1 interpreter from Expert Systems International. A version of d-Prolog also runs under LM-Prolog on the Lambda from Lisp Machines Incorporated.

[4] A formal system and a formal semantics for defeasible reasoning can be found in Nute (1986). This formal system and formal semantics do not exactly correspond to the

analysis of defeasible reasoning presented here although they are very close. This raises an important question about the method described in this paper. Did I really have a formal system or a formal semantics already in mind as I developed my program? The answer to this question is a definite 'No!' There certainly would be nothing wrong with such a procedure, but it is not the way that the proposed analysis evolved.

My original plan was to build a theorem prover for some suitable conditional logic and use this to represent defeasible reasoning. This plan had to be abandoned almost immediately since all the conditional logics I know about allow you to detach the consequent of a conditional when the antecedent is satisfied regardless of available evidence to the contrary. The conditionals of contemporary conditional logics are falsifiable, but they are not defeasible.

At this point, I began to develop a program that mirrored my observations of the way people resolve conflicts involving the use of rules of thumb like 'Birds fly.' This work was all done in Prolog. The initial version was very simple, but it captured for the simplest cases the basic idea that when two rules conflict we use the rule that is better informed. A complex method for handling bad chains of rules was developed then later abandoned in favor of the treatment included here. I also changed the ways that the antecedents for competing rules could be satisfied during the process of trying to defeat a rule. At some point in the process, the importance of defeaters was noticed and they were added to the program. Throughout the development of the analysis, the only test of adequacy was direct comparison with ordinary use and intuition as represented in the opinions of readily available subjects (usually my colleagues and myself).

[5] It would be strange to say that a conditional were true if its antecedent were true and its conclusion were false. But to say that a rule is defeasible is just to admit the possibility that, at least in some instances, its antecedent will be true and its conclusion false. Rather than trying to discover truth conditions for such rules, it makes more sense to try to provide conditions under which we would be justified in accepting such a rule. This is one kind of semantics that we might try to develop for defeasible rules, although I will not pursue this line of investigation here. Pollock (1974) and Stalnaker (1984) contain relevant discussions.

[6] Other attempts to model non-monotonic reasoning without the use of numerical confidence factors or probabilities are reported in McDermott and Doyle (1980), Reiter (1980), McCarthy (1980), Glymour and Thomason (1984), and elsewhere. Some of these are discussed and compared with the present analysis in Nute (1985) and Nute (1986).

REFERENCES

Clocksin, W. F., land C. S. Mellish: 1981, *Programming in Prolog*, Springer-Verlag, Berlin.

Covington, Michael, Donald Nute and Andre Vellino: 1987, *Prolog Programming Techniques*, Scott-Foresman, Glenview, Illinois. (To appear.)

Fulton, Ken: 1984, *The Light-Hearted Astronomer*, AstroMedia. Milwaukee.

Glymour, Clark, and Richmond Thomason: 1984, 'Default reasoning and the logic of theory perturbation', *Proceedings of the AAAI Workshop on Non-Monotonic Reasoning*, October 17—19, New Paltz, New York.

McCarthy, John: 1980, 'Circumscription — a form of non-monotonic reasoning', *Artificial Intelligence* **13**, 27—39.

McDermott, Drew, and Jon Doyle: 1980, 'Non-monotonic logic I', *Artificial Intelligence* **13**, 41—72.

Nute, Donald: 1984, 'Non-monotonic reasoning and conditionals', ACMC Research Report 01-0002, University of Georgia, Athens.

Nute, Donald: 1985, 'Non-monotonic logic based on conditional logic', ACMC Research Report 01-0007, University of Georgia, Athens.

Nute, Donald: 1986, 'LDR: a logic for defeasible reasoning', ACMC Research Report 01-0013, University of Georgia, Athens. A short version of this report will appear in *Proceedings of the 20th Hawaiian International Conference on System Science*, University of Hawaii, 1987.

Nute, Donald and Michael Covington: 1986, 'Implicature, disjunction, and non-monotonic logic', ACMC Research Report 01-0015, University of Georgia.

Nute, Donald and Michael Lewis: 1986, 'd-Prolog: a users manual', ACMC Research Report 01-0017, University of Georgia, Athens.

Pollock, John: 1974, *Knowledge and Justification*, Princeton University Press, Princeton.

Pollock, John: 1976, *Subjunctive Reasoning*, D. Reidel, Dordrecht, Holland.

Reiter, Raymond: 1980, 'A logic for default reasoning', *Artificial Intelligence* **13**, 81—132.

Stalnaker, Robert: 1968. 'A theory of conditionals.' *American Philosophical Quarterly*, monograph series 2 (edited by Nicholas Rescher), 98—112.

Stalnaker, Robert: 1984, *Inquiry*, MIT, Cambridge.

Toulmin, Steven, Richard Rieke and Allan Janik: 1984, *An Introduction to Reasoning*, Macmillan, New York.

Department of Philosophy
Advanced Computational Methods Center
University of Georgia
Athens, GA 30602, U.S.A.

TERRY L. RANKIN

WHEN IS REASONING NONMONOTONIC?

Recent advances in Artificial Intelligence (AI) emphasize the crucial importance of nonmonotonic reasoning in any adequate scheme of knowledge representation.[1] As Nute (1984) observes, humans notoriously rely upon nonmonotonic reasoning, and any "automated reasoning system should also reason nonmonotonically in a way which people can easily understand". Conclusions that can be inferred on the basis of a given set of premises may often be withdrawn or even overruled when new evidence is provided in the form of additional premises. Told only that a match has been struck, for example, humans will typically infer that that match did light and burn. But if they are also told that the match in question was wet when struck, they will tend to revise their inference and conclude that the match did not light. If told further that the match was wet but coated in paraffin when struck, most (reasonable) humans would infer that the match did light and burn after all, thus revising their inference once more — unless, of course, told no oxygen was present when the match was struck, in which case still another revision is called for. And this is precisely the character of nonmonotonic reasoning, as Nute suggests, i.e., "people draw conclusions based on incomplete information, but these conclusions are open to revision as better information becomes available".[2]

But is reasoning of this kind deductive or inductive? What of reasoning in which warranted conclusions do not vary no matter how many further premises are obtained: is it deductive or inductive? Just how does the distinction between deductive and inductive modes of inference relate to the distinction between monotonic and nonmonotonic reasoning? Is there an exact correspondence between them or are they only incidentally found together? The analysis and examples which follow will attempt to show that an argument is deductively valid only if it is monotonic, and similarly that an argument is inductively proper only if it is nonmonotonic. This correspondence appears to be one of direct *parallelism*, therefore, insofar as any deductively valid inference will be monotonic, just as any inductively proper inference will be nonmonotonic. It will be important to observe, however, that, while all

289

James H. Fetzer (ed.), Aspects of Artificial Intelligence, 289—308.
© 1988 *by Kluwer Academic Publishers.*

monotonic inferences are *deductively valid*, not all nonmonotonic inferences are *inductively proper*.[3]

DEDUCTION AND INDUCTION

The distinguishing features of valid deductive arguments, of course, are comparatively well-known and widely recognized, and they can be relatively easily summarized. In Fetzer (1981) and (1984), for example, any deductive argument is characterized as being *valid* just in case (a) its conclusion could not be false if all its premises were true; (b) its conclusion contains no more content than is already provided in its premises; and (c) the addition of further premises can neither strengthen nor weaken this argument, which is already maximally strong". These criteria identify valid deductive inferences as being essentially (a) demonstrative, (b) nonampliative, and (c) additive. Thus, a deductive argument is *valid* if and only if it satisfies these criteria, and it is *sound* just in case it is both valid and its premises are true. Fetzer further identifies the primary aim, purpose, or goal of deductive inference as being that of truth-preservation.[4]

By contrast, the distinguishing features of proper inductive inference are not generally agreed upon. Fetzer (1981) and (1984), nevertheless, also suggest that the basic features of induction can be partially identified, at least to the extent that the above criteria for deduction are explicit complements of corresponding criteria for induction. Hence, an inductive argument will be (what Fetzer calls) *proper* only if (a′) its conclusion could be false even though all its premises were true, (b′) its conclusion contains more content than is provided in its premises, and (c′) the addition of further premises could either strength or weaken this argument — even to the extent of converting an (originally) inductive argument into a (subsequently) deductive argument. These criteria reflect the general assumption that inductive inferences are supposed to be knowledge-extending, where *proper* inductive arguments are (a′) nondemonstrative, (b′) ampliative, and (c′) nonadditive, while *correct* inductive arguments are both proper and have true premises as well.

Notice, however, that (a′), (b′), and (c′) only identify necessary but not sufficient conditions for an inference to be inductively acceptable. Any argument intended to be deductive that is either nondemonstrative, ampliative, or nonadditive — and therefore *invalid* — satisfies these conditions. Even if an argument is for these reasons regarded as being inductive, it remains to be shown that that argument is not fallacious, since sufficient conditions for inductive propriety are required in addition to the necessary conditions identified by (a'), (b'), and (c'). Fetzer (1984) thus characterizes this problem as follows:

The principal difficulty, . . . , is distinguishing inductive reasoning from inferential fallacies, for innumerable fallacious arguments seem capable of satisfying the conditions of being (i) ampliative, (ii) nondemonstrative, and (iii) nonadditive. Even if the corresponding conditions are sufficient to identify (acceptable) deductive principles of inference, (i), (ii), and (iii) are not sufficient to identify (acceptable) principles of inductive inference. If we adopt the standard terminology for deductive arguments possessing the demonstrative property as "valid" arguments, where those that are both valid and have true premises are also "sound," then inductive arguments possessing the ampliative property may be referred to as "proper" and those that are both proper and have true premises as "correct"; yet there is no generally accepted account of the conditions for an argument to be proper.[5]

Perhaps an example may help to illustrate these differences between deductive and inductive modes of inference. Suppose that one match after another is removed from a box of matches and that each match is struck and does indeed light and burn. If some number, say the first twenty-five, of those matches has lit and burned when struck, one might be tempted to infer that the next (or the twenty-sixth) match selected from the box will also light and burn if struck. This inference will be deductively invalid, clearly, since the conclusion might turn out to be false even though each of the premises is true: i.e., each one of the first twenty-five matches might actually light and burn when struck, but the next (twenty-sixth) match might fail to light or burn when struck, for any of a number of reasons, e.g., depletion of the available oxygen, that particular match might have no head, and so on. Such an argument would be nondemonstrative, ampliative, and nonadditive, thus failing to satisfy the criteria for a valid deductive inference, but apparently fulfilling the criteria for acceptable induction, at least to the extent that those criteria specify *necessary* conditions of proper inductive inference. Yet the inference in question still might not qualify as being inductively proper insofar as some further *sufficient* conditions for proper inductive inference remain unspecified and possibly unfulfilled.

As Fetzer (1981) also points out, various attempts have been made to identify the sufficient condition(s) for an inductive inference to qualify as being proper: one approach has sought to combine the ampliative aspect of induction with the demonstrative function of deduction, for example, in an attempt to generate a class of 'ampliative-

demonstrative' inferences and a set of rules for regulating those inferences. Fetzer rejects this approach — and rightly so — insofar as "rules
of inference that satisfy one of these conditions . . . do so at the expense
of the other, which dictates the result that this contradictory conception
of the inductive program cannot possibly succeed". With respect to
scientific knowledge, Reichenbach (1949) and Salmon (1967), in particular, adopt an alternative conception of the aim, purpose, or goal of
induction, namely, "to ascertain the limiting frequencies with which
different attributes occur within different reference classes during the
course of the world's history . . . ." According to Fetzer, this approach
also confronts problems of definitional relevance and theoretical significance to the extent to which actual scientific practice fails to conform to
this interpretation. The underlying issues are summarized by Fetzer as
follows:

Not the least of the reasons for objecting to this conception, . . . , is that actual scientific
practice appears to pursue the discovery of scientific laws rather than ascertaining
limiting frequencies, to search for principles of prediction for the past and the present
as well as for the distant future, and to value explanations for "the single case" in
addition to those accessible for long and short runs. The point, let me emphasize, is not
that the case for one conception rather than the other is clear-cut and beyond debate,
but rather that there is an issue here which requires explicit attention and careful
argumentation for its tentative resolution, namely: *the problem of defining the inductive
program.* And insofar as there appear to be objective standards of logical consistency,
of definitional relevance, and of theoretical significance which are applicable to such a
dispute, the defense of some specific conception of the inductive program is an
indispensable ingredient of attempts to provide an adequate justification for induction,
an aspect we may refer to as *the process of exoneration.*[6]

Despite the emphasis here upon the scientific desideratum of an
adequate justification of the program of induction, it should be clear,
nonetheless, that the example provided earlier — of matches that light
and burn (or not) when struck — illustrates the problem of securing an
appropriate conception of inductive inference within ordinary as well as
within scientific contexts. And it should also be clear that the criteria of
deduction and of induction recommended by Fetzer and outlined above
establish necessary and sufficient conditions for deductive validity and
necessary conditions for inductive propriety. What remains to be
considered, therefore, is the manner in which these criteria can be
applied in distinguishing between monotonic and nonmonotonic inferences and arguments together with their validity or invalidity (if deductive) or their propriety or impropriety (if inductive).

MONOTONICITY AND NONMONOTONICITY

Let us begin with the following straightforward example:

(I)   All bachelors are unmarried.
        John is a bachelor.
        _____
        John is unmarried.

Obviously, this is a valid deductive inference: if the premises were true, then the conclusion could not be false, on pain of contradiction. Moreover, the claim that 'John is unmarried' contains no information that is not already conveyed — implicitly or explicitly — by the premises that 'All bachelors are unmarried' and that 'John is a bachelor'. The introduction of further information about John will in no way compromise the validity of any inference from those premises to the conclusion that 'John is unmarried'. That is, the inference in question is explicitly demonstrative, nonampliative, and additive, and it is therefore deductively valid. If the individual named by 'John' in this example is in fact a bachelor, then we may validly infer that that individual is in fact unmarried, regardless of any further information about that individual that might or might not come to light. It is this property, in particular, that illuminates the *monotonicity* of the inference, i.e., that the same conclusion could be reached using any other set of premises that contained the two premises, namely, 'All bachelors are unmarried' and 'John is a bachelor'.[7]

This example, of course, tacitly appeals to logical, analytic, or tautological truth by definition, insofar as "a bachelor" is *defined* as being an unmarried person. Precisely how such a definiendum is so related to such a definiens is not at issue here; indeed, our general premise could be contingent without affecting the validity of our example. Suppose a slightly expanded example is used:

(II)   If John is still married to Mary, then he has been married for
        ten years.
        John is still married to Mary.
        _____
        John has been married for ten years.

We may assume it is not logically, analytically, or tautologically true that John's marriage to Mary is a marriage of at least ten years' duration, nor is it true by definition that John is in fact married to

Mary. Nevertheless, if these claims are given as premises, then once again our conclusion follows from them validly, i.e., it could not be false so long as these premises were true; it recapitulates part or all of the content of these premises; and the introduction of further information into an expanded premise set would not alter the validity of this specific inference.

This example is a typical instance where the well-known deductive rule of *modus ponens* can be applied. Clearly, the additive feature of valid deductions surfaces as the primary grounds for classifying such an inference as monotonic. On the other hand, consider the following as a further extention of our earlier examples (I) and (II):

(III)   John was married to Mary for five years and then they divorced.

John was married to Jane for five years and then they divorced.

John was married to Joan for five years and then they divorced.

John was married to Doris yesterday.

John will be married to Doris for five years and then they will divorce.

Now this obviously is *not* a valid deductive argument. Even if each of its premises were true, its conclusion could still turn out to be false, so it represents a *non*demonstrative inference. And the conclusion clearly contains more information than is given in its premises, insofar as it predicts the occurrence of a particular outcome on the basis of past events, which renders it *ampliative* with respect to those prior events. And finally, there are many premises which could, if added to the given premises, undermine the evidential strength of the argument as it is stated above: John or Doris might not live another five years or another five days; John might have vowed never to endure another divorce proceeding; or, in fact, John and Doris might remain married happily ever after, as the story unfolds.[8]

The salient point is that this version of our example now satisfies the necessary conditions for a proper inductive inference, insofar as it is nondemonstrative, ampliative, and nonadditive. Whether or not it satisfies the sufficient conditions of proper inductive argument, however, depends upon the specific criteria of evidential support one might adopt for inductions. Criteria of this sort would specify the nature and extent of evidence that must be given in a set of premises in order to warrant the sort of ampliative conclusion illustrated by our example to some specific degree.

In this example, the *non*additive aspect of our inference similarly reflects the *non*monotonicity that is characteristic of inductive argumentation, much as the additive aspect of the prior examples emphasized the monotonicity of deductive argumentation. It thus appears to be the case that an inference will be deductively valid if and only if it is monotonic; but an inference that is nonmonotonic may be either inductively proper or — inductively or deductively — fallacious, depending upon the appropriate conditions of inductive propriety and evidential support and the purpose it was intended to serve. Clearly, inductive arguments may be deductively invalid without being *therefore* fallacious, by contrast with arguments that are *intended* to be valid but fail to be, which are always fallacious.

FORMAL SYSTEMS

In order to formally explore the characteristics of monotonic inference, let us adopt the following definition:

> Φ follows from K *monotonically* =df If a proposition Φ is either syntactically derivable from, or semantically entailed by, some set of propositions K, then that proposition Φ will also be syntactically derivable from, or semantically entailed by, any set of propositions containing K.

Now in Nute (1981), formal systems are formally defined as being inherently deductive and monotonic. Nute explicitly incorporates the property of monotonicity into his definition of a (deductive) *formal system* as follows, where F is the set of formulae of the language L, K and Γ are subsets of F, and $D(K)$ and $D(\Gamma)$ are sets of formulae derivable from K and from Γ, respectively:

M1. $K \subseteq \Gamma \subseteq F \rightarrow D(K) \subseteq D(\Gamma)$; i.e., if K is contained in (or identical to) Γ, and Γ is contained in (or identical to) F, then the syntactically derivable consequences of K are contained in (or identical to) the syntactically derivable consequences of Γ.

On Nute's view, therefore, *any* deductive formal system must be monotonic, entailing the result that a deductive inference is valid only if it is monotonic — or, in other words, implying that *non*monotonic inference is either nondeductive or else invalid — while permitting the possibility that an argument might be inductively proper if it is not monotonic, as already suggested above. While Nute (1981) does not elaborate the formal characteristics of induction, it is nonetheless clear that any argument will be deductively valid on this view only if it is monotonic.

The validity of monotonic inference may be determined either syntactically or semantically, in fact, although M1 specifically refers to the feature of syntactic derivability. In particular, for every world in which all of the propositions contained in Γ are true, all of the propositions contained in K must also be true, since $K \subseteq \Gamma$; and, if Φ is true in all worlds in which every proposition contained in K is true (i.e., K semantically entails Φ), then Φ must be true in every world in which all propositions contained in Γ are true. The semantic counterpart of a monotonic syntactic derivability relation in which the rules or the axioms of a formal system will validate monotonic inferences (apart from any considerations of specific atomic or molecular truth values) can thus be obtained using semantics accommodated by a possible worlds model of that formal system. Since Nute's conception of formal systems will validate every monotonic inference as being *deductively* valid, and since nonmonotonic inferences cannot be deductively validated, as we have seen, it follows that formal systems in which nonmonotonic inferences are acceptable must be *non*deductive systems, necessarily.[10]

Recall our initial example involving matches being struck and then either lighting and burning or not, which we will now reformulate more explicitly as follows:

(IV)   a. Match no. 1 lit and burned when struck.
        b. Match no. 2 lit and burned when struck.
        c. Match no. 3 lit and burned when struck.
        .
        .
        .
        y. Match no. 25 lit and burned when struck.
        z. Match no. 26 is now being struck.
        ───
        Match no. 26 will light and burn.

As we initially observed, this inference ought to be acceptable so long as the same conditions are in effect when match no. 26 is struck, as expressed in (IV-z): no. 26 must resemble nos. 1—25 in in the appropriate ways (wood or paper stem, head that ignites under friction against certain surfaces, etc.), sufficient oxygen must be present for combustion to occur, sufficient striking force must also be applied to match no. 26 (as applied to nos. 1—25), humidity and wind conditions must remain very nearly the same in (IV-z) as they were in (IV-a) through (IV-y), and so on. The acceptability of the inference in question also depends upon the assumption, moreover, that the prior twenty-five outcomes under these conditions constitute sufficient evidential support to infer that that same outcome will also occur on the twenty-sixth trial, *under those same conditions*. Hence, there appear to be *two* kinds of conditions determining the acceptability of the inference in question: (a) relevant condition specifications that describe the test conditions; and, (b) criteria of evidential support relating the given premises to the stated conclusion.

Notice, especially, that the "specification of relevant conditions" identifies distinctively *nonlogical* criteria for accepting the truth of the premises in question, whereas the criteria of evidential support identify distinctively *logical* conditions for accepting the argument in question. Being nonlogical, the specification of relevant conditions can only be regarded as being *neither* deductive *nor* inductive in kind; as criteria of evidential support, however, the genuinely logical conditions that should apply are clearly *inductive* in character. The distinction between the two kinds of conditions that must be used to determine the acceptability of the argument expressed in (IV) is even more clearly revealed when the argument is modified, allowing the relevant properties of each trial to vary as follows:

(V)   a. Match no. 1 lit and burned when struck.

          b. Match no. 2 had no head, so it did *not* light and burn when struck.

          c. Match no. 3 had a head, so it did light and burn when struck.

          d. Match no. 4 had a head, but it was wet, so it did *not* light and burn when struck.

          e. Match no. 5 had a head and was dry, so it did light and burn when struck.

(V)   f. Match no. 6 had a head and was dry, but the striking
          surface has worn down, so it did *not* light and burn
          when struck.

            .

            .

            .

       y. Match no. 26 has a head and is dry, and a new striking
          surface is available.
       z. Match no. 26 is now being struck.
       ───

       Match no. 26 will light and burn.

What one *wants* to say, of course, is that 'All other things equal, a
match will light and burn when struck'. But as these examples vividly
reveal, inferences of this kind are simply not that straightforward.
Whenever unspecified relevant conditions are included (e.g., by means
of *ceteris paribus* clauses) as nonlogical grounds upon which the
acceptability of an inference depends, that inference will be *either*
deductively or inductively fallacious, *or else* inductively proper. In no
case will it be deductively valid, insofar as its acceptability relies upon
the *nonlogical* considerations discounted in the *ceteris paribus* conditions! Whether such an inference turns out to be inductively proper or
not, moreover, thus depends upon the criteria of evidential support
identified as a set of sufficient *logical* conditions for its acceptability as
an inference which is clearly (a′) nondemonstrative, (b′) ampliative,
and (c′) nonadditive (where these conditions, as discussed earlier, are
necessary features of proper induction). *Ceteris paribus* clauses thus
can fill in for unspecified relevant conditions, but this involves an
ampliative assumption which rules out the possibility of deductive
validity, although it does leave open the question of inductive propriety.
    Somewhat surprisingly, Gabbay (1982), Nute (1984), and Nute and
Gabbay (1984) attempt to introduce *non*monotonic inferences into
deductive formal systems by suggesting that subjunctive conditionals
can be used to accommodate them. As a framework for *non*monotonic
inference, Nute (1984) recommends a (partial) axiomatization of Lewis'
(1973) conditional logic VW, which he has implemented in micro-
PROLOG on an IBM Personal Computer. Elsewhere, however, Nute
identifies VW as a conditional logic which satisfies the criteria of formal
systems, including M1 above: in Nute (1981), for example, the same

axiomatization of VW establishes it as a formal system of conditional logic whose derivability relation exactly conforms to M1 above![11] In conversations involving examples resembling (IV) and (V), Nute invariably incorporates the kinds of *ceteris paribus* clauses just discussed, as in the following:[12]

(IV) a. All other things equal, if a match were struck it would light.
   b. All other things equal, if a wet match were struck it would *not* light.
   c. All other things equal, if a wet match coated in paraffin were struck it would light.
   d. All other things equal, if a wet paraffin-coated match were struck in a vacuum, it would *not* light.

   .
   .
   .
   .

These examples are intended by Nute to illustrate his view that a conditional such as "if a match were struck then it would light" might have to be revised or even withdrawn as new premises are introduced which would *prevent* the match from lighting if it were struck — or, as in the case of the paraffin coating, ensuring that it would light if it were struck even though it were wet. But the real effect of the disclaimer, i.e., "All other things equal", is to completely *discount* all those premises which have not yet been added to the initial set. The crucial point is that those discounted premises might in fact be relevant to the acceptability of the inference in question: as we have seen, any *ceteris paribus* conditions of this sort can only serve to conceal the underlying problem as to whether that inference is either deductively or inductively fallacious, or else inductively proper according to criteria for evidential support that specify sufficient condition(s) of inductive acceptance. While crucial distinctions between subjunctive and material conditionals and between conditionals and inferences *per se* should obviously not be conflated, they are equally (if differently) vulnerable to this ambiguity with respect to the acceptability of inferences that involve *ceteris paribus* disclaimers.

Separately, Gabbay (1982) recommends that the nonmonotonic logics constructed by McDermott and Doyle (1980) and by McCarthy

(1980), for example, can be significantly improved through the use of intuitionistic logic based (once again) upon subjunctive conditionals. Gabbay suggests that a subjunctive conditional such as "$A > B$" can be interpreted as "B is expected on the basis of A", and proposes its use in place of the unary operator "M" in the logic advanced by McDermott and Doyle (where "Mp" is read as "p is consistent with what we currently know"). Gabbay also suggests that McCarthy's logic of circumscription is (partially, at least) contained in the logic he proposes and claims further that his intuitionistic analysis of subjunctive conditionals closely resembles the default reasoning system developed by Reiter (1980).[13]

Yet each of these analyses apparently involves an (implicit or explicit) appeal to nonlogical *ceteris paribus* conditions which, as we have already discovered, completely obscure the genuine problem to be solved. Insofar as "consistency with what we currently know", for example, or "expecting B on the basis of knowing A", moreover, can both be said to completely *discount* the relevance of what is *not* currently known or of what is *not* currently being considered, to that extent neither the unary "M" operator nor the binary subjunctive could be said to address the underlying problem of disclosing the deductive or the inductive principles involved, much less the inductive propriety of the inference in question.

In collaboration, Nute and Gabbay (1984) recommend that an alternative to M1 can be incorporated into a broadened definition of deductive formal systems which would accommodate nonmonotonic inferences. Their proposed alternative can be stated as follows, using the same notation appearing in M1 above:[14]

NM1. $(\Phi \& \Theta) \in D(K) \rightarrow \Theta \in D(K \cup \{\Phi\})$, and, $(\Phi \& \Theta) \in D(K) \rightarrow \Phi \in D(K \cup \{\Theta\})$; i.e., if the conjunction of two propositions is a member of the syntactically derivable consequences of some set of formulae K, then either of those two propositions can be derived from a set formed by uniting K with the other proposition.

But surely the question can be raised as to whether or not NM1 is a principle of *non*monotonicity at all: does it explicitly violate the definition of monotonicity stated above? Notice, especially, that $K \cup \{\Phi\}$ and $K \cup \{\Theta\}$ are not just *any* supersets of K; they are *special cases* of supersets of K, namely, those that contain K and one of its syntactically derivable consequences. But if Φ & Θ is derivable from K, they cannot be contradictory (i.e., neither $\Phi \leftrightarrow \neg\Theta$ nor $\Theta \leftrightarrow \neg\Phi$), unless K is inconsistent. If Φ and Θ are not contingently related (i.e., neither $\Phi \rightarrow \Theta$ nor $\Theta \rightarrow \Phi$), then neither Φ nor Θ has any *relevance* to the kinds of inferences under consideration: if neither Φ nor Θ describes the properties of matches or the conditions under which a match would light and burn, while K (either partially or completely) describes a reference class of matches under test conditions for their being struck, then the inclusion of Φ or Θ in K can have no effect upon the acceptability of inferring an occurrence of the outcome under consideration, namely, lighting and burning. And finally, even if $\Phi \leftrightarrow \Theta$ or $\Phi \rightarrow \Theta$ or $\Theta \rightarrow \Phi$ *and* $K \vdash (\Phi$ & $\Theta)$, then NM1 only reveals the *nonampliative* aspect, in particular, of *deductive* inference: material equivalence or contingency involving Φ and Θ only emphasizes the (implicit or explicit) recapitulation which characterizes deductive inference, namely, that the information contained in Φ or in Θ is already contained in K as a premise set.

Odd as it might seem, however, both Nute and Gabbay have endorsed the view that NM1 can only ensure that "if a pair of propositions is derivable from some set K, then either member of that pair can be derived from K whenever K is expanded to include the other member" and that "*all* of the relevant premises can also be derived". Yet both of them regard NM1 as an acceptable principle of *nonmonotonic* inference! What NM1 does ensure is the *specificity* of any premise set involved in inferences of this kind, i.e., that the premise set must contain all premises that are actually *relevant* to the (deductive) inference in question. But this is quite different from establishing the *nonmonotonicity* of an (inductive) inference, as we shall see.

SPECIFICITY AND NONMONOTONICITY

Nute's definition of a formal system is too restrictive to accommodate nonmonotonic inference, clearly, since condition M1 of that definition imposes monotonicity and thus fails to provide for induction. As suggested by Fetzer's characterization of the various modes of inference, Nute's definition adequately captures the desideratum of monotonicity for deductive formal systems, but the equally desirable nonmonotonicity of inductive inference is not thereby established. Nor can NM1 succeed as an alternative to M1 for accommodating 'nonmonotonic deduction', as we have seen. A fully comprehensive conception of formal systems should support both deduction and induction as distinct and complementary modes of inference, ensuring their monotonicity and nonmonotonicity, respectively.

The most promising solution to the problem of monotonicity seems to be Fetzer's implicit position (with respect to additivity and nonadditivity as features of inference) that deductive inference is inherently monotonic, as it should be, and that induction is inherently nonmonotonic, as it should be. On this view, examples (III)–(V) are best seen as being either inductive (in which case their deductive invalidity arises from their nonmonotonicity), or deductive (in which case their invalidity stems from a failure to adequately fulfill *The Requirement of Maximal Specificity* shown below).[15] From this point of view, the goal of 'nonmonotonic deduction' is neither desirable nor attainable, and it should not be included in the definition of "formal systems". Hence, in order properly to accommodate both deductive and inductive modes of inference, formal systems should encompass distinct classes of monotonic and of nonmonotonic inference. Such a system would specify complementary rules of inference (appropriately monotonic or nonmonotonic in kind) for each of those classes, respectively. Consideration of the situation encountered with respect to scientific reasoning may further illuminate these issues.

For deductive arguments involving scientific conditionals, in particular, Fetzer (1981) introduces his conception of *The Requirement of Maximal Specificity* (*RMS*, for short), which clarifies the significance of *ceteris paribus* clauses that may occur whenever all nomically relevant conditions have not been specified. *RMS* also reveals that NM1 above is itself a kind of specificity requirement, insofar as NM1 is equivalent to the following formulations — themselves equivalent — of *RMS* itself:

> *The Requirement of Maximal Specificity*: If a nomically relevant predicate is added to the reference class description of a scientific conditional S which is true in L^*, then the resulting sentence S^* is such that either S^* is no longer true in L^* (by virtue of the fact that its antecedent is now self-contradictory), or S^* is logically equivalent to S in L^* (by virtue of the fact that that predicate was already entailed by the antecedent of S).

An alternative but equivalent formulation of this requirement (which some might find more intuitively appealing), moreover, may be advanced as follows: if 'p' is a true scientific conditional, 'K' is the reference class description of 'p' and 'F' is any predicate which is nomically relevant to the truth of 'p', then either $(x)(t)(Kxt \rightarrow Fxt)$ or $(x)(t)(Kxt \rightarrow -Fxt)$ is a logical truth; that is, for every nomically relevant predicate 'F', relative to 'p', either it or its negation must be entailed by the reference class description of 'p' if 'p' is true.[16]

RMS explicitly requires that the antecedent of any scientific conditional that happens to be true must already contain *all* of the nomically relevant predicates and conditions that will be required to infer that a sentence describing the occurrence of a specific outcome event will be true, given a description of test conditions and a reference class description as that antecedent.

RMS is a truth-condition for lawlike sentences, which are subjunctive generalizations, and for nomological conditionals, which are instantiations of lawlike sentences. Taken together, these make up the class of (what Fetzer (1981) refers to as) scientific conditionals. By contrast, validity and invalidity are properties of deductive arguments that may or may not contain scientific conditionals. When the validity of an argument depends upon one or more conditional premises, it can be sound only when those premises are true. The crucial point to observe is that a deductive argument could contain scientitfic conditionals that did not conform to *RMS* and still remain valid as such, but the argument could not be sound, insofar as any such conditionals that fail to satisfy *RMS* could not possibly be true.

RMS ensures that the addition of further predicates to scientific conditionals that occur in valid deductive arguments results in either (a) an inconsistent antecedent, in which case the argument will remain deductively (but vacuously) valid, or (b) an antecedent that is logically equivalent to the original, in which case the argument retains the deductive validity it already displayed. Hence, the definition of monotonicity provided at the beginning of the last section clearly applies, and the arguments in question will be explicitly *monotonic*: the maximal specificity of the scientific conditionals involved supports that monotonicity, in fact, and effectively rules out the possibility of 'nonmonotonic deduction' altogether. Whenever *RMS* is not satisfied, of course, the problem remains of ascertaining whether the argument is (deductively or inductively) fallacious or inductively proper, which (as we have repeatedly emphasized) will depend upon the criteria of evidential support that ought to be applied, the rules of inductive inference that have been adopted, and the purpose, aim, or goal that the inference in question is intended to fulfill.

Keep in mind that it is very easy to confuse questions of validity with questions of soundness when dealing with arguments that contain scientific conditionals. The reason for this is that, in order to appraise the validity of an argument, one must assume hypothetically that its premises are true and determine whether or not its conclusion could be false given that hypothetical assumption. But in assuming the truth of a scientific conditional one thereby also assumes that that conditional is maximally specific, since a scientific conditional that is not maximally specific cannot be true. The validity of such arguments, therefore, is, as always, a purely formal question, independent of the actual truth or falsity of its premises. The soundness of these arguments, however, naturally requires that those premises be true, which in turn requires that they be maximally specific, not as a question of logical form, but as an empirical characterization of the physical world, which is not a hypothetical matter at all.

To summarize, every *argument* has both premises and conclusions, whereas every *conditional* has both an antecedent and a consequent. In formal discourse, premises must be explicitly stated; and if the premise set includes conditionals, then their antecedents and consequents must also be explicitly stated. Otherwise, some of the relevant conditions pertaining to the inference in question will remain unspecified. If the conditionals are subjunctive, lawlike, or causal, moreover, then their antecedents must be *maximally specific* (as required by Fetzer's *RMS*) in order for the conditionals to be true. As we have seen, any conditional which invokes a *ceteris paribus* clause leaves its antecedent only partially specified, i.e., it is not maximally specific in this crucial sense, and it could well be false! Any argument or inference based upon a conditional of this sort may be unsound, therefore, where any putative (deductive) validity that is claimed for such an inference will depend upon the presumed truth of those *ceteris paribus* disclaimers which — let us note — cannot be empirically falsified. For the presumption of their truth, after all, reflects what may appropriately be viewed as an appeal to ignorance, i.e., the assumption that conditions of which we are ignorant will not defeat our inference. Failure to account adequately for these important distinctions only conflates the additivity of monotonic inference and the specificity of conditional propositions, and thus obscures the underlying problem of distinguishing between those modes of inference which are deductively valid or sound, and those which, by contrast, are inductively proper or correct.

Ceteris paribus disclaimers presume that all conditions not explicitly mentioned in the premises of an argument will be either irrelevant or 'equal' (in the sense of being relevant, perhaps, but somehow offset or balanced out). Depending upon how they are formally invoked, the resulting arguments clearly fail to qualify as being deductively valid, since these arguments either (a) will not be deductively well-formed, in leaving out a necessary premise, or else, (b) will be inductively ampliative as a result of begging the question of the truth of the premises involved, as the following examples are intended to illustrate:

(VII)   *Ceteris paribus*, if a match were struck, it would light.

A match is now being struck.

The match will light.

In (VII), if detachment is to proceed with deductive validity, then the truth of the *ceteris paribus* disclaimer appearing in the antecedent of the major premise must be explicitly asserted in another premise. Since it is not, detachment is blocked and the argument can easily be shown to be deductively invalid. Alternatively, however, the argument could be viewed as an ampliative inference whose deductive invalidity would not compromise its inductive propriety, but on this interpretation, of course, the argument should be schematized with a double line separating premises from conclusion to indicate its inductive character. By contrast, consider the following:

(VIII)   *Ceteris paribus*, if a match were struck, it would light.

Ceteris paribus (i.e., all other conditions are either irrelevant or 'equal').

A match is now being struck.

The match will light.

In (VIII), an explicit premise is included which does assert the truth of the clause, formally permitting detachment to proceed. But how is the truth of that premise to be established? If it could be shown, indeed, that all other conditions were irrelevant or 'equal' in the sense suggested above, then the premise would not qualify as a disclaimer at all, the major premise could then be regarded as possessing an antecedent which satisfies *RMS*, and the inference in question would be deductively valid. As it stands, however, the *ceteris paribus* clause begs that very question, and it should better be viewed as a premise in an inductively proper inference, if indeed it is acceptable at all. Finally, if *ceteris paribus* clauses are not understood as being assertions at all but rather as a logical feature of conditional assertions that are in some sense 'defeasible', i.e., capable of being defeated — or shown to be false — under certain conditions, then the metalinguistic rule(s) of 'non-monotonic deduction' through which that detachment proceeds would clearly beg the very same question of the truth of that conditional, which, of course, is not maximally specific.

In conclusion, therefore, valid deductive inferences clearly are *never* nonmonotonic but *always* monotonic, where the problem remains of ascertaining the sufficient conditions under which any nonmonotonic inference may still be acceptable. This will depend, of course, upon the sufficient condition(s) for proper induction that should be endorsed in addition to the necessary conditions already specified for induction. Hence, it seems that Gabbay, McCarthy, McDermott and Doyle, Nute, and others have only succeeded in obscuring the underlying problem of defining and of exonerating the program of induction (in Fetzer's sense) in misconstruing nonmonotonicity as an aspect of deduction by confusing it with nonspecificity as a feature of subjunctive premises: the former cannot be ascribed to any valid deductive inference, while the latter is an equally undesirable feature of subjunctive argumentation, especially in relation to scientific discourse, generally. Insofar as AI science seeks to achieve an adequate representation of scientific knowledge, therefore, distinctions between *monotonic deduction* and the necessary and sufficient conditions of proper *nonmonotonic induction* are clearly required.

NOTES

[1] See McDermott and Doyle (1980) and McCarthy (1980) and (1984), for example.
[2] The quotes appearing in this paragraph are contained in Nute (1984), p. 1.
[3] Fetzer (1981), p. 177, and Fetzer (1984), p. 8, recommend the use of "proper" and "correct" as inductive counterparts for "valid" and "sound", respectively, as the latter are typically applied to deductive arguments.
[4] Fetzer (1981), p. 178, and Fetzer (1984), p. 6.
[5] Fetzer (1984), p. 8.
[6] Fetzer (1981), p. 179.

[7] The example presumes, of course, that divorce reinstates bachelorhood. The implicit difference between being *unmarried* and *never having been married* at all is not a crucial factor in these concerns since a term could easily be introduced into the lexicon to explicitly capture the distinction. The same approach can also apply concerning the gender of bachelors, if necessary.

[8] Notice that on some relative frequency interpretations of this example, moreover, it might be 'certain' (probability = 1) that John will divorce Doris after a marriage lasting five years!

[9] As shown here, M1 is identical to condition (c) of the definition of formal systems appearing in Nute (1981), pp. 39—40.

[10] The semantic counterpart of M1, therefore, would simply substitute another symbol for the '⊢' that typically denotes the kind of derivability relation appearing in M1 as "$D(K)$". Various alternative possible worlds models that could accomplish this are discussed in Nute (1980) and (1981).

[11] Nute (1981), p. 131 (definition of logistic systems and theorem 75), p. 152 (definition of a conditional logic), pp. 160—164 (the conditional logic VW, esp. theorem 129).

[12] The same examples appear in Nute (1984), but with no explicit mention of the nonlogical *ceteris paribus* clauses being discussed here.

[13] See Gabbay (1982), pp. 262, 266, 268, and 271.

[14] NM1 was proposed by Gabbay and Nute during a seminar on Artificial Intelligence at the University of Georgia's Advanced Computational Methods Center, August, 1984.

[15] Notice that examples (III) and (IV) involve homogeneous reference classes and test conditions, whereas example (V) does not, and example (VI) is no argument at all, but merely a list of conditionals whose reference classes are also heterogeneous.

[16] Fetzer (1981), pp. 49—51. According to Fetzer, Nute himself recommended the alternative *RMS* formulation shown here as Fetzer and Nute (1979) was being prepared. Yet with Gabbay, the same notion seems to have emerged as a basis for nonmonotonic inference, contrary to the character of *RMS* as a truth condition for lawlike sentences and nomological conditionals. Nute is apparently of the opinion that nonmonotonic deduction is a justifiable mode of inference, which Fetzer would deny. See also Gabbay (1982).

REFERENCES

Fetzer, James H.: 1981, *Scientific Knowledge: Causation, Explanation, and Corroboration*, D. Reidel, Dordrecht, Holland.

Fetzer, James H.: 1984, 'Philosophical Reasoning', in *Principles of Philosophical Reasoning*, James H. Fetzer, editor. Rowman & Allanheld, Totowa, New Jersey.

Fetzer, James H., and Donald E. Nute: 1979, 'Syntax, Semantics, and Ontology: A Probabilistic Causal Calculus', *Synthese* **40**, 453—95.

Fetzer, James H., and Donald E. Nute: 1980, 'A Probabilistic Causal Calculus: Conflicting Conceptions', *Synthese* **48**, 241—246.

Gabbay, Dov M.: 1982, 'Intuitionistic Basis for Non-Monotonic Logic', *Proceedings of the Conference on Automated Deduction*, Springer Lecture Notes in Computer Science, No. 6, Springer-Verlag, Berlin, Germany.

Gabbay, Dov M. and Donald E. Nute: 1984, Seminar on Artificial Intelligence, sponsored by the University of Georgia Advanced Computational Methods Center, August 1984.

Lewis, D. K.: 1973, *Counterfactuals*. Library of Philosophy and Logic, P. T. Geach, P. F. Strawson, D. Wiggins, and P. Winch (eds.), Basil Blackwell, Oxford.

McCarthy, John: 1980, 'Circumscription: A Form of Non-monotonic Reasoning', *Artificial Intelligence* **13**, 27—39.

McCarthy, John: 19--, 'Applications of Circumscription to Formalizing Common Sense Knowledge' (forthcoming).

McDermott, Drew and Jon Doyle: 1980, 'Non-monotonic Logic I', *Artificial Intelligence* **13**, 41—72.

Nute, Donald E.: 1980, *Topics in Conditional Logic*, D. Reidel, Dordrecht, Holland.

Nute, Donald E.: 1981, *Essential Formal Semantics*, Rowman and Littlefield, Totowa, New Jersey.

Nute, Donald E.: 1984, 'Non-monotonic Reasoning and Conditionals', University of Georgia Advanced Computational Methods Center, Research Report No. 01—0002.

Reichenbach, Hans: 1949, *The Theory of Probability*, University of California, Berkeley.

Reiter, R.: 1980, 'A Logic for Default Reasoning', *Artificial Intelligence* **13**, 81—132.

Salmon, Wesley C.: 1967, *The Foundations of Scientific Inference*, University of Pittsburgh Press, Pittsburgh, Pennsylvania.

IBM AI Support Center
1501 California Avenue
Palo Alto, CA 94304, U.S.A.

KEVIN T. KELLY

ARTIFICIAL INTELLIGENCE AND
EFFECTIVE EPISTEMOLOGY

1. INTRODUCTION

Most philosophical interest in artificial intelligence arises in the areas of philosophy of mind and philosophy of psychology. The idea is that artificial intelligence is an attempt at an existence proof for the computational theory of mind. If one can write a computer program that has the input-output behavior of a being generally agreed to have a mind (e.g. a human), then we have evidence that cognition is somehow nothing more than computation.

But much of artificial intelligence makes more sense as a kind of epistemological inquiry. In the second and third sections of this paper I propose a computational approach to the study of epistemology. The second section overturns some standard objections to the philosophical interest in hypothesis generation procedures. The third section shows how computational considerations can undercut the motivation behind standard theories of hypothesis evaluation.

In the fourth and fifth sections of this paper, I relate *effective epistemology* (the study of epistemic norms for computational agents) to psychology and to artificial intelligence. I conclude that much artificial intelligence practice (in contrast to its rhetoric) is construed better as an approach to the study of effective epistemology than as an attempt to duplicate human cognitive processes.

2. HYPOTHESIS GENERATION PROCEDURES

Prior to the nineteenth century, many philosophers, scientists, and methodologists were interested in finding procedures that generate or discover knowledge. In his *Posterior Analytics*, Aristotle attempted to provide an account of how to discover causes by means of partially specified syllogisms. Francis Bacon envisioned something like an industry for generating scientific knowledge. And many subsequent methodologists, including the likes of Newton, Whewell, Herschell, and Mill, all

309

James H. Fetzer (ed.), Aspects of Artificial Intelligence, 309–322.

believed that there exist good, effective methods for making causal discoveries.

In the early twentieth century, however, the study of hypothesis generation procedures was largely abandoned in epistemology. There were some good reasons for this shift in attitude. For one thing, philosophers had to contend with the new edifice of modern physics. It was quite natural for them to analyze the riches at hand rather than to study procedures for generating more.

Moreover, many philosophers had become familiar with mathematical logic and probability theory. These disciplines provide excellent tools for the study of the structure and justification of scientific theories. But if hypothesis generation methods are to be of use to beings like us, they should be explicit procedures. And the proper setting for the study of procedures is computation theory. But computation theory was largely unavailable to epistemologists in the first half of this century. Hence the formal tools available to epistemologists during this period directed them to study logical and probabilistic relations of evidential support rather than effective methods for generating hypotheses from evidence.

But epistemologists did not appeal to these practical reasons for postponing the study of hypothesis generation methods. Rather, they attacked the philosophical study of such methods as a kind of conceptual error. Three distinct strategies of argument were pursued:

(A)   there are no good discovery methods to be found;
(B)   even if there were, we should not use them; and
(C)   even if it were not reprehensible to use them, such methods would be of interest to psychologists, not to epistemologists.

(A) was seriously proposed by Rudolph Carnap, Karl Popper and Carl Hempel; (B) is supported by the philosopher Ron Giere and many frequentist statisticians; and (C) was proposed by Popper and Hempel, and has been revived recently by the philosopher Larry Laudan.

2.1. *You Can't*

So far as (A) is concerned, it is a nontrivial matter to prove that a problem cannot be solved by a computer. And this assumes that one can state with mathematical precision what the problem to be solved is. Traditional advocates of (A) did not state the discovery problem they had in mind with any precision; nor did they have the theoretical wherewithal to prove a problem uncomputable. So (A) originated as a bluff, and has not advanced beyond this state in the philosophical literature.[1]

2.2 *You Shouldn't*

There is, at least, a plausible argument for position (B).

1. Evidence only supports an hypothesis if it can possibly refute the hypothesis.
2. But if a sensible procedure looks at the available evidence to construct an hypothesis to fit it, then the evidence looked at cannot possibly refute the hypothesis constructed to fit it.
3. Therefore, the evidence input to an hypothesis generator does not support the hypothesis generated.
4. But an hypothesis guessed without looking at the data could possibly be refuted by the available evidence.
5. So if one wants hypotheses supported by the current evidence, one *should not* use a generation procedure that looks at the data.

But the plausibility of this argument hinges on an equivocation over the sense of "possibility" in premises (1) and (2).

Consider what it means in premise (1) for evidence to "possibly refute an hypothesis." To say that some evidence can possibly refute a given hypothesis is not to say that the relation of inconsistency possibly holds between the evidence and the hypothesis. For notice that inconsistency is a logical relation, which either holds necessarily or necessarily fails to hold. So if it possibly holds, then it holds necessarily. But then the first premise says that evidence confirms an hypothesis only if this evidence (necessarily) refutes the hypothesis, which is absurd.

I propose, instead, that the sense of "possibly refutes" in premise (1) is this: our actual evidence supports our hypothesis only if *we might possibly have sampled* evidence that (necessarily) refutes the hypothesis, where our evidence collection procedure is fixed over all possible worlds. Under this construal, premise (1) is plausible. For suppose I test my hypothesis that "all ravens are black" by using a flying scoop that collects only black things. There is no possible world in which this procedure collects evidence that refutes my hypothesis, and the evidence collected does seem unpersuasive. The same can be said about a procedure that can collect only uninformative or tautologous evidence. Such evidence is consistent with any hypothesis, so our procedure cannot possibly produce evidence that contradicts our hypothesis. So when we talk about evidence possibly refuting an hypothesis, we are really talking about dispositions of our evidence gathering procedure.

But now consider premise (2). When we say that the output of a hypothesis generator that uses the evidence to patch together an hypothesis cannot possibly be refuted by the input evidence, what we mean is that the procedure is sure never to output an hypothesis refuted by its input, regardless of what the input is. That is, the procedural architecture of the procedure necessitates the consistency of its output with its evidential input. The same can be said of premise (4).

So the argument equivocates between properties of evidence-gathering procedures and properties of hypothesis-generation procedures. Considering a procedure that produces an unrefuted hypothesis for any consistent evidence may clarify this point. Collect some evidence and feed it to the procedure. Eventually, an hypothesis pops out. Suppose further that the evidence-gathering procedure is unbiased with respect to the hypothesis produced: that is, it can collect evidence that refutes the hypothesis in any world in which there is such evidence. In this case we can say that the evidence might have refuted the hypothesis produced because we might have gathered evidence that refutes it, despite the fact that the generation procedure never produces an output that is refuted with respect to its input. Hence, an hypothesis produced by a generation procedure that peeks at the data can nonetheless be supported by this data, at least so far as premise (1) is concerned.

There are other arguments against the reliance on discovery methods. These include the familiar claim the test statistics are "biased" if one looks at the data used in the test and the claim that such methods are too unreliable. The response to the first objection is that test statistics mean the same thing (from a frequentist point of view) when a discovery procedure employs them as when a human uses them. In neither case is the test *actually* repeated forever. In both cases, a new hypothesis is conjured as soon as the first is rejected. Of course, it would be a mistake to confuse the test significance level with the mechanical procedure's probability of conjecturing the truth, but it would be equally mistaken to confuse the human's reliability with the significance level of his tests. The response to the second objection is similar. The reliability of humans dreaming up hypotheses and employing statistical tests is unknown, but it can probably be exceeded by the reliability of well-designed formal methods.[2]

2.3. *Who Cares?*

Finally, consider claim (C), that theory-generation procedures are of interest to psychologists but not to epistemologists. The usual argument for this position is that the epistemologist is interested only in the justification of hypotheses, not in their discovery. How an hypothesis happens to be discovered or constructed is a mere matter of fact, whereas one's *justification* in believing it is a normative, philosophical issue [Popper, 1968].

But it is equally true that what people happen to believe is a mere matter of fact, while deciding which discovery method one *ought* to choose is a normative, philosophical issue. In general, what people ought to believe and the scientific methods they ought to choose are both normative, epistemological questions, whereas how they happen to dream up conjectures or come to believe them are psychological questions.

Of course, it would be another matter if one could show that no interesting normative issues arise in the choice of a method for generating good theories. But of course there are many such issues. What sorts of inputs should an hypothesis generator receive? Should an adequate method converge to the truth on complete evidence? If so, how shall we construe convergence? Must the method know when convergence is achieved? Should the method's conjectures always be confirmed with respect to the input evidence? Should its conjectures always be consistent with the evidence? Should it maintain coherence among one's beliefs? Should a finite being be capable of following it? On the face of it, these are all interesting normative questions concerning hypothesis generation methods.

Proponents of (C) might respond that the answers to such questions are all parasitic on finding an adequate theory of confirmation. Hence, the study of theory-generating procedure is not irrelevant to epistemology, but it is nonetheless redundant.[3]

But this position is also implausible, for there are obvious criteria of evaluation for hypothesis generators that cannot be maximized jointly with the aim of producing only highly probable or highly confirmed hypotheses. One thing we might desire in a method is that it converge to the truth in a broad range of possible worlds [Putnam, 1963]. Another is that it be effective so that it can be of use to us. But Osherson and Weinstein [Osherson, 1986] have demonstrated that there is an effective discovery method that can converge to the truth in more possible worlds[4] than any effective method which generates an hypothesis of maximal posterior probability on any given evidence.[5] That is, producing probable hypotheses is at odds with the joint goals of effectiveness and convergence to the truth in a broad range of possible worlds. It is also known that there are classes of possible worlds such that a method that sometimes conjectures hypotheses inconsistent with the input evidence can converge to the truth in each world in the class, but no method whose conjectures are always consistent with the evidence can. It is clear from these examples that theories of rational belief can be at odds with other criteria of evaluation for discovery methods. Hence, not all interesting normative questions concerning discovery methods are reducible to questions concerning standard theories of hypothesis evaluation.

3. RATIONAL BELIEF

So far, I have argued that computation theory is essential to the study of theory-generation methods and that the study of theory generation is an abstract, normative topic suitable for philosophical investigation. But computation theory also has direct relevance for standard theories of hypothesis evaluation. It can show that standard norms of hypothesis evaluation are in conflict for beings with limited cognitive resources. For example, some Bayesians like to issue the following Three Commandments to the teeming multitude:

1. Thou shalt be coherent (i.e. thou shalt distribute thy degrees of belief as a probability measure on Ye algebra of propositions).
2. Thou shalt modify thy degrees of belief by conditionalization on Ye evidence.
3. Thou shalt not be dogmatic (i.e. thou shalt not believe Ye contingent propositions to Ye degree 1 or to Ye degree 0).

The advantage of the first commandment is that it prevents one from taking bets he must lose, and the advantage of the second two is that anyone who follows them will (in his own opinion anyway) converge to the truth on increasing, complete evidence. But Haim Gaifman (Gaifman, 1982) has shown the following. Assuming that one can talk about arithmetic, and assuming that one is coherent, one's degrees of belief become *increasingly impossible* to compute as one becomes less dogmatic. That is, for any computational agent, avoiding hopeless bets is at odds with convergence to the truth. This is a computational result that should be of interest even to epistemologists who ignore questions of theory generation. What might have been taken to be mutually reinforcing reasons to be a Bayesian are actually conflicting desiderata for real agents.

But some philosophers may insist that epistemology concerns norms for *ideal* agents rather than for merely actual ones. And unlike every system of interest to man — including himself, robots, animals, and even aliens from other planets — an "ideal agent's" logical and mathematical ability is unlimited by the results of computation theory. It is fair to ask why epistemology should be for such peculiar agents, who are limited like all ordinary agents in one respect (observationally) but not in another (cognitively).

One proposal is that ideal agents are an abstraction, and that abstractions are a practical necessity in all formal reasoning. Messy and ill-understood factors may be disregarded in favor of elegant, formal principles. But if every system in which we are interested seems to have a limitation, and if we have an elegant formal theory about this limitation, as we do in the case of computability, then the abstraction argument suggests more of sloth than of virtue.

A stronger negative position is that epistemology is the study of justified belief for any agent whose observational power is limited. Since the limitations of ideal agents transfer to any agent with other limitations, epistemological results for ideal agents are more general, and therefore more interesting, than those for computationally limited agents.

While it is true that the limitations on ideal agents are limitations on all more limited agents, it is false that any solution to the problems of an ideal agent is a solution to the problems of a real one. In Gaifman's theorem, for example, we see that the Bayesian Commandments cannot all be observed by a real agent. So generality of applicability of epistemological principles is no reason to focus on ideal agents.

A third position is that although ideals are not themselves achievable by real agents, they are normative in the sense of promoting the realizable actions that better *approximate* them. This is fair enough, but notice that it is no argument for *ignoring* computation in epistemology. Rather, it is an invitation to characterize the sense in which a proposed ideal can be effectively approximated.

An adequate theory of approximating an ideal must have at least two parts. First, there must be a concept of distance from the ideal so that some acts can be determined to be further from the ideal than others. Second, there must be a motivation for achieving better approximations of the ideal that is a natural generalization of the motivation for achieving the ideal itself. If there is no well-motivated account of effective approximation for a proposed ideal, then the ideal is not normative in the sense of guiding action, for either there is no way to tell which actions are better than others, or there is no motivation for calling the better actions better.

For example, consider the Bayesian ideal of coherence. It turns out that there are hypothesis spaces over which there are countably additive probability distributions, but no *computable*, countably additive probability distributions.[6] So coherence is an unachievable ideal when one entertains certain classes of hypotheses. The question is then how to approximate coherence in such a case. And whatever the metric of approximation, it had better be the case that better effective approximations of coherence yield more benefits analogous to immunity to Dutch book. So we also need to invent something like "degrees of Dutch book", and to show that better degrees of approximation to coherence achieve lower degrees of Dutch book. Without such a theory, coherence over rich hypothesis spaces is a fatuous, nonnormative ideal

The moral is this: if you *insist* on advising the public to hitch its wagon to a star, you ought to provide a way to characterize which broken wagons came closer to the stellar rendezvous than others and to explain why trying and failing is better than never trying.

4. EFFECTIVE EPISTEMOLOGY AND PSYCHOLOGY

Since humans believe, reason, and attempt to justify their beliefs, and since psychologists intend to explain these facts with computational models, effective epistemology and cognitive psychology can appear quite similar. But similarity is not identity. The epistemologist's aim has typically been to *improve* the human condition rather than to describe it.[7] It is easy to see that epistemologists can evaluate lots of possible methods for testing and generating hypotheses or for altering degrees of belief even if no human ever uses them.

But when we move from the question of ends to the question of means, the relationship between psychology and epistemology becomes more tangled. For if past success is any evidence of a heuristic principle's soundness, then to the extent that humans are successful in science, it would seem fruitful to codify and to recommend the heuristics that successful humans use. And these principles may be covert, in the sense that no human could "find" them in a simple introspective exercise. So to find them, the epistemologist would have to resort to the usual experimental and theoretical techniques of cognitive psychologists. On this approach, epistemology can look just like empirical psychology, even though its *aims* are quite distinct. Herbert Simon is a notable proponent of this conception of epistemological method.

There is nothing wrong with beginning effective epistemology with the study of covert human methods, provided the empirical problems involved are not too onerous. There are weak evolutionary and empirical arguments in their favor as a place to start in the search for good techniques. But these arguments are just the starting point for an epistemological analysis of these techniques. For example, one might plausibly expect Einstein's implicit search heuristics to be good ones for generating scientific theories, for he did, after all, invent relativity theory. But it is possible that his commitments to field theory and to symmetry conditions made him a very inflexible theory generator. Perhaps he would have been a crank in any possible world with a messier space-time structure than ours.

It does not suffice to reply that success in the actual world is all that counts. First of all, one only uses a theory generator when one does not know which possible world he is in. Hence, a method with strong *a priori* commitments to one world in the set is undesirable from the point of view of the user, even if its commitment happens to be true. Moreover, each new subject matter in our universe presents us with an entirely new hidden structure to contend with. To succeed in theorizing about these different subject matters, we must essentially succeed in decoding many distinct possible structures. The structure of cognition is a world apart from the structure of a cell's chemical pathways, which in turn is a world apart from the dynamical structure of a galaxy. Hence, past success in one subject matter is not persuasive evidence of future

KEVIN T. KELLY

success in others, unless one knows on formal grounds that one's method is general.

A lucky crank can look like a genius in his subject without being one. One difference between the crank and the genius is that the latter is more *flexible* than the former. A more flexible agent can *succeed* in a broader range of possible circumstances. Success in a world involves several factors, some of which are in conflict. Would the generator's conjectures fit the input evidence that might be encountered in this world, or are there inputs for which its hypotheses are vacuous, irrelevant, or false? Can the procedure converge to the truth in the world in question? Another aspect of rational method is efficiency, for an inefficient method costs more to yield the same benefit. So another important question is whether there exist much faster methods that are as flexible as the method in question.

Questions of the efficiency, generality, and correctness of algorithms are just the sorts of questions computation theory can address in a mathematical manner. If a discovery algorithm appears to work pretty well over a class of trials, epistemologists should feel under some obligation to prove some facts about its scope and limits, its strong suits and blind spots. Psychologists are under no such obligation, although such analyses may facilitate the explanatory and predictive power of psychological theory.[8]

5. EFFECTIVE EPISTEMOLOGY AND ARTIFICIAL INTELLIGENCE

Artificial intelligence has emerged from its academic honeymoon. Theoretical computer scientists already view it as a kind of clean witchcraft, in which eyes of newts are *symbolically* mixed with warts of toads. A little pseudo-probability here, a little Aristotelian metaphysics there, and a good deal of unintelligible hacking to hold it all together, and *voila*! Frankenstein's monster surprises its creator as it clanks through its unexpected behaviors.

When pressed regarding the lack of theoretical depth in the field, many AI proponents slide into the posture of cognitive modelling. Since the brain's procedures may be messy or poorly motivated, why shouldn't cognitive models be the same way? But on this view, the absence of psychological evidence in most AI articles may raise the questioning eyebrows of cognitive psychologists, whose models are usually less detailed, but whose empirical arguments are often very sophisticated.

But AI is a more creditable study when it is interpreted as effective epistemology. Like epistemologists (but unlike psychologists), AI technicians can pursue interesting studies in blissful ignorance of how humans actually work. An AI programmer would be overjoyed to develop a super-human discovery engine that can demonstrably satisfy various methodological criteria. So would an epistemologist. And AI enthusiasts like epistemologists, have no qualms about saying that you *ought* to use their methods.

And like epistemologists (but unlike computation theorists), AI technicians rarely have clear conceptions of the problems their procedures are to solve. This is not so much a shortcoming in AI practice as a fundamental fact about the vague subject matter of the field. In computation theory, a problem is just a mathematical function. A program solves a problem (function) just in case it computes it. So the natural order of business in computation theory is to define a function extensionally, and to decide whether it is computable or not, how impossible it is to compute if it cannot be computed, and how long it takes to compute if it can be computed.

The usual technique in artificial intelligence is quite different. A "problem" is a vaguely specified area of human competence — say *learning*. Once a problem area is specified, the AI programmer typically begins to play around with data structures and procedures that seem to do what humans take themselves to be doing when they address problems in the relevant area. And if the resulting procedures take a lot of time to run, the offending subroutines are altered until they run quickly — even if this alters the input-output behavior of the overall system. The final result is a program that runs large cases on a real computer and that goes through steps that seem reasonable to humans, but whose overall input-output behavior may be quite unknown to the designer.

From the computation theorist's point of view this way of proceeding appears underhanded. Whenever the problem to be solved by his current program becomes too difficult, the AI programmer changes the program until it solves an easier problem — and then calls the result progress. The computation theorist accepts no progress other than solving the *same* (mathematically precise) problem in a less costly way. And finding a moderately efficient algorithm for an easy problem should never be confused with finding a very efficient solution to a given, difficult problem.

But if AI programmers do not usually make mathematical progress, their approach can make a kind of epistemological progress. If the view propounded in this essay is correct, epistemologists should strive for rational, *effective* methods for discovering, testing, and maintaining theories. And in searching for such methods, one may either examine effective methods that seem plausible to see whether they are rational, or one may first propose constraints on rational behavior and then analyze the computational difficulties of these behaviors. At best, the AI approach of starting out with procedures focuses attention on principles that can possibly be normative for real robots and people. At worst, it can lead to a rambling, unintelligible program that runs on a computer but which carries our understanding of the kinematics of rational belief not one whit further. On the other hand, the standard, philosophical approach focuses attention on the motivation for a method rather than on getting it to run on a computer. But this "abstract" approach has its own dark side. At worst, it can lead to nonnormative, inapproximable ideals, which in the phrase of Hilary Putnam, are "of no use to anybody".

6. PROSPECTS

I maintain that effective methods of discovery and hypothesis evaluation are not only acceptable objects of epistemological study, but are its proper objects — so far as physically possible beings are concerned. One can begin the study in various ways. One can attempt to discover the human methods as a benchmark and then evaluate and improve them; one can design computer programs that seem to perform intelligently and then analyze them; or one can define abstract criteria of adequacy and subsequently search for procedures that satisfy them. The first approach involves the techniques of cognitive psychology, the second is the approach of artificial intelligence, and the third is the standard approach of epistemologists.

Some logicians and computer scientists have already been busy bridging the gap between disciplines in the pursuit of effective epistemology. For example, there is an extensive interdisciplinary literature spanning the fields of recursion theory, linguistics, statistics and logic that focuses on the ability of discovery methods to converge to the truth in various classes of possible worlds.[9] These studies juggle the desideratum of convergence to the truth with those of effectiveness, complexity, verisimilitude and confirmation. Other research centers on the effectiveness of probabilistic norms, as we have seen in the case of Gaifman's paper.

The cross-fertilization of computer science and epistemology is still in its infancy, and the prospects for discoveries are still good. My own research concerns computational issues in the discovery of complete universal theories. Another natural project would be to find a well-motivated theory of effectively approximable Bayesian coherence. Still another would be to investigate the computational complexity of nondogmatic coherence over propositional languages. Only an artificial disdain for practicality prevents a revolutionary, computational reworking of epistemology. The tools are available.

NOTES

[1] An exception is Hilary Putnam's (1963). My response to this argument may be found in (Kelly, 86).

[2] For a more thorough discussion, see (Glymour, 1986).

[3] Larry Laudan has made roughly this point in (Laudan, 1980).

[4] in the sense of set inclusion.

[5] A method is said to be Bayesian if its conjecture always has maximal posterior probability on the input evidence.

[6] Proof: Let propositions be partial recursive functions, and let hypotheses be indices drawn from an acceptable numbering [Rogers, 1967] of the partial recursive functions. Hence, any two indices of the same function are equivalent, and must therefore be assigned the same probability values. Suppose for *reductio* that P is a countably additive probability distribution on the partial recursive functions. P cannot be uniform, for there is a countable infinity of partial recursive functions, and if P were to assign each function f the same value r, the countable sum over all functions would be unbounded. Let i be an index, and let φ_i be the ith partial recursive function. Let $[i]$ be the set of all j such that $P(j) = P(i)$. Since P is not uniform, $[i]$ is neither the set of all indices nor the empty set. Since P is computable, $[i]$ is a recursive set (on input k, just compute $P(k)$ and $P(i)$ and see whether the results are identical). But notice that $[i]$ is the set of all indices of some non-universal and non-empty subset of the partial recursive functions. But by Rice's theorem (Rogers, 1967), no such set of indices is recursive. Hence, P is not effective. Q.E.D.

[7] Quine's views being a notable exception.

[8] As a case in point, consider the application of the theory of learnability to linguistics by Kenneth Wexler (Wexler, 1983).

[9] For a good survey, see (Angluin, 1980).

REFERENCES

Angluin, Dana: 1980, 'Finding Patterns Common to a Set of Strings', *Journal of Computer and System Sciences*, **21**, 46-62.

Gaifman, Haim and Snir, Marc: 1982, *Journal of Symbolic Logic*, **47**, 495-548.

Glymour, C., Kelly, K., Scheines, R., and Spirtes, P.: 1986, *Discovering Causal Structure: Artificial Intelligence for Statistical Modelling*, Academic Press, New York.

Kelly, Kevin T.: 1986, 'The Automated Discovery of Universal Theories', Ph.D. thesis, University of Pittsburgh.

Laudan, Larry: 1980, 'Why was the Logic of Discovery Abandoned?', in *Scientific Discovery, Logic, and Rationality*, D. Reidel, Dordrecht and Boston.

Osherson, D.N., Stob, M., and Weinstein, S.: 1986, 'Mechanical Learners Pay a Price for Bayesianism', Unpublished Manuscript.

Popper, Karl R.: 1968, *The Logic of Scientific Discovery*, Harper and Row, New York.

Putnam, Hilary: 1963, '"Degree of Confirmation" and Inductive Logic', In *The Philosophy of Rudolph Carnap*, Arthur Schilpp (ed.) Open Court, Illinois.

Rogers, Hartley: 1967, *Theory of Recursive Functions and Effective Computability*, McGraw-Hill, New York.

Wexler, Kenneth, and Culicover, Peter W.: 1983, *Formal Principles of Language Acquisition*, MIT Press, Cambridge, MA.

Department of Philosophy
Carnegie-Mellon University
Pittsburgh, PA 15213 U.S.A.

RICHARD A. VAUGHAN

MAINTAINING AN INDUCTIVE DATABASE

Many "state-of-the-art" expert computer systems function through the application of a finite set of deductive production rules to a static database of knowledge. For some well-studied fields such a system is sufficient, as there are no significant changes being made in the fundamental theory behind the knowledge represented in the database. An example of such a system is MYCIN, a medical diagnosis system. MYCIN's database is made up of the associations between symptoms and diseases.[1] This system works because these associations are well-documented and supported by centuries of medical study.

Such an expert system would be inadequate, however, in a new field of study in which the associations are not completely known, or in any area where much new information is being accumulated. How can these databases be kept up-to-date when conclusions must be drawn from incomplete data? Is there any feasible way of measuring the support for these conclusions? The static database of an expert system could be updated through some form of batch processing, or perhaps infrequently by unloading, updating, and reloading the database, but such a system still lacks any dynamic measure of the reliability of the conclusions drawn from this incomplete data. Thus, for applications requiring additions or modifications to database information, methods of dynamically updating the database and of measuring the support for conclusions drawn from the database information must be devised. Note that such a system would be ampliative, nonadditive and nondemonstrative, where these attributes are characteristic of induction.[2]

By implementing the system's measures of evidential support on a zero-to-one scale with some sort of probabilistic scheme, a number of assumptions can be made. The probability of a statement or hypothesis always lies between zero and one, inclusive. The probability of a tautology is one. The probability of a contradiction is zero. The sum of the probabilities of a statement or hypothesis and its negation is one. The probability of the disjunction of two mutually exclusive statements or hypotheses is the sum of their individual probabilities.[3] Thus implemented, the system would have the ability to test hypotheses and to

323

James H. Fetzer (ed.), Aspects of Artificial Intelligence, 323—335.
© 1988 *by Kluwer Academic Publishers.*

respond to inquiries about the probability of some statement or hypothesis being true, given a set of circumstances. Probabilities in the system would be computed with respect to all relevant evidence available in the database.

One technique that has been suggested for implementing such a system is Bayesian conditionalization. It can be represented by the following formula:

$$P(h/d) = \frac{P(d/h) * P(h)}{P(d)} \text{ given nonzero } P(d).^4$$

where: $P(h/d)$ = probability of hypothesis h given data d as true; = posterior probability of hypothesis h, having conditionalized on data/evidence d; $P(d/h)$ = probability of data d given hypothesis h as true; = likelihood of hypothesis h given data d; $P(h)$ = prior probability of hypothesis h; and, $P(d)$ = prior probability of data/evidence d.

Note that at any time t, the probability currently associated with hypothesis h is the probability that would be assigned to prior probability $P(h)$, if it were to be conditionalized given some new datum d. This method could function as follows: newly acquired facts (e.g. from experimental results) are translated into a computer-readable form; a system utility then reads this data and updates all hypotheses to which each such datum is relevant.[5]

To evaluate the effectiveness of the Bayesian conditionalization technique, each of the three terms in the formula must be analyzed: the likelihood $P(d/h)$, and the two prior probabilities $P(d)$ and $P(h)$. The prior probability of the data can be replaced by the following sum of products:

$$P(d) = P(d/h) * P(h) + P(d/-h) * P(-h).$$

Since $P(-h) = 1 - P(h)$, an equivalent formula is:

$$P(h/d) = \frac{P(d/h) * P(h)}{P(d)}$$

$$= \frac{P(d/h) * P(h)}{P(d/h) * P(h) + P(d/-h) * P(-h)}$$

$$= \frac{P(d/h) * P(h)}{P(d/h) * P(h) + P(d/-h) * (1 - P(h))}$$

which is made up of two likelihoods and a prior probability.

The likelihood $P(d/h)$ is the probability of the data d given the hypothesis h. This is a calculable quantity,[6] provided that the hypothesis and the data are stated in the database in an acceptable manner. Similarly the likelihood $P(d/-h)$, i.e., the probability of the data d given the negation of the hypothesis h, is calculable under the same requirements. The prior probability of the hypothesis, $P(h)$, however, is not so easily determined.

$P(h)$ is the current probability associated with the hypothesis, not considering the new information present in the data of the active conditionalization. Note that the value resulting from the active conditionalization would be assigned to be the prior probability of the hypothesis for the next conditionalization involving this hypothesis. Thus this process requires only an initial assignment for the prior probability of the hypothesis in the first conditionalization involving the hypothesis. The choice of an initial value becomes somewhat arbitrary, though. An initial assignment of zero or one is unacceptable because the value associated with $P(h)$ becomes fixed; it cannot then be altered by any number of Bayesian conditionalizations (see Figure 1).[7]

Whenever a hypothesis is plausible (i.e., noncontradictory), Salmon suggests the assignment of a low prior probability to $P(h)$ because "even a very small prior probability is compatible with a very high posterior probability,"[8] especially since, with a large number of conditionalizations, the effect of any nonextreme prior probability becomes negligible. He explains that a high posterior probability can occur even with a low prior probability when the likelihood $P(d/h)$ is high and the likelihood $P(d/-h)$ is very near zero:[9]

$$P(h/d) = \frac{P(d/h) * P(h)}{P(d/h) * P(h) + P(d/-h) * P(-h)}$$

$$= \frac{\text{high} * \text{low}}{\text{high} * \text{low} + \text{very-near-zero} * \text{high}}$$

$$= \frac{\text{medium}}{\text{medium} + \text{very-low}}$$

$$= \text{high}$$

Note that the same cannot be claimed for a low posterior probability resulting from a high prior probability when the likelihood $P(d/h)$ is low and likelihood $P(d/-h)$ is very near one:

Initially: $P(h) = 0$:

Later conditionalizations:

$$P(h/d) = \frac{P(d/h) * P(h)}{P(d/h) * P(h) + P(d/-h) * (1 - P(h))}$$

$$= \frac{P(d/h) * 0}{P(d/h) * 0 + P(d/-h) * (1 - 0)}$$

$$= \frac{0}{0 + P(d/-h) * 1}$$

$$= \frac{0}{P(d/-h)}$$

$$= 0$$

Initially: $P(h) = 1$:

Later conditionalizations:

$$P(h/d) = \frac{P(d/h) * P(h)}{P(d/h) * P(h) + P(d/-h) * (1 - P(h))}$$

$$= \frac{P(d/h) * 1}{P(d/h) * 1 + P(d/-h) * (1 - 1)}$$

$$= \frac{P(d/h)}{P(d/h) + P(d/-h) * 0}$$

$$= \frac{P(d/h)}{P(d/h)}$$

$$= 1$$

Fig. 1. Prior probabilities of zero or one are permanent under the scheme of Bayesian conditionalizations

$$P(h/d) = \frac{P(d/h) * P(h)}{P(d/h) * P(h) + P(d/-h) * P(-h)}$$

$$= \frac{\text{low} * \text{high}}{\text{low} * \text{high} + \text{very-near-one} * \text{low}}$$

$$= \frac{\text{medium}}{\text{medium} + \text{low}}$$

$$= \text{medium-high}$$

A low prior probability is desirable because it requires more extreme values to reduce a high prior probability to a low posterior probability than it does to increase a low prior probability to a high posterior probability (due to mathematical principles). Even if low prior probabilities are known to be advantageous, however, there remains the arbitrariness of what low prior probability to initially associate with each hypothesis.

The initial prior probability might be calculated from the first set of data against which the hypothesis would be tested, where conditionalizations could be performed against all other relevant sets of data. The likelihood of the hypothesis given this first set of data would then become the initial prior probability for the first conditionalization. A problem could arise only when the data fully supports the hypothesis, inducing an irrevocable associated probability of one. A simple solution to the problem might be to place in effect some limiting maximum so that this condition could not arise.

One must also consider the effects of an especially high or low initial prior probability $P(h)$. A high $P(h)$ at or near the limiting maximum, indicating high plausibility of the hypothesis, makes the hypothesis h more acceptable by artificially inducing a high associated probability. Similarly, a low $P(h)$, near zero, indicating low plausibility of the hypothesis, makes the hypothesis h more falsifiable.[10]

These effects become negligible after a large number of Bayesian conditionalizations on new data,[11] so some sort of counter might be associated with each hypothesis indicating the number of conditionalizations that have been performed on it. This count would represent the amount of data against which the hypothesis has been condition-alized. Such an indicator could be expanded to include some form of weighting to stress the need for gathering data from a variety of sources under a variety of conditions.[12]

With a counter and a probability now associated with each hypoth-esis in the database, some criteria could be developed for acceptance and rejection of hypotheses. A hypothesis might be accepted when (1) the counter of conditionalizations associated with it is greater than the minimum number of conditionalizations for acceptance (some specified large integer weighted as discussed previously); (2) its associated probability is greater than the minimum probability for acceptance (some real number near one); and (3) the data used in the condition-alizations is known to be chosen from various sources under various conditions. The third criterion, although reasonable, is not one deter-minable by the system as presented here.

A hypothesis could be rejected when (1) its associated counter is greater than the minimum number of conditionalizations for rejection (some specified large integer weighted as above) and (2) its associated probability is less than the maximum probability for rejection (some real number near zero). A third criterion is not necessary for rejection: if the hypothesis does not hold in one region of the relevant sample space, then it does not hold in all of them. Note that an accepted or rejected hypothesis can become unaccepted or unrejected if new evidence, when conditionalized on, reduces the probability of an accepted hypothesis below its acceptance limit or increases the pro-bability of a rejected hypothesis above its rejection limit. Remember that this unaccepting and unrejecting cannot take place when one or zero probability, respectively, has been associated with a hypothesis.

Since the most significant difficulty with the Bayesian conditionaliza-tion approach is choosing the prior probability of the hypothesis, a method based upon likelihoods that dispenses with prior probabilities suggests itself. The higher the likelihood $P(d/h)$ of a hypothesis h with respect to evidence d, the better the hypothesis is supported by that evidence. Such a likelihood method could function, for example, as follows: given a hypothesis entered into the database, a set of mutually exclusive and exhaustive hypotheses is generated; a likelihood is associated with each of the hypotheses; and the highest associated likelihood among these hypotheses indicates the best supported hypoth-esis.[13] Although the likelihood is calculable and thus a convenient measure of the support for a hypothesis, it alone is not an adequate measure for accepting the hypothesis. As Fetzer, for example, has explained,

According to precepts [of likelihood], the data d supports an hypothesis [i] better than it supports an hypothesis [j] so long as the likelihood [P(d/i)] exceeds that of [P(d/j)]; thus, the best supported members of a specified set of incompatible hypotheses for their common domain will always be those with maximum likelihood in relation to the available evidence d, i.e., those whose likelihood on that evidence is not exceeded by that of any other member of that set. It should not be difficult to discern, however, that more than one hypothetical alternative may possess maximum likelihood relative to d at any particular time: If innumerable hypotheses . . . all entail that data, then they will all be supported by that evidence with likelihoods of one, which, of course, confer upon each of them not only maximum likelihood as *the best supported hypotheses* . . . but maximal likelihood as receiving *the best possible support* . . . , in relation to the law of likelihood itself. Since not more than one of these hypotheses could possibly be true, therefore, it should be obvious that likelihood relations, at best, establish necessary but not sufficient conditions for resolving the problem of acceptance.[14]

Thus, to differentiate between such equally likely hypotheses, some additional factor — such as confidence in the data — requires consideration. Fetzer proposes using the degree of divergence of the data from a statistically normal distribution as an objective measure of confidence that the sample is random.[15] When the data is distributed more normally (i.e., with a smaller degree of divergence), the level of confidence is correspondingly higher.

This proposal eliminates the need for prior probabilities, but while doing so it also eliminates the system's ability to associate a probability with each hypothesis. Instead a measure of support is associated with a hypothesis for each set of data against which it is tested. Since these measures of support are separate for each set of data and do not consider the measure of support previously associated with the hypothesis, revisions to this value, due to newly accumulated data, seem to require a prohibitive amount of computation. Two possibilities present themselves: (A) integrating the data relevant to each such hypothesis and considering it as one large set of data; or, (B) integrating the measures of support calculated individually against each set of relevant data in the database.

The former suggestion (A) would necessitate either a tremendous amount of redundancy in the database keeping data associated with more than one hypothesis in several places or a list of pointers for each hypothesis in several places to indicate which data is relevant to it.

Using some form of set theory might reduce the significant amount of space required for these redundancies. There still remains the cost, in time and money, to recompute — against all the relevant data associated with each hypothesis, every time more information is added to the database — the measure of support for all hypotheses to which that new information is relevant. This expensive recomputation tends to exclude this suggestion from practical consideration.

The latter suggestion (B) would necessitate a method of determining the weight of support from a single set of data relative to the weight of support from all the previously used sets of data. This measure of relative significance could be based on the sample size of data. The likelihood 0.7400 due to some new set of data (of sample size 300), for example, might be integrated with the previously integrated likelihoods of 0.9200 (with sample size 9700). The new integrated likelihood would be calculated as follows:

$$\begin{matrix} \text{new integrated} \\ \text{sample size} \end{matrix} = \begin{matrix} \text{old integrated} + \text{new sample size} \\ \text{sample size} \end{matrix}$$
$$= 9700 + 300$$
$$= 10\,000$$

$$\begin{matrix} \text{new} \\ \text{integrated} = \\ \text{likelihood} \end{matrix} \dfrac{\begin{matrix}\text{old} & \text{old} & \text{likelihood} & \text{new} \\ \text{integrated} * \text{integrated} + \text{from} * \text{sample} \\ \text{likelihood} & \text{sample size} & \text{new data} & \text{size}\end{matrix}}{\text{new integrated sample size}}$$

$$= \dfrac{(0.9200 * 9700) + (0.7400 * 300)}{10\,000}$$

$$= (8924.00 + 222.00) / 10\,000$$
$$= 9146.00 / 10\,000$$
$$= 0.9146$$

Then some measure of confidence could be introduced into this calculation for each set of data. If Fetzer's proposed degree of divergence of the data from a normal distribution were used, then this value would have to be associated with each set of data in the database. Since a low degree of divergence indicates a high degree of confidence, and a high degree of divergence indicates a low degree of confidence, an inverse relationship is quite appropriate with respect to likelihood. Simply dividing by the degree of divergence of the data is impossible, however, because the boundary value of zero — for the degree of divergence of data which is perfectly normally distributed — would yield an undefined integrated likelihood. Values near zero, similarly, would give near-infinite integrated likelihoods. It can be shown that by adding one to the degree of divergence before performing this division, however, the boundary value of zero leaves the likelihood unchanged, as is appropriate. Given a degree of divergence of 0.4 for the new data, for example, the equations above would be altered to:

$$\text{new likelihood with confidence} = \frac{\text{likelihood from new data}}{1 + \text{degree of divergence of new data from a normal distribution}}$$

$$= \frac{0.74}{1 + 0.4}$$

$$= \frac{0.74}{1.4}$$

$$= 0.5286 \text{ [16]}$$

$$\text{new integrated likelihood} = \frac{\begin{array}{c}\text{old} \\ \text{integrated} \\ \text{likelihood}\end{array} * \begin{array}{c}\text{old} \\ \text{integrated} \\ \text{sample size}\end{array} + \begin{array}{c}\text{new likeli-} \\ \text{hood with} \\ \text{confidence}\end{array} * \begin{array}{c}\text{new} \\ \text{sample} \\ \text{size}\end{array}}{\text{new integrated sample size}}$$

$$= \frac{(0.9200 * 9700) + (0.5286 * 300)}{10\,000}$$

$$= (8924.00 + 158.58) / 10\,000$$

$$= 9082.58 / 10\,000$$

$$= 0.9083$$

Acceptance and rejection under this likelihood technique can be similar to that under Bayesian conditionalization.[17] A hypothesis is acceptable when the integrated sample size is greater than the system's lower sample size limit for acceptance (some specified large integer); its associated integrated likelihood is greater than the system's lower likelihood limit for acceptance (some real number near one); and the data used in the integration of likelihoods is known to be gathered from various sources under various conditions. Again, this third criterion is undeterminable by the system as presented here. Rejection would occur only when the integrated sample size has reached the system's lower limit for rejection (some specified large integer), and its associated integrated likelihood is less than the system's upper likelihood limit for rejection (some real number near zero). Unlike the Bayesian conditionalization method, a likelihood technique such as that described here allows integrated likelihoods of zero and one to be altered given appropriate data. Unacceptance and unrejection also occur in a straightforward manner when associated integrated likelihoods drop below the lower likelihood limit for acceptance or increase above the upper likelihood limit for rejection, respectively.

To summarize the two methods presented here, the Bayesian technique consists of:

Production rules:

1. for initial prior probability of a hypothesis, consisting of the likelihood of the first evidence conditionalized on or the arbitrary maximum for initial prior probability, whichever is smaller;
2. for the updating of the probability associated with a hypothesis, given some new evidence, consisting of the Bayesian conditionalization technique:

$$P(h/d) = \frac{P(d/h) * P(h)}{P(d/h) * P(h) + P(d/-h) * (1 - P(h))}$$

where the prior probability for the hypothesis $P(h)$ in this current conditionalization is the posterior probability of the previous conditionalization on this hypothesis;
3. for acceptance and rejection of hypotheses when the counter associated with a hypothesis exceeds the necessary number of conditionalizations, tentatively accepting when the associated probability is greater than the minimum limit for acceptance and tentatively rejecting when the associated probability is less than the maximum limit for rejection; and,

Database:

1. hypotheses:
   a. hypotheses stated in appropriate form;
   b. associated probabilities;
   c. associated counters;
2. evidence/data.

The likelihood technique consists of:

Production rules:

1. for calculating the integrated likelihood:

   a. $\text{new integrated sample size} = \text{old integrated sample size} + \text{new sample size};$

   b. $\text{new likelihood with confidence} = \dfrac{\text{likelihood from new data}}{1 + \text{degree of divergence of new data from a normal distribution}};$

   c. $\text{new integrated likelihood} = \dfrac{\text{old integrated likelihood} * \text{old integrated sample size} + \text{new likelihood with confidence} * \text{new sample size}}{\text{new integrated sample size}}$

2. for acceptance and rejection of hypotheses when the sample size associated with a hypothesis exceeds the necessary quantity of data, tentatively accepting when the associated integrated likelihood is greater than the minimum limit for acceptance and tentatively rejecting when the associated integrated likelihood is less than the maximum limit for rejection; and,

Database:

1. hypotheses:
   a. hypotheses stated in appropriate form;
   b. associated integrated likelihoods;
   c. associated sample sizes;
2. evidence/data.

The Bayesian technique's major difficulty, the arbitrary choice of an initial prior probability, of course, becomes less important as more conditionalizations improve the accuracy of the probability associated with a hypothesis. There does not appear to be any easy way to eliminate this problem. To remain aware of the number of conditionalizations that have been performed, a counter is associated with each hypothesis. To ease the arbitrariness of initial prior probability assignment, no extreme values should be initially assigned, particularly values of zero or one, which become permanent.

The use of a likelihood scheme provides a method with no initial arbitrariness. However, integrated likelihood is not a probability, and there is an underlying assumption that the data should be distributed normally. The suggestion here for integrating the idea of confidence into the likelihood scheme reduces the likelihood calculated from evidence which is not normally distributed.[18]

Both techniques have acceptance and rejection limits which can be tailored to almost any environment by adjusting the system's boundary values on the number of conditionalizations and upper and lower probabilities in the Bayesian technique, and on the cumulative sample size and the upper and lower integrated likelihoods in the likelihood scheme. To work most effectively, both techniques require the data to come from samples acquired from various sources under various conditions.

Neither technique seems to have a complete advantage over the other under all circumstances. The Bayesian method is most effective in situations where many conditionalizations occur due to large influxes of new data. Because there is no initial arbitrariness to eliminate, the likelihood scheme functions well with smaller amounts of new information. Both techniques could be implemented as discussed here in a working database environment. When a hypothesis is added, it can be evaluated against all relevant data in the database, and when a set of data is added, all hypotheses to which it is relevant can be evaluated against it. In accordance with procedures of either kind, therefore, it should be possible to implement the types of production rules that would be required to maintain an inductive data base for an expert system.

NOTES

[1] For more information about MYCIN and other expert systems, see Webber and Nilsson (1981).
[2] For a fairly concise overview of the arguments for and against the use of inductive inference, see Salmon (1967), pp. 11—54.
[3] Skyrms, (1975), pp. 168—169.
[4] Barnett, (1982), pp. 192—193.
[5] Determining which hypotheses need to be updated might require supplementary principles and procedures from mathematical set theory.
[6] Hacking, (1965), p. 56.
[7] Fetzer, (1981), pp. 207—208.
[8] Salmon, (1967), p. 128.
[9] Salmon, (1967), p. 129.
[10] Salmon, (1967), pp. 118—119.
[11] Salmon, (1967), pp. 128—129.
[12] Fetzer (1981), pp. 216, 253.
[13] Hacking (1965), pp. 70—71.
[14] Fetzer, (1981), p. 231.
[15] Fetzer, (1981), pp. 252—253.
[16] Note that the values here are rounded to the fourth decimal place. In practice, accuracy would depend on the implementation.
[17] Fetzer (1981), esp. pp. 248—254; see pp. 255—263 for further discussion.
[18] For the rationale underlying this conception, see Fetzer, (1981), esp. pp. 250—251.

REFERENCES

Barnett, Vic.: 1982, *Comparative Statistical Inference*, John Wiley & Sons, New York.
Fetzer, James H.: 1981, *Scientific Knowledge*, D. Reidel, Dordrecht.
Hacking, Ian: 1965, *Logic of Statistical Inference*, Cambridge University Press, Cambridge.
Nilsson, Nils J.: 1980, *Principles of Artificial Intelligence*, Tioga Publishing Co., Palo Alto, Calif.
Salmon, Wesley C.: 1967, *The Foundations of Scientific Inference*, University of Pittsburgh Press, Pittsburgh.
Skyrms, Brian: 1975, *Choice and Chance*, Dickenson Publishing Co., Inc., Encino, Calif.
Webber, Bonnie Lynn and Nilsson, Nils J. (eds.): 1981, *Readings in Artificial Intelligence*, Tioga Publishing Co., Palo Alto, Calif.

Department of Computer Science
Stanford University
Stanford, CA 94305. U.S.A.

EPILOGUE

RICHARD SCHEINES

AUTOMATING CREATIVITY*

1. INTRODUCTION

Critics of the very idea of artificial intelligence[1] fall into two main classes. There are those who argue that it is impossible to build a machine that truly mimics human cognitive behavior, and there are those who argue that even if a machine could mimic *cognitive* behavior, it would never "really" have the *emotional* experiences we humans do. Critics in the first class usually argue their case by pointing to a cognitive task that humans accomplish with ease but which a computer could not, even *in principle*, do. Because creativity is mysterious, seemingly impossible to teach, and not subject to any easily discernible set of "rules", it is a popular candidate for such a cognitive task.

Many people are especially doubtful about 'automating' creativity, freedom, and the like. No computer, they suppose, could ever be truly inventive . . . because 'it can only do what it's programmed to do'. (Haugeland, 1985, p. 9).

John Haugeland suggests two ways to defeat this type of argument. One is to show that humans are, on some important level, only capable of doing what we were designed or programmed (by evolution) to do. This assumes that we are strongly analogous to computers and therefore begs the question. The other is to agree on a definition of creativity and provide an example of a machine that satisfies this definition. In what follows I intend to pursue the second strategy.

Unfortunately, few who are skeptical about artificial intelligence are willing to offer a clear account of just what it is they are claiming cannot be automated. Those of us more optimistic are forced to either wait for an account to refute or to create our own targets. To the optimist the debate is essentially a very slow retreat by the skeptics. With each clear formulation of some task that cannot be automated comes a frontier to push back. If the task turns out to be possible, then the skeptic retrenches into a different formulation of the task.

A philosopher of science no less distinguished than Carl Hempel recently argued specifically against the possibility of automating *crea-*

James H. Fetzer (ed.), Aspects of Artificial Intelligence, 339—365.

tive science (Hempel, 1985). Hempel argued that the limitations on scientific discovery by computer are more severe than are those on scientific discovery by humans. He proposed a relatively clear criterion for when a theoretical term in science is novel and argued that while humans have obviously introduced such terms, computers could not be made to do so. I argue: (1) that new terms are introduced to provide a theory with explanatory power it cannot possess without such terms, (2) that a clear criterion of relative explanatory virtue can indicate when to introduce such terms and when not to, and (3) that in delimited scientific domains, both the criterion and a heuristic procedure for introducing terms are easily automated. Finally, I give two examples from empirical social science, one in which a computer program helped to suggest the introduction of terms which are novel by Hempel's definition, and one in which it helped conclude that no such terms should be introduced.

2. HEMPEL'S CASE

In 'Thoughts on the Limitations of Discovery by Computer' (Hempel, 1985), Hempel argued that there are limits to what a computer could be programmed to discover. Hempel's remarks were in some part a response to a paper preceding his by Bruce Buchanan on an existing discovery program, Meta-DENDRAL. In the introduction to the collection, Schnaffner provided the context for Hempel's remarks by quoting Buchanan and then Hempel. Here are Schnaffner's quotes slightly extended.

Buchanan:
> The Meta-DENDRAL program is designed to aid chemists find and explain regularities in a collection of data .... Although the most revolutionary discoveries involve postulating new theoretical entities (i.e., developing a new theory), finding general rules is also a creative activity within an existing theory. It is at this level of scientific activity that the Meta-DENDRAL program operates. It does not postulate new terms but tries to find new regularities and explain them with rules written in the predefined vocabulary. (Schaffner, 1985, p. 102)

Hempel:
> The formulation of powerful explanatory principles, and especially theories, normally involves the introduction of a novel conceptual and terminological apparatus. The explanation of combustion by the conflicting theories of dephlogistication and of oxidation illustrates the point.

The new concepts introduced by a theory of this kind cannot, as a rule, be defined by those previously available; they are characterized, rather, by means of a set of theoretical principles linking the new concepts to each other and to previously available concepts that serve to describe the phenomena to be explained. Thus the discovery of an explanatory theory for a given class of occurrences requires the introduction both of new theoretical terms and of new theoretical principles. It does not seem clear at all how a computer might be programmed to discover such powerful theories.

It was no doubt this kind of consideration that led Einstein to emphasize that there are no general systematic procedures which can lead the scientist from a set of data to a scientific theory, and that powerful theories are arrived at by the exercise of the scientist's free creative imagination. (Hempel, 1985, p. 120)

Buchanan refers to creativity on Meta-DENDRAL's *level of scientific activity*. Hempel answers by referring to a different and higher level of scientific activity. Implicit in their discussion is a range of such levels. At the bottom is something like Kuhnian normal science: Discovering new patterns in data and explaining them with already extant theoretical apparatus. At the top is something like Kuhnian revolutionary science: Totally **replacing** the old theoretical apparatus with new theoretical apparatus. In the ignored but most commonly practiced middle of this range is at least the following type of activity. Explaining patterns (old or new) in the data by adding to the existing theoretical apparatus. Hempel, even if he agrees with Buchanan that Meta-DENDRAL exhibits creativity on its level of scientific activity, wants to argue that creativity on certain levels cannot be automated. His particular point is this: creativity that demands the introduction of new theoretical terms not explicitly definable from old ones, and the introduction of relations in which these terms stand to each other, are tasks as that cannot be automated. He argues that it is for this reason that a computer could not do science on the highest level. Actually he is making an even stronger claim. He is asserting that it is not possible to automate scientific discovery on the highest level *or* on the middle level, where we explain by adding to the existing theoretical apparatus.

It is possible that Hempel is right that science on the highest level cannot be automated, but not for the reason he gives. It *is* possible to automate the introduction of theoretical terms not definable only from old ones. I do not intend to argue that doing science automatically on the highest level (e.g., replacing the phlogiston theory of combustion with the modern theory from only the chemical data available in the eighteenth century and the phlogiston theory) is within our reach. I do intend to show that in automating the introduction of novel theoretical apparatus, an important form of automatic scientific discovery involving true creativity is possible.

3. NOVELTY

Suppose we restrict ourselves to discussing a certain class of scientific theories, one in which a theory is formulated as a set of equations along with some interpretation of the terms that occur in the equations. A **term** is any member of an equation that is not a number or a standard mathematical operator or relation, e.g., an algebraic variable is a term and "$>$" is not. The set of all terms that occur in the equations can be thought of as the **terminological apparatus** of the theory. **Theoretical terms** are those which, for whatever reason, cannot currently be measured directly. **Observable terms** are those which can be assigned a value by performing a measurement. A **theoretical principle** is an equation in which at least two distinct terms occur. The **theoretical apparatus** with which explanations may be offered includes the mathematical theory and any of its consequences. To add to the theoretical apparatus one must add to the terminological apparatus or to the set of theoretical principles.

Suppose we are given a set of equations E that contains a set of terms T, and a set of equations E' that contains a set of terms T'. A **term** t' from E' **is novel** to E iff t' is not an element of T and there does not exist an equation in which t' occurs alone on one side and all terms that occur on the other side are elements of T. A **theoretical principle** from E' **is novel** to E iff the equation that represents the principle is not deducible from E or any equivalent to E.

If we start with a theory T1 and introduce a novel term t' to arrive at T2, then any theoretical principle in T2 involving t' is novel to T1. It is not true that a theoretical principle in T2 that does not involve t' is never novel. In other words, you can extend a theory in two ways in this scheme of things. Either you can introduce a new term (along with new theoretical principles that involve that term), or you can introduce just a new relation among already extant terms.

This scheme of novelty may be illustrated with an example from Newtonian mechanics. Suppose we have before us a theory about force, mass, and energy. Suppose that we are given Newton's three laws and the conservation of energy. Since acceleration is actually expressed as a function of position and time, and since energy is a function of mass and position and time, the terms that occur are force, mass, time, and position. If we wanted to explain an empirical regularity such as colliding billiard balls, we might do so by citing the law of conservation of momentum. We would not, however, be adding to our terminological apparatus at all. Clearly momentum, which is just mv, is not truly a *new* term. It is definable with reference to only mass, time and position. We would not be extending the set of theoretical principles either, because the conservation of momentum is deducible from the second law: $f = ma$, and the third law: every application of force from body 1 to body 2 produces an equal and opposite application of force from 2 to 1.

Universal gravitation seems a natural candidate for an example of a term novel to Newtonian Mechanics, but under my scheme it is not. Gravitation can be expressed as an equation involving only mass and position terms. The equation expressing universal gravitation is certainly a novel *theoretical principle*, however. It claims that gravitation is an instance of a Newtonian force, and it describes how that force is related to mass and distance.

Although this depiction of one kind of scientific theory is far too rough and ready, I think it captures the spirit of Hempel's scheme and can therefore serve as a framework with which to examine his claims. My task now is to show how to automate the introduction of plausible explanations that extend the theoretical apparatus of a scientific theory.

4. EXPLANATION

The workhorse of automatic discovery is a precise formulation of what it means for one theory to explain empirical regularities better than another theory does. If we have no desire to explain regularities, and all we want is to adopt a theory that is empirically adequate, then there is nothing to prevent us from simply restating the data we have observed. Hempel himself made this point in "The Theoretician's Dilemma." (Hempel, 1965). He recognized the relevance of his earlier argument to computer aided discovery and used it as further support of his general claim.

... an appropriate hypothesis should explain all or most of the given data. But this alone surely is not enough, for otherwise a satisfactory hypothesis for any given set of data could be constructed trivially by conjoining the data: from that conjunction any one of the given data would be deducible. What is wanted is evidently a hypothesis that goes beyond the data . . . .

A computer program for the discovery of acceptable hypotheses would therefore have to include criteria of preference to govern the choice among different hypotheses each of which covers the available evidence. . . . There are, indeed, several such criteria of preference that are frequently invoked — among them, simplicity and large range of potential applicability — but these notions have not, so far, received a precise characterization. . . . no exact and generally acknowledged formulation of corresponding criteria is presently available or likely to be forthcoming. (Hempel, 1965, p. 121)

Formulating a *precise* characterization of explanatory virtue that favors hypotheses that in some way go *beyond the data* is a serious challenge. A characterization of explanatory virtue should help accomplish two crucial tasks in creative scientific discovery. It should indicate when the existing theoretical apparatus is insufficient to furnish an adequate explanation for a class of empirical regularities and it should guide in constructing new theoretical apparatus that *will* provide the kind of explanation sought. Before I formulate a principle of relative explanatory virtue I need to say a little about scientific theories.

Scientific equations can often be divided into **structural terms** and **free parameters**. Structural terms are those that refer crucially to the structure of the natural system they are intended to represent. Free parameters are said to be free because their value does not affect the structural terms or principles of the theory, and therefore they are free to take on any value, although they are thought to have a unique value in nature.

To illustrate, in the early Nineteenth Century there was a vital debate between two conceptions of light, the corpuscular theory and the wave theory. In the corpuscular theory, among the structural terms were those that described light as a particle having certain properties, in the wave theory were those that described light as a wave. Theoretical principles involving light particles might be that they travel in straight lines and obey the laws of Newtonian mechanics. A theoretical principle involving light waves might be that they exhibit interference phenomena. The different *speeds* of light in various media are free parameters. There is nothing about the particular value of the speed of sunlight in air or water that is crucial to either theory's structural components. In this example the free parameter is measurable, but that is not always so. In the later part of the nineteenth century, when the wave theory was dominant, theorists decided that a wave must have a medium, leading to the concept of the luminiferous ether. Ether theorists would tell you that the particular value of the elasticity of the luminiferous ether was a free parameter, even though they could not hope to measure it.

With this distinction between structural terms and free parameters in hand, formulating principles of relative explanatory virtue is now feasible. Other things being equal, an explanation that does not in any essential way depend on the particular values of free parameters is better than one that does.[2]

Principle 1:

> If, other things being equal, theory T1 explains regularity R for all values of its free parameters, and theory T2 explains R only for particular values of its free parameters, then T1 is a better explanation of R than is T2.

To illustrate this principle, consider the Ptolemaic and Copernican theories of the solar system. Ptolemy's theory fixed the Earth at the center and had the planets and the Sun revolving around it, while Copernicus had the Sun at the center and the planets revolving around it. These claims, plus the order of the orbits, constituted a portion of the structural part of their theories. Among the many empirical regularities each theory had to account for was the relationship between Mercury and the Sun. Mercury always appeared close to the Sun, oscillating back and forth behind it or in front of it. Ptolemy's account postulated that the center of Mercury's main epicycle was on the direct line between the Earth's center and the Sun's center. Further, he postulated that the orbital speed of the Sun and the center of Mercury's epicycle were identical. Copernicus' explanation was far more plausible. Because Mercury was the innermost planet, and because the radius of Earth's orbit was much larger than Mercury's, the regularity was revealed to be a necessity. In Ptolemy's explanation, the parameters representing the starting position of the Sun, Mercury and the Earth, and those representing the orbital speed of the center of Mercury's epicycle and the Sun's center *did the explanatory work.* If the orbital speed of Mercury's main epicycle and the Sun's center around the Earth were not identical, then the observed regularity should not have persisted. In Copernicus' explanation, however, the parameters representing the orbital speed of Mercury and the Earth made virtually no difference. The explanation would have worked regardless of their values.

Another principle of relative explanatory virtue concerns the *number* of free parameters involved in an explanation.

Principle 2:

> If, other things being equal, T1 explains R and T2 explains R, and T1 makes essential reference to fewer parameters than does T2, T1 is a better explanation of R.

The same famous example provides an illustration of this principle as well. Due to assuming the Earth was stationary and at the center of the solar system, Ptolemy had to add an epicycle (essentially just a free parameter that corresponded to the Earth's actual motion) to every other planet's motion. Copernicus, by assuming the Sun is stationary and the Earth in orbital motion, eliminated each of these free parameters and thus reduced the number of epicycles by five.[3] Copernicus managed to postulate approximately twenty fewer epicycles in his system than did Ptolemy. To reduce the number of epicycles was to improve the explanatory virtue of the theory.

Bas van Fraassen argues that better explanations do not compel one to take a theory's unobservable ontological assertions seriously, although he is perfectly willing to admit that explanatory power is a legitimate criterion for choosing among competing theories, given that they are both equally empirically adequate (Van Fraassen, 1980). It should be clear that I am not committed to arguing that a theory which provides a better explanation is likely to be closer to the truth than its competitor. The issue is whether criteria of relative explanatory virtue can lead to automatic creativity that mimics good scientific creativity. Whether good scientific creativity leads us closer to the truth is, at least in this context, beside the point.

No one would seriously deny that scientists argue for theories by showing that they provide better explanations of known empirical regularities than do competitors. The question becomes: are the principles above a plausible rendering of the intuitions scientists have about better explanation, and if they are, can they enable the automatic introduction of novel terms when they are needed for better explanations? That these principles capture scientific practice can be seen from both the history and philosophy of science.

I will assume without argument that the more free parameters a theory has to work with, the better the chances that it can, by some contorted wiggling of these parameters, be made to explain a range of empirical regularities. If Principle 2 were not descriptive of scientific practice, then there would be a straightforward strategy for finding good theories. Simply multiply the number of free parameters until the phenomena can be saved no matter what they are. But the scientific community does not allow this unless there is no competitor that is even remotely empirically adequate. Examples of this strategy abound when there is no competitor, and Ptolemy with his epicycles is the paradigm case. But when other criteria among competing theories are even remotely equal, scientists will choose the theory that has fewer free parameters.

Dalton used a version of Principle 1 in arguing for his atomic theory of matter, as did Kepler in arguing that the Copernican model of the solar system was to be preferred (Glymour, 1980). I have already remarked that Copernican theory was an improvement over Ptolemic on Principle 2.

In short, the history of science is replete with appeals for an inference to the best explanation that amounts to one of the principles above. On the philosophical front, Robert Causey, in *The Unity of Science*[4] argues that scientific explanation amounts to reducing empirical regularity to necessity. A theory that explains a regularity for all values of contingent parameters may be said to render it necessary. The Bayesian account of scientific methodology can also be used to justify both principles.[5]

5. CAUSAL MODELS

Causal models are a class of mathematical theories used widely in sociology, econometrics, psychology, political science, epidemiology, educational research and many other non-experimental sciences. The goal in causal modelling is to create mathematical models of "social systems" that will allow us to understand the causal structure of those systems and that will therefore tell us how we can best influence them. Causal models are not grandiose social theories like Marxian economics. Rather a causal model is meant to apply to a specified population of individuals, usually limited in scope. The population may be made up of countries, counties, people who are on welfare, children under the age of 10, or any other purportedly meaningful social group.

We measure the value of certain variables (e.g. a person's score on an IQ test) for as many individuals in the population as is feasible, and we usually express these measurements in terms of statistical properties they possess, e.g. their means, variances and covariances.

Mathematically, causal models fit very neatly into the scheme laid out above. When formulated as systems of linear equations, they are typically called **structural equation models**. Structural equation models include regression models, factor analysis models, path models, simultaneous equation models, and multiple indicator models. They consist of two different kinds of equations. The first kind are **structural equations**, so called because they are intended to capture the causal relations among the variables in the model. A structural equation is routinely written in a canonical form in which a dependent variable (the effect) stands alone on the left side of the equation, and it is set equal to a linear combination of all variables which directly cause it. Two kinds of variables, or terms, occur in structural equations, **measured variables** and **latent variables**. Measured values are typically recorded as a point on a discrete scale (e.g., a person's IQ test score). A latent (unmeasured) variable is a member of one of two categories. It is either a **theoretical construct** or an **error term**. The theorist specifies the set of theoretical constructs and measured variables, and also the causal relations among them. An **exogenous** variable is one that is only a cause and not an effect. Any variable that *does* have a causal predecessor among the set of measured variables and theoretical constructs is **endogenous**. Each endogenous variable is given its own unmeasured cause, or error term. Intuitively, a theorist cannot hope to capture *all* the causes of variables he or she specifies. An error term is added in order to represent the "other causes" of variables explicitly defined in the system. The linear coefficient representing the amount of dependence any variable has on its error term is always assumed to be 1.

The second kind of equation in causal models is a **stochastic equation**. These equations express statistical constraints, e.g., that the covariance between two variables is vanishing. It is routine to assume that the covariance between explicitly specified causes of a variable and the error term for that variable is 0. A further set of mathematical assumptions the theorist makes involve the forms of probability distributions over the variables in the model.

To illustrate with the simplest example, suppose we theorize that variable x causes variable y. We can represent this situation with a causal picture, a structural equation and a stochastic equation. The variable u is y's error term. The term b_{yx} is the linear coefficient of y on x. *The linear coefficients in causal models are* **free parameters** *to be estimated from the data.*

Causal Picture

$$x \to y \overset{u}{\swarrow}$$

Structural Equations

$$y = b_{yx}x + u$$

Stochastic Equations

$$E(u) = 0^6$$
$$E(xu) = 0$$

Fig. 5.1.

Herb Simon attempted to formulate a criterion with which to disambiguate the causal relations from just the structural equations, but his attempt is inadequate for a number of reasons (Simon, 1953). The stochastic equations are essential in specifying the *causal asymmetry* that the structural equations alone do not. Take the first case, where x causes y. The structural equation could just as easily be written as:

$$x = 1/b_{yx}y - 1/b_{yx}u$$

If we substitute c_{xy} for $1/b_{yx}$, and v for $-1/b_{yx}u$, then we have:

$$x = c_{xy}y + v$$

It appears that this equivalent equation indicates that y is a cause of x with v as a disturbance term. But if we add the initial model's stochastic equations, then we can prove that $E(vy) \neq 0$, hence this representation is not consistent with the usual stochastic assumptions we would make about it.

The most natural representation for a causal model is a directed graph. A **directed graph**, $G = \langle V, E \rangle$, is a set of vertices V and a set of edges E which are ordered pairs of vertices. We can define a labelling function $L(e)$ that assigns a unique label to each edge. A directed graph is easy to represent as a causal picture. For example:

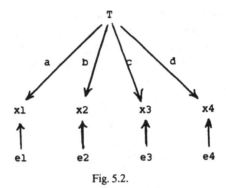

Fig. 5.2.

This directed graph corresponds to the following structural equation model.

Structural Equations

$x1 = aT + e1$
$x2 = bT + e2$
$x3 = cT + e3$
$x4 = dT + e4$

Stochastic Equations

$E(e1, T) = E(e2, T) = E(e3, T) = E(e4, T) = 0$
$E(e1, e2) = E(e1, e3) = E(e1, 4) = 0$
$E(e2, e3) = E(e2, e4) = E(e3, e4) = 0$
$E(e1) = E(e2) = E(e3) = E(4) = 0$

The natural interpretation of a directed edge is as a direct cause. That is, if there is a directed edge from x to y, we say that x is a direct cause of y. When higher order relations are defined on them, directed graphs have a tremendous amount of structure that naturally corresponds to causal intuitions and to formal parts of other representations of causal systems.

An example of a higher order relation is a **path**. A path is a sequence of edges (ordered pairs):

$$\langle\langle v1, v2\rangle, \langle v2, v3\rangle, \ldots, \langle vi, vj\rangle, \langle vj, vk\rangle\rangle$$

such that the second member of an ordered pair $p1$ is identical to the first member in the ordered pair $p2$ that follows $p1$ in the sequence.

A **trek** between two variables x and y is either a path from x to y, or a path from y to x, or a pair of paths from a third variable w, to x and to y such that the two paths have exactly one variable, w, in common. For example, in Figure 5.2, the trek between $x1$ and $x2$ is:

$$\langle\langle T, x1 \rangle, \langle T, x2 \rangle\rangle$$

A path is just one special sort of trek.

In causal models we want to explain covariation when we observe it. We interpret the vertices in our directed graphs as the variables that occur in the structural equations. The assertion our model makes about the sources of covariation between two variables is made explicit and clear by a directed graph. **Any trek between two variables will contribute to the covariation of those two variables.** Therefore, if two variables have no trek connecting them, we may assume their covariance is 0.

With a few simple rules we can move unambiguously from a directed graph to a causal model:

1. Take every vertex in the graph which has an arrow into it and express it as a linear combination of all the vertices that are at the head of those arrows, with the edge labels as the linear coefficients.
2. If two vertices $v1$ and $v2$ in the graph have no trek between them then Covariance$(v1, v2) = 0$.

Above I distinguished between the structural part of a scientific theory and its free parameters. The *directed graph alone* comprises the structural part of the causal model. When the linear coefficients, or edge labels, or free parameters, are estimated from the data, I call the model a **statistical model**. A causal model is either the directed graph with undetermined linear coefficients, or it is a statistical model.

6. NOVEL THEORETICAL APPARATUS IN CAUSAL MODELS

What does it mean to introduce novel theoretical apparatus in the context of causal models? As above, there are two ways we might add to the existing apparatus. we may introduce a new term and/or we may introduce a new theoretical principle.

Recall that a new term is one that cannot be defined from only old ones. "Cannot be defined" means that there is no equation such that the new term appears on the left and only old terms appear on the right. Consider the following two causal models:

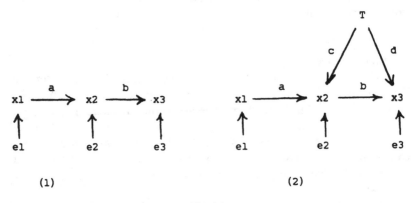

Fig. 6.1.

It would appear that Model 2 introduces a novel term to Model 1, namely the latent variable T. Whether or not it does depends on whether we count free parameters as terms or not. Model 1 can be represented by the following set of structural equations:

$$x2 = ax1 + e2$$
$$x3 = bx2 + e3$$

and Model 2 by:

$$x2 = ax1 + cT + e2$$
$$x3 = bx2 + dT + e3$$

From this, T can be defined:

$$T = (x2 - ax1 - e2)/c$$

Except for "c", every term on the right side of the equation occurs in the set of equations that represent Model 1. To make my case more persuasive I am willing to suffer the slight perversity of not counting T as a novel term. That is, I will *not* allow free parameters to count as terms[7] and therefore T can be defined by terms already extant in Model 1.

Model 2 does introduce a *novel theoretical principle*, however. Intuitively, it postulates a different *relationship* between variables $x2$ and $x3$ than Model 1 does. Expressing $x3$ in terms of $x2$ in Model 1:

$$x3 = bx2 + e3$$

and in Model 2:

$$x3 = bx2 + d((x2 - ax1 - e2)/c) + e3$$

Clearly the last equation is not deducible from the two structural equations that represent Model 1. Consider a third model:

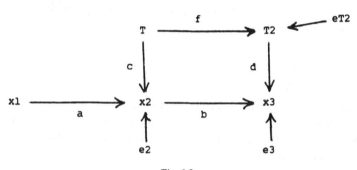

Fig. 6.2.

Here T2 *is* a novel term to the system of Model 1. Its definition cannot avoid the error term $eT2$ and thus it is cannot be defined with only the vocabulary of Model 1. This model of course introduces new theoretical principles as well.

7. EXPLANATION WITH CAUSAL MODELS

7.1. *Empirical Regularities*

Researchers employing causal models measure data and calculate sample correlations, both multiple and partial. In the measured data for causal models there are many kinds of potential empirical regularities that we might attempt to explain if supported by the data. Two that occur commonly are tetrad equations and vanishing partial correlations.

The first order[8] **partial correlation** of $x1$ and $x3$ with respect to $x2$ is defined to be (see Anderson, 1984):

$$\rho_{13.2} = \frac{\rho_{13} - \rho_{12}\rho_{23}}{(1 - \rho_{12}^2)^{1/2}(1 - \rho_{23}^2)^{1/2}}$$

THEOREM 1: *The* directed graph *of a causal model* alone *determines a set of first order vanishing partial correlations that the causal model implies for all values of the linear coefficients.*[9]

Tetrad equations involve products of correlations among a set of four measured variables. There are three possible inequivalent products of two correlations which involve all and only four variables. E.g., if the four variables are w, x, y, and z, then the three possible correlation products are

$$\rho_{wx}\rho_{yz}$$
$$\rho_{wy}\rho_{xz}$$
$$\rho_{wz}\rho_{xy}$$

Tetrad equations say that one product of correlations (or covariances) equals another product of correlations (or covariances). There are three possible tetrad equations, any two of which are independent, in a set of four variables. They are:

$$\rho_{wx}\rho_{yz} = \rho_{wy}\rho_{xz}$$
$$\rho_{wy}\rho_{xz} = \rho_{wz}\rho_{xy}$$
$$\rho_{wx}\rho_{yz} = \rho_{wz}\rho_{xy}$$

THEOREM 2: *The* directed graph *of a causal model* alone *determines a set of tetrad equations that the causal model implies for all values of the linear coefficients.*

7.2. *Two Forms of Explanation*

There are statistical tests to determine if the sample tetrad difference or the sample partial correlation are close enough to vanishing to be counted as being 0 in the population.[10] If they are vanishing, then they constitute empirical regularities in need of explanation. There are two ways a causal model can be used to explain such regularities.

A causal model may entail a tetrad equation or a vanishing partial correlation from *the directed graph alone*. To illustrate, consider the following causal model, where a, b, and c are undetermined linear coefficients.

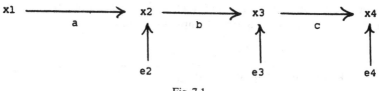

Fig. 7.1.

There is an algorithm to calculate whether a directed graph implies a tetrad equation or vanishing partial by substituting the sum of the products of all the edge labels on a trek between variables 1 and 2 for the correlation between variables 1 and 2.[11] Using the graph in Figure 7.1 to substitute into the numerator of the above formula for the partial correlation of $x1$ and $x3$ on $x2$ gives us:

$$\rho_{13.2} = \frac{ab - ab}{(1 - \rho_{12}^2)^{1/2}(1 - \rho_{23}^2)^{1/2}}$$

Thus, no matter what the values of the unspecified parameters a and b are, this causal model implies that $\rho_{13.2} = 0$.

To illustrate the algorithm for tetrad equations, consider the graph in Figure 5.2 again:

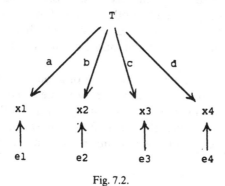

Fig. 7.2.

Substituting for the correlations in the equation:

$$\rho_{x1,x3} * \rho_{x2,x4} = \rho_{x1,x4} * \rho_{x2,x3}$$

gives us:

$$ac * bd = ad * bc$$

which is a simple algebraic identity.

The graphs in Figure 7.1 and Figure 7.2 above both imply tetrad equations. The graph in Figure 7.1 implies only one:

$$\rho_{x1, x3} * \rho_{x2, x4} = \rho_{x1, x4} * \rho_{x2, x3}$$

The graph in Figure 7.2 implies all three:

$$\rho_{x1, x3} * \rho_{x2, x4} = \rho_{x1, x4} * \rho_{x2, x3}$$
$$\rho_{x1, x3} * \rho_{x2, x4} = \rho_{x1, x2} * \rho_{x3, x4}$$
$$\rho_{x1, x2} * \rho_{x3, x4} = \rho_{x1, x4} * \rho_{x2, x3}$$

A different directed graph may entail the same vanishing partial correlation or tetrad equations, but only for *particular values of the free parameters*. For example, the graph in Figure 7.2 does not alone imply $\rho_{13.2} = 0$. In this case the numerator in the substituted expression for $\rho_{13.2}$ is:

$$ac - ab * bc$$

Unless $b = 1$, the partial correlation will not vanish.

These two different explanations of $\rho_{13.2} = 0$ comprise straightforward instances of the two kinds of explanations described in Principle 1 of relative explanatory virtue formulated above. If *after* estimating the free parameters, both the causal model in Figure 7.1 and the causal model in Figure 7.2 entail and therefore explain the empirical regularity $\rho_{13.2} = 0$, the model in Figure 7.1 provides a *better* explanation.

If one causal model is an elaboration of another there is no way for the two principles of relative explanatory virtue to conflict. In directed graphs, one graph $G1 = \langle V, E \rangle$ is a subgraph of another graph $G2 = \langle V', E' \rangle$ iff V is a subset of V' and E is a subset of E'. It is straightforward that $G1$ cannot have a greater number of free parameters than $G2$.

THEOREM 3: *If G' is a subgraph of G, then G' implies every vanishing tetrad difference and every vanishing partial correlation implied by G.*

Therefore any graph $G1$ that is an elaboration of G has a greater number of free parameters than G and also can never explain more empirical regularities[12] than G.

7.3. *Explanations That Demand Novel Theoretical Apparatus*

As you might suspect, implying tetrad equations is not independent of implying vanishing partials.

THEOREM 4: *A directed graph G implies the tetrad equation*:

$$\rho_{i,j} * \rho_{k,l} = \rho_{i,k} * \rho_{j,l}$$

if for some variable v it implies the four conditions:

$$\rho_{i,j} - \rho_{i,v}\rho_{v,j} = 0$$
$$\rho_{k,l} - \rho_{k,v}\rho_{v,l} = 0$$
$$\rho_{i,k} - \rho_{i,v}\rho_{v,k} = 0$$
$$\rho_{j,l} - \rho_{j,v}\rho_{v,l} = 0$$

To illustrate the theorem, look again at the model in Figure 7.1. The reader can verify that in this model:

$$\rho_{x1,x3} - \rho_{x1,x3}\rho_{x3,x3} = 0$$
$$\rho_{x2,x4} - \rho_{x2,x3}\rho_{x3,x4} = 0$$
$$\rho_{x1,x4} - \rho_{x1,x3}\rho_{x3,x4} = 0$$
$$\rho_{x2,x3} - \rho_{x2,x3}\rho_{x3,x3} = 0$$

The converse of the theorem is unfortunately not generally true, although the only cases in which it is false involve five or more variables.[13] In most acyclic cases examined the converse of the theorem is true. Thus we can postulate a strong heuristic guide to the relation between the implications of tetrad and partial correlation constraints in causal models expressed as directed graphs.

Heuristic 1:

> If a tetrad equation is implied by a directed graph then there is a variable v in the graph such that the graph implies that the partial correlation of each variable pair appearing in the tetrad equation on v vanishes.

Suppose that we are exploring the causal relations among four measured variables, $x1 - x4$. Suppose further that the data we measure strongly support one tetrad equation:

$$\rho_{x1,x3} * \rho_{x2,x4} = \rho_{x1,x4} * \rho_{x2,x3}$$

and that they do *not* support any vanishing partial correlations of the pairs occurring in this tetrad equation partialled on $x1$, $x2$, $x3$ or $x4$. *Any* directed graph involving *only* variables $x1-x4$ and their error terms which implies the above tetrad equation will also imply vanishing partial correlations that are, by hypothesis, empirically false. Explanatory Principle 1 tells us that we want to explain the above tetrad equation (by implying it) with only the directed graph, but we cannot do this with only the existing theoretical apparatus at our disposal. The only solution to our dilemma is to add a latent variable to our terminological apparatus, and then form a causal model that implies the above tetrad equation and does not imply any of the vanishing partials that are empirically false. Put another way, we need to add a latent variable in such a way that *it* becomes the variable v in Heuristic 1.

Examples like these lead to a second heuristic principle that is a combination of explanatory Principle 1 and Heuristic 1 above:

Heuristic 2:

> If a set of tetrad equations hold among measured variables, and no vanishing partial correlations hold among the measured variables that imply the tetrad equations, and no cyclic model without latent variables is acceptable, then, *ceteris paribus*, a model with latent variables provides the best explanation of the data.[14]

Suppose again that we are exploring the causal relations among measured variables $x1-x4$, and that the above tetrad equation holds. Suppose this time that the data also strongly support vanishing partial equations for the pairs involved in the tetrad equation when they are partialed on $x2$ and $x3$. Suppose that the model pictured in Figure 7.1 implies the tetrad equation and it implies the vanishing partials that hold. In this case we have the best explanation possible according to explanatory Principle 1. In fact we have something stronger. For if we add a latent variable to the model in such a way as to preserve the implication of the tetrad equation that holds, then we *cannot* preserve the implication of the vanishing partial correlations. This leads to a third heuristic principle for causal models.

Heuristic 3:

> If a set of vanishing partial correlations among a set of measured variables hold in the data and imply tetrad equations that also hold, then do **not** introduce a latent variable as a common cause of the measured variables unless there is good substantive reason to do so.

We have arrived at two heuristic guides to creativity that rely on principles of explanatory virtue and the mathematical properties of directed graphs. They are not toy heuristics, in fact they have already guided model building in real cases, as I will now show.

8. EMPIRICAL CASES

8.1. *Automatically Introducing Latent Variables*

In the manual to a statistical package, LISREL IV (Joreskog 84), Joreskog and Sorbom present data on the performance of 799 school girls on the Scholastic Aptitude Test. The variables they measure are the girls' scores in 5th, 7th, 9th, and 11th grades. The model they arrive at, shown in Figure 8.3, is highly plausible. Starting with an initial model, shown in Figure 8.2, that has no latent variables but which is identifiable, I will show how a computer can be creative enough to discover the final model that Joreskog and Sorbom offer.

Using the TETRAD program,[15] we compute, on the assumption that the population difference is 0, the probability that each first order partial correlation and each tetrad difference observed is as great as it is. Here are the results:

Tetrad Equation[16]	Residual	P(diff.)
$q5\,q7, q9\,q1 = q5\,q9, q7\,q1$	0.0953	0.0000
$q5\,q7, q1\,q9 = q5\,q1, q7\,q9$	0.0917	0.0000
$q5\,q9, q1\,q7 = q5\,q1, q9\,q7$	0.0036	0.7580

Partial[17]	Residual	P(diff.)
$q5\,q7 \cdot q9$	0.4806	0.0000
$q5\,q7 \cdot q1$	0.4542	0.0000
$q5\,q9 \cdot q7$	0.2931	0.0000
$q5\,q9 \cdot q1$	0.2655	0.0000
$q5\,q1 \cdot q7$	0.3177	0.0000
$q5\,q1 \cdot q9$	0.3379	0.0000
$q7\,q9 \cdot q5$	0.4595	0.0000
$q7\,q9 \cdot q1$	0.3208	0.0000
$q7\,q1 \cdot q5$	0.4742	0.0000
$q7\,q1 \cdot q9$	0.3819	0.0000
$q9\,q1 \cdot q5$	0.6347	0.0000
$q9\,q1 \cdot q7$	0.5763	0.0000

Fig. 8.1.

You can see that no vanishing partial correlations hold in the data and that one tetrad equation holds very closely. Suppose that we had an initial theory that suggested the following plausible model with which to explain the above data.

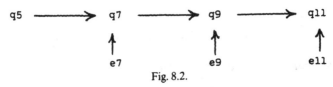

Fig. 8.2.

This model implies one tetrad equation, the one that holds, so it provides a good explanation of the equation. Unfortunately, the model implies the equation *because* it implies the partial correlations that correspond to the pairs occurring in the tetrad equation partialled on $q7$ or $q9$. As we already saw from TETRAD's output, these partial correlations are strongly rejected by the data. We are in the exact situation I used to illustrate the heuristics formulated in the last section. The only way to explain the tetrad equation that holds closely in the data, and not also imply partial correlation equations that are false in the data, is to introduce new theoretical apparatus and reformulate our model. Heuristic 2 is especially appropriate in this case because the variables are time-ordered, thus no cyclic model among only $q5$, $q7$, $q9$, and $q11$ is even remotely reasonable.

The automatic construction of such a model can proceed as follows. The set of models whose members entail the tetrad equation that holds and do not entail the partial correlation equations that do not hold can be partially ordered by the number of free parameters they contain. Rule out any which violate plausibility constraints (e.g., they posit backward causation) and then choose the one that is the most plausible among those with the fewest free parameters.

This kind of procedure easily leads to the following model:

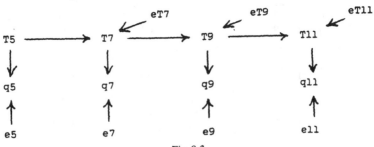

Fig. 8.3.

The model implies the tetrad equation that holds in the data but it does not imply the partial correlation constraints that do not hold in the data. The latent variables T7, T9 and T11 are novel to the model in Figure 8.2, and a number of theoretical principles involving them are novel as well.

8.2. Automatically Preventing the Introduction of Latent Variables

A series of articles about the causal relations between constituent attitudes and congressional roll call voting behavior appeared in the *American Political Science Review* in the 1960s.[18] The models discussed dealt with four measured variables:

Attitudes of the Constituents in a Representative's District = *d*

Representative's Attitudes = *a*

Representative's Perception of Constituents' Attitudes = *p*

Congressional Roll Call Vote = *r*

On a sample of 116 congressmen, the data for voting on issues concerning civil rights showed the following support for tetrad and partial correlation constraints:

The sample size is: 116

Partial	Residual	P(diff.)
$dp \cdot a$	0.6291	0.0000
$dp \cdot r$	0.4718	0.0000
$da \cdot p$	0.0454	0.6253
$da \cdot r$	0.0570	0.5470
$dr \cdot p$	0.1086	0.2485
$dr \cdot a$	0.4825	0.0000
$pa \cdot d$	0.4708	0.0000
$pa \cdot r$	0.1261	0.1776
$pr \cdot d$	0.6701	0.0000
$pr \cdot a$	0.6772	0.0000
$ar \cdot d$	0.6030	0.0000
$ar \cdot p$	0.4409	0.0000

Tetrad Equation	Residual	P(diff.)
$dp, ar = da, pr$	0.1222	0.0136
$dp, ra = dr, pa$	0.1148	0.0040
$da, rp = dr, ap$	0.0075	0.8388

Fig. 8.4.

Like the SAT case, the data strongly support one of three possible tetrad equations, but unlike that case they also support vanishing partial correlations involving the pairs occurring in the tetrad equation partialled on r or p. If the causal model implies either the vanishing partials on r or the ones on p, then it also implies the tetrad equation that holds.[19] Thus we have a straightforward instance of Heuristic 3 above. We should only consider models which add novel theoretical apparatus if we have good substantive reasons to do so. The data and principles of good explanation tell us not to introduce any latent variables to whatever causal model we devise for this system. Although a number of models were suggested in the literature that only dealt with the four variables I mentioned above, no author gave an argument for why the set of alternatives that included latent variables was excluded from consideration.

9. CONCLUSION

Hempel argued that scientific discovery on the highest level could not be automated because it involves the introduction of novel theoretical apparatus. To Hempel, the introduction of novel theoretical apparatus meant the introduction of *terms* not explicitly definable from those previously available and it meant the introduction of *relationships* among a theory's entities that were not somehow already contained in the original theory. I have given an interpretation of these notions in the context of scientific theories expressable as systems of equations. Using principles of relative explanatory virtue common to all quantitative science, instantiated in the class of mathematical theories known as causal models, I have argued that we can automatically determine when we should or should not introduce novel theoretical apparatus, and I have also shown *how* such apparatus can be introduced.

What is now called artificial intelligence ranges from the development of "expert systems", some of which are really just immense look-up tables, to projects that aim to be as independent as possible of domain specific knowledge. Projects like these latter ones must find heuristics or principles that are general, so that they do not rely essentially on the particular features of a prescribed domain, and that can be specified precisely, so that they can be automated. Douglas Lenat's work on automatic discovery in mathematics (Lenat, 1983), Pat Langley and Herb Simon's work on the BACON programs (Langley,

1983), and Jaime Carbonell's work on analogical reasoning (Carbonell, 1983), are all good examples of the latter type of work. Projects that are even more ambitious attempt to rely on the same sort of general principles of preference to guide them (Larkin, 1983). The work dealing with causal models that I described above is not as general as these are, but it shares one of their most important virtues. It employs principles of relative explanatory virtue that are *general* to good science. When these principles are instantiated in the limited mathematical domain of causal modelling, they actually help to automate the discovery of novel scientific theoretical apparatus.

The level of creativity for which I demonstrate the possibility of automating is a far cry from the level that, say, Einstein employed in revolutionizing modern physics. The frontier of "impossible to automate" tasks is only slightly pushed back by showing that novel theoretical apparatus can be automatically introduced into causal models.

Nevertheless there are important similarities. In his famous 1905 paper, the first point Einstein made is that there was a flaw in the way traditional electrodynamics *explained* the phenomenon of inducing a current in a coil with a magnetic field (Einstein, 1905). He pointed out that two situations — the coil still and the magnet moving, and the coil moving and the magnet still — were perfectly symmetric physically, yet the theoretical explanations of them are asymmetric. In effect Einstein invoked a principle of explanatory adequacy to argue that the traditional theory needed to be changed, and then a principle of relative explanatory virtue to argue for the theoretical apparatus he introduced as an alternative to Maxwell's and Lorentz's theory. Einstein placed constraints on the type of explanations he was searching for. They had to have certain *symmetry properties*, they had to be *simple*, and of course they had to account for a wide range of already established empirical phenomena.

By no means did these constraints automatically pick out the Special Theory of Relativity for Einstein. Instead this example shows that even the most revolutionary episodes in science, those that seem to be almost mystical leaps of the creative mind, are indeed *guided* by strong methodological principles and constraints. If even vague methodological principles can be abstracted from science in general, which is partly the task of historians and philosophers of science, and these can be made precise in more local domains, then automating discovery on many levels will be within our reach.

NOTES

* I gratefully acknowledge Joel Smith for help with the examples and for catching a number of silly mistakes. I especially thank Clark Glymour, who I have had the good fortune to work closely with over the past several years. The ideas in this paper are as much his as anybody's.

[1] I take this terminology from John Haugeland (Haugeland, 1985).

[2] This is by no means the first statement of such a principle. See Glymour, 1979, Glymour, 1980, Garfinkel, 1981, and Causey, 1977.

[3] I am indebted to Joel Smith for this example.

[4] See Causey, 1977.

[5] See Rosenkrantz, 1977.

[6] $E(u)$ means "the mathematical expectation of u."

[7] Free parameters occur as algebraic unknowns in the initial formulation of a theory. They are interpreted as simply constants of nature, and therefore have a different ontological status from structural variables.

[8] We say "first order" because the pair is only partialled on one other variable. The partial correlation of $x1$ and $x3$ w.r.t. $x2$ and $x4$ is well defined, but I do not use it.

[9] For the proof of this and every other theorem I use, see *Discovering Causal Structure*, Glymour *et al.*, 1987.

[10] See DCS.

[11] For the proof of the correctness of this algorithm, see DCS, chapter 10.

[12] That is, the empirical regularities of tetrad equations and vanishing partials.

[13] See DCS.

[14] This and heuristic 3 below are quoted from DCS.

[15] See DCS.

[16] The notation: "$q5\ q7, q9\ q1$" stands for: "$\rho_{q5,q7} * \rho_{q5,q7}$".

[17] The notation: "$q5\ q7 \cdot q9$" stands for: "$\rho_{q5,q7 \cdot q9} = 0$".

[18] See Miller, 1963, Cnudde, 1966, and Forbes, 1968.

[19] Look again at Theorem 4 above.

REFERENCES

Anderson, T.: 1984, *An Introduction to Multivariate Statistical Analysis*, Wiley, New York.

Carbonell, J. G.: 1983, 'Learning by Analogy: Formulating and Generalizing Plans from Past Experience', in R. S. Michalski, J. G. Carbonell and T. M. Mitchell (eds.), *Machine Learning: An Artificial Intelligence Approach*, Tioga Press, Palo Alto.

Causey, Robert L.: 1977, *Unity of Science*, D. Reidel Publishing Company, Dordrecht, 1977.

Cnudde, C., McCrone, D.: 1966, 'The Linkage Between Constituency Attitudes and Congressional Voting Behavior: A Causal Model', *American Political Science Review* **60**, 66—72.

Einstein, Albert: 1952, 'On the Electrodynamics of Moving Bodies', in *The Principle of Relativity*, chapter 3, Dover, New York.

Forbes, H., Tufte, E.: 1968, 'A Note of Caution in Causal Modelling', *American Political Science Review* **62**, 1258—1264.

Garfinkel, A.: 1981, *Forms of Explanation*, Yale University Press, New Haven.

Glymour, C.: 1979, 'Explanations, Tests, Unity, and Necessity', *Nous* **14**, 31—50.

Glymour, C.: 1980, *Theory and Evidence*, Princeton University Press, Princeton.

Glymour, C., Scheines, R., Spirtes, P., and Kelly, K.: 1987, *Discovering Causal Structure: Artificial Intelligence, Philosophy of Science, and Statistical Modelling*, Academic Press, Orlando, FL, forthcoming.

Haugeland, John: 1985, *Artificial Intelligence: The Very Idea*, MIT Press, Cambridge, MA.

Hempel, Carl G.: 1965, *Aspects of Scientific Explanation*, The Free Press, New York.

Hempel, Carl G.: 1985, 'Thoughts on the Limitations of Discovery by Computer', in Kenneth F. Schaffner (ed.), *Logic of Discovery and Diagnosis in Medicine*, chapter 5, University of California Press, Berkeley.

Joreskog, K. and Sorbom, D.: 1984, *LISREL VI User's Guide*, third edition, Scientific Software, Inc., Mooresville, Indiana.

Langley, P., Bradshaw, G., and Simon, H. A.: 1983, 'Rediscovering chemistry with the BACON system', in R. S. Michalshki, J. G. Carbonell, and T. M. Mitchell (eds.), *Machine Learning: An Artificial Intelligence Approach*, Tioga Press, Palo Alto.

Jaime G. Carbonell, Jill H. Larkin and F. Reif: 1983, *Toward a General Scientific Reasoning Engine*, Joint Computer Science and Psychology technical report C.I.P. 445, Carnegie Mellion.

Lenat, D. B.: 1983, 'The Role of Heuristics in Learning by Discovery: Three Case Studies', in R. S. Michalski, J. G. Carbonell and T. M. Mitchell (eds.), *Machine Learning, An Artificial Intelligence Approach*, Tiogo Press, Palo Alto.

Miller, Warren, and Stokes, Donald: 1963, 'Constituency Influence in Congress', *American Political Science Review*, 45—56.

Rosenkrantz, R.: 1977, *Inference, Method and Decision*, D. Reidel, Dordrecht, Holland.

Schaffner, K. F. (ed.): 1985, *Logic of Discovery and Diagnosis in Medicine*, University of California Press, Berkeley.

Simon, H.: 1953, 'Causal Ordering and Identifiability', in *Models of Discovery*, D. Reidel, Dordrecht, pp. 53—80.

Van Fraassen, Bas C.: 1980, *The Scientific Image*, Clarendon Press, Oxford.

History and Philosophy of Science
University of Pittsburgh
Pittsburgh, PA 15260
U.S.A.

INDEX OF NAMES

INDEX OF SUBJECTS